ELECTROCHEMICAL BIOSENSORS

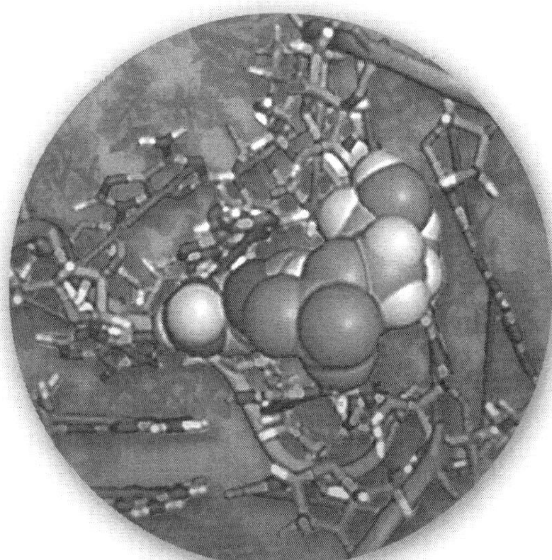

Pan Stanford Series on the High-Tech of Biotechnology

Robert S. Marks
Series Founding Editor

Avram and Stella Goldstein-Goren Department of Biotechnology Engineering
National Institute for Biotechnology Engineering
Ilse Katz Institute for Nanoscale Science & Technology
Ben Gurion University of the Negev
Israel

Titles in the Series

Published

Vol. 1

Nanoantenna: Plasmon-Enhanced Spectroscopies for Biotechnological Applications

Marc Lamy de la Chapelle and Annemarie Pucci, eds.

2013

978-981-4303-61-3 (Hardcover)
978-981-4303-62-0 (eBook)

Vol. 2

Viral Diagnostics: Advances and Applications

Robert S. Marks, Leslie Lobel, and Amadou Alpha Sall, eds.

2015

978-981-4364-43-0 (Hardcover)
978-981-4364-44-7 (eBook)

Vol. 3

Electrochemical Biosensors

Serge Cosnier, ed.

2015

978-981-4411-46-2 (Hardcover)
978-981-4411-47-9 (eBook)

Forthcoming

Vol. 4

Nanomaterials for Water Management: Signal Amplification for Biosensing from Nanostructures

Ibrahim Abdulhalim and Robert S. Marks, eds.

978-981-4463-47-8 (Hardcover)
978-981-4463-48-5 (eBook)

Vol. 5

Luminescent Biosensors

Gerald Thouand and Robert S. Marks, eds.

Vol. 6

Fiber-Optic Immunosensors and Biosensors

Robert S. Marks, ed.

Pan Stanford Series on the High-Tech of Biotechnology
Volume 3

ELECTROCHEMICAL BIOSENSORS

edited by
Serge Cosnier

PAN STANFORD PUBLISHING

Published by

Pan Stanford Publishing Pte. Ltd.
Penthouse Level, Suntec Tower 3
8 Temasek Boulevard
Singapore 038988

Email: editorial@panstanford.com
Web: www.panstanford.com

British Library Cataloguing-in-Publication Data
A catalogue record for this book is available from the British Library.

Electrochemical Biosensors

ISBN 978-981-4411-46-2 (Hardcover)
ISBN 978-981-4411-47-9 (eBook)

Printed in the USA

Contents

Preface xiii

1 Electrochemical Detection of Exocytosis: A Survey from the Earliest Amperometry at Carbon Fiber Ultramicroelectrodes to Recent Integrated Systems **1**

Christian Amatore, Manon Guille Collignon, and Frédéric Lemaître

1.1 Monitoring the Exocytotic Secretion at the Single-Cell Level 2

1.1.1 Principles of Exocytosis 2

1.1.1.1 Biological mechanism 2

1.1.1.2 Analytical requirements: necessity of real-time monitoring for understanding exocytosis 4

1.1.2 Analytical Techniques for Real-Time Detection of Exocytosis 5

1.1.2.1 Optical techniques 5

1.1.2.2 Electrophysiological techniques 8

1.1.2.3 Electrochemistry at ultramicroelectrodes: the semiartificial synapse configuration 9

1.2 Electrochemical Detection of Exocytosis: from Basic Carbon Fiber Ultramicroelectrodes to Combinatory Systems 12

1.2.1 Properties of Ultramicroelectrodes for Biological Measurements 12

1.2.2 Typical Single-Cell Models 15

1.2.3 1990–2000: Carbon Fibers' Years for Apex Detection 17

1.2.3.1 Types of ultramicroelectrodes 17
1.2.3.2 Voltammetry vs. amperometry: advantages and drawbacks 18
1.2.3.3 Applications of amperometry to the investigation of biological or/and physicochemical parameters 19
1.2.4 Beginning of the 2000s: Development of Coupled Analytical Approaches 25
1.2.4.1 Limitations of basic single UMEs with apex detection 25
1.2.4.2 Coupling amperometry with other techniques 26
1.2.4.3 Decreasing the electrode size 29
1.2.4.4 Integrating methods in microsystems 31
1.3 Conclusion 40

2 Adsorptive Stripping Voltammetric Determination of Metabolites 53
Xiaoquan Lu, Hui Li, and Yaya Li
2.1 Introduction 53
2.2 Classification, Mechanism, and Theory of Adsorptive Stripping Voltammetry 56
2.2.1 Mechanism and Characteristics of the First Adsorptive Stripping Voltammetry 57
2.2.2 Mechanism and Characteristics of the Second Adsorptive Stripping Voltammetry 59
2.3 Experimental Techniques of Adsorptive Stripping Voltammetry 61
2.3.1 Subtraction 62
2.3.2 Media Exchange 63
2.4 Electrodes 65
2.4.1 Hanging Mercury Drop Electrode 65
2.4.2 Mercury Film Electrode 65
2.4.3 Ultramicroelectrodes 66
2.4.4 Carbon Paste Electrode and Platinum Electrode 66
2.5 Representative Examples 67

3 Electrochemical Nucleic Acid Aptamer–Based Biosensors **75**
Ilaria Palchetti and Marco Mascini
3.1 Introduction 75
3.1.1 Nucleic Acid Aptamers 76
3.2 Electrochemical Aptasensors 81
3.3 Conclusion: Criticisms and Future Perspectives 87

4 Amperometric Enzyme Electrodes **93**
Serge Cosnier and Gareth P. Keeley
4.1 Introduction 93
4.2 Functioning Principles of Amperometric Enzyme Electrodes 95
4.2.1 Enzyme Electrodes Based on Mediated Electron Transfer 99
4.2.2 Enzyme Electrodes Based on Direct Electron Transfer 102
4.3 Bioelectrodes Based on Multienzyme Configurations 103
4.3.1 Extension of the Substrate Range via Enzyme Association 104
4.3.2 Improvement of the Selectivity of the Amperometric Transduction Step 105
4.4 Amplification of the Biosensor Response 107
4.5 New Concept of Bienzyme Sensors Displaying a Positive Amperometric Response to the Inhibitive Effect 109
4.6 DNA Sensors and Immunosensors Based on Enzyme Labeling for Amperometric Transduction of the Biological Sensor Response 111
4.6.1 Immunosensors Based on the Electroenzymatic Response 111
4.6.2 DNA Sensors Based on the Electroenzymatic Response 114
4.7 Future Directions and Challenges 117

5 Electrochemical Biosensors Based on Carbon Nanotubes and Nanohybrids: From Fundamental to Biological Architectures **125**
Youxing Fang and Erkang Wang
5.1 Introduction 125

5.2 Synthesis and Functionalization of CNTs 128
5.3 Fabrication of CNT-Based Biosensors 129
5.4 Biomolecular Sensing Monosaccharides 130
5.4.1 Glucose 130
5.4.1.1 Glucose sensors with enzymes 131
5.4.1.2 Glucose sensors without enzymes 134
5.4.2 Other Monosaccharides 134
5.4.3 Dihydronicotinamide Adenine Dinucleotide 135
5.4.4 Neurotransmitters/Neurochemicals 136
5.4.4.1 Dopamine 136
5.4.4.2 Other important neurochemicals 137
5.4.5 Proteins 138
5.4.5.1 Amino acids 138
5.4.5.2 Some important proteins 139
5.4.6 DNA Sensors 141
5.5 Cells and Microorganism Sensing 144
5.6 Conclusion and Prospects 145

6 Gold and Silver Nanoparticles for Electrochemical Immunosensors 161
Barry G. D. Haggett
6.1 Introduction 161
6.1.1 Nanoparticles 161
6.1.2 Electrochemical Immunoassays 162
6.1.3 Use of Gold and Silver Nanoparticles 165
6.2 Modification of the Working Electrode 165
6.2.1 Use of Gold or Silver Nanoparticles 165
6.3 Use of Antibodies and/or Other Materials 172
6.3.1 Antibodies Indirectly Attached to the Base Electrode 173
6.3.2 Antibodies on Nanoparticles Attached to Intermediate Layers on the Base Electrode 173
6.3.3 Antibodies on Nanoparticles Incorporated into Composite Layers on the Base Electrode 173
6.4 Nanoparticle Modification of Antibodies/Antigens not Bound to the Working Electrode 174
6.5 Antibodies Bound to Magnetic Composite Nanoparticles 174

6.6 Antibodies/Other Biomaterials Bound to Gold or Silver Nanoparticles 175
6.7 Generation of the Analytical Signal at the Working Electrode 176
6.8 Gold Nanoparticles 177
6.9 Gold Nanoparticles with Silver Deposition 177
6.10 Silver Nanoparticles 179
6.11 Conclusion 180

7 Electrochemical DNA Sensors Based on Nanoparticles 195
Claudio Parolo, Alfredo de la Escosura-Muñiz, and Arben Merkoçi
7.1 Introduction 195
7.2 General Aspects of DNA and Nanoparticles 196
7.2.1 DNA Structure 196
7.2.2 DNA Amplification Techniques 197
7.2.3 What Nanoparticles Can Bring to DNA Sensors? 199
7.2.4 Designs of Electrochemical DNA Biosensors 200
7.3 Applied Technologies 202
7.3.1 Gold Nanoparticle–Based Technologies 202
7.3.1.1 Gold nanoparticles as electroactive and catalytic labels 202
7.3.1.2 Gold nanoparticles as signal amplifiers and carriers of other labels 206
7.3.1.3 Gold nanoparticles as modifiers of electrotransducers 208
7.3.2 Other Nanoparticles 210
7.3.2.1 Silver nanoparticles 210
7.3.2.2 Semiconductor nanoparticles 211
7.3.2.3 Other nanoparticles 213
7.4 Conclusion and Future Perspective 216

8 Electroenzymatic Labeling for Immunosensors and DNA Sensors 223
Dianping Tang
8.1 Introduction 223
8.2 Electroenzymatic Labeling for Immunosensors 225
8.2.1 Electroenzymatic Labeling for Potentiometric Immunosensors 226

8.2.2 Electroenzymatic Labeling for Amperometric Immunosensors 228
8.2.3 Electroenzymatic Labeling for Impedimetric Immunosensors 231
8.2.4 Electroenzymatic Labeling for Conductometric Immunosensors 233
8.3 Electroenzymatic Labeling for DNA Sensors 235
8.3.1 Electroenzymatic Labeling for Potentiometric DNA Sensors 236
8.3.2 Electroenzymatic Labeling for Amperometric DNA Sensors 237
8.3.3 Electroenzymatic Labeling for Impedimetric DNA Sensors 238
8.3.4 Electroenzymatic Labeling for Conductometric DNA Sensors 240
8.4 Conclusion 242

9 Conductometric Enzyme Biosensors **247**
Sergei Dzyadevych and Nicole Jaffrezic-Renault
9.1 Introduction 248
9.2 Conductometry in Enzyme Catalysis 249
9.2.1 Conductometric Enzyme Biosensors Based on Direct Analysis 250
9.2.1.1 Glucose biosensors 250
9.2.1.2 Urea biosensors 252
9.2.1.3 Biosensors for arginine determination 255
9.2.1.4 Biosensor for testing protein denaturation 257
9.2.1.5 Biosensors for formaldehyde determination 258
9.2.1.6 Biosensors for proteins determination 261
9.2.1.7 Biosensors for phosphate determination 263
9.2.1.8 Biosensors for nitrite determination 263
9.2.1.9 Biosensors for nitrate determination 266
9.2.1.10 Biosensors for sucrose determination 268
9.2.1.11 Biosensors for lactose determination 269
9.2.1.12 Biosensors for maltose determination 270

9.2.2 Conductometric Enzyme Biosensors Based on Inhibition Analysis 273

9.2.2.1 Biosensor for organophosphorous pesticides determination 273

9.2.2.2 Biosensor for heavy metal ion determination 276

9.2.2.3 Biosensor for surfactant determination 277

9.2.2.4 Biosensor for detection of diuron and atrazine 278

9.2.2.5 Biosensor for nitrite determination 280

9.2.2.6 Biosensor for cyanide determination 280

9.3 Conclusion 281

10 Impedance Immunosensors **289**

Frank Davis and Séamus P. J. Higson

10.1 Introduction 289

10.2 The AC Impedance Method 291

10.3 Impedance Immunosensors 295

10.4 Immunosensors Containing Nanomaterials 298

10.5 Immunosensor Arrays 300

10.6 Conclusion and Outlook 304

11 Transduction of Biochemical Reactions by Use of Quantum Dots and Photocurrent Detection **309**

Fred Lisdat and Daniel Schäfer

11.1 Introduction to Quantum Dots 310

11.2 Concepts of Using Quantum Dots in Electrochemical Bioanalysis 313

11.3 Immobilization of Quantum Dots on Electrodes 317

11.4 QD Electrodes for Detection of Small Molecules in Solution 319

11.5 QD Electrodes and Enzymatic Signal Chains Based on First Sensor Generation 322

11.6 QD Electrodes and Enzymatic Signal Chains Based on Mediated Electron Transfer 327

11.7 QD Electrodes and Enzymatic Signal Chains Based on Direct Protein Electron Transfer 327

11.8 Summary 331

12 Biosensors Based on Electrochemiluminescence **337**
Huangxian Ju
12.1 Introduction 338
12.2 Principles of ECL Biosensing 339
12.2.1 Ion Annihilation Mechanism 340
12.2.2 Coreactant ECL 341
12.2.3 ECL from NPs 343
12.3 ECL Biosensor and Its Applications 345
12.3.1 Immobilization of $Ru(bpy)_3{}^{2+}$ and Its Derivatives for Biosensing 346
12.3.2 Enzymatic ECL Biosensing with Immobilized Emitters 347
12.3.3 Labels with $Ru(bpy)_3{}^{2+}$ and Its Derivatives for Biosensing 348
12.3.4 Biosensing Based on ECL Emission from QDs 351
12.3.4.1 Generation or consumption of coreactants 351
12.3.4.2 Quenching effect of analytes 355
12.3.4.3 Electron transfer kinetics 357
12.3.4.4 ECL biosensing using QDs to label binding partners 358
12.4 Conclusion 361

13 The Self-Powered Biosensors Based on Biofuel Cells **371**
Liu Deng and Shaojun Dong
13.1 Introduction 372
13.2 The Self-Powered Biosensor Based on the Enzymatic Biofuel Cell 373
13.3 The Self-Powered Biosensor Based on Mitochondrial Biofuel Cells 378
13.4 The Self-Powered Biosensor Based on Microbial Fuel Cells 379
13.5 Outlook 381

Index 389

Preface

Since biological sensors were first reported several decades ago, electrochemistry, being one of the few known transduction methods for biological processes and biosensors, has played a pivotal role in the development of these devices. Electrochemical transduction, a generic concept involving techniques such as amperometry, conductimetry, potentiometry, square-wave voltammetry impedance, and electrochemiluminescence, is currently one of the main strategies used in the biosensor field. The latter includes enzyme electrodes, immunosensors, protein sensors, DNA sensors, and biochips. The explosion of activity in nanoscience and nanotechnology and their huge success have profoundly affected the biosensor field, opening new avenues of research for materials and biomaterials for electrodes. Electrochemical biosensors are rapidly gaining traction in new applications with increasingly adjustable biomaterials with superior electrical, mechanical, biocompatible, and physical properties. In particular, the nano-structuring of electrode biomaterials provides higher surface area and better porosity. Involvement of nano-objects such as quantum dots, carbon nanotubes, and nanohybrids in the development of bioarchitectures yields fascinating properties and constitutes a strategic booming field in the development of biosensors.

Thanks to their adaptability, ease of use in relatively complex samples and their portability, interest in biosensors is growing exponentially and now constitutes one of the mainstays of analytical chemistry.

In this context, the aim of this book is to provide an overview of different types of electrochemical biosensors as well as to present the use of electrochemistry for the detection of metabolites and the exploration of physiological processes with ultramicroelectrodes.

The most common electrochemical biosensors are classified as electrochemical DNA sensors, amperometric enzyme electrodes, conductimetric biosensors, impedance immunosensors, and electrochemiluminescent biosensors. Areas of particular interest are the use of silver and gold nanoparticles for signal amplification, photocurrent transduction, and aptamer design. Moreover, advanced insights in the innovative concept of self-powered biosensors derived from biofuel cells are also discussed.

This book compiles contributions written by world-recognized scientists known for their expertise in different fields of electrochemical transduction for biological phenomena and biosensors. This was in order to cover major aspects of electrochemical biosensors from fundamentals to various applications. Special emphasis was given to the latest trends and innovations. We believe this book will appeal to a wide scientific audience as well as graduate students.

Serge Cosnier

Chapter 1

Electrochemical Detection of Exocytosis: A Survey from the Earliest Amperometry at Carbon Fiber Ultramicroelectrodes to Recent Integrated Systems

Christian Amatore, Manon Guille Collignon, and Frédéric Lemaître

UMR 8640, Département de Chimie, Ecole Normale Supérieure, 75231 Paris Cedex 5, France

christian.amatore@ens.fr

The intrinsic properties of ultramicroelectrodes make them particularly suitable for electrochemical detection of exocytosis at the single-cell level. During the nineties, carbon fiber ultramicroelectrodes (5–10 μm in diameter) used in the "semiartificial synapse" configuration were essential for improving the understanding of the mechanism of exocytotic release not only by addressing the precise detection and monitoring of small effectors fluxes but also by unraveling multiple factors governing release, multiple vesicular populations, fusion pore kinetics, etc. Over the past 10 years, new questions appeared, thus requiring global monitoring of a release event, enhancing the spatial resolution of the detection or avoiding all the experimental problems related to cell variability. Therefore,

Electrochemical Biosensors
Edited by Serge Cosnier

ISBN 978-981-4411-46-2 (Hardcover), 978-981-4411-47-9 (eBook)
www.panstanford.com

the early simple microelectrode composed of a fiber with possible surface treatment could not solve these issues and was required to be adapted, modified, or even integrated in multicomponent microassemblies. In this chapter, we wish to present how exocytosis may be electrochemically investigated in real time at the single-cell level and how the seminal carbon fiber electrode method evolved over 20 years. The examples produced to illustrate this chapter are obviously not exhaustive, taking into account the tremendous number of studies dealing with this expanding domain. The reader of this chapter is thus encouraged to consult the last comprehensive reviews that have blossomed during the past five years [1–8].

1.1 Monitoring the Exocytotic Secretion at the Single-Cell Level

1.1.1 *Principles of Exocytosis*

1.1.1.1 Biological mechanism

Vesicular exocytosis is a ubiquitous secretory pathway used for an organism to control the delivery of biochemical or chemical messengers. Intercellular communication in multicellular organisms is based on the release of acting molecules into the extracellular medium, from an emitting cell to a target cell [9]. When transmission is achieved by vesicular exocytosis, the molecules are stored beforehand in secretory vesicles. These act as cargos ensuring both storage and transport from the site at which the messengers are biosynthesized to the cell membrane at which they release their contents to the outside. Specialized secretory cells, such as endocrine cells or neurons, require fast secretion on demand. In regulated secretory pathways, release is triggered by extracellular signals, a specificity that allows the control of the release of required messengers or effectors (hormones, peptides, proteins, etc.) [10, 11]. For instance, exocytosis is a key mechanism in chemical synapse transmission, thereby transferring an electrical signal from an emitting neuron to the receiving neuron or muscle receptor via the intermediate release of chemical messengers called neurotransmitters in the restricted volume defined by the two

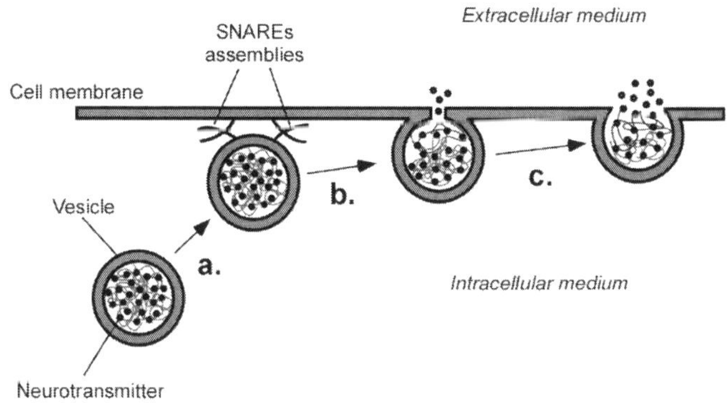

Figure 1.1 Main phases of vesicular dense-core exocytosis. After appropriate stimulation, vesicles that are primed to undergo exocytosis dock to the cell membrane by the means of SNARE complexes (step a). The cell and vesicular membrane mixing provokes fusion pore formation, through which neurotransmitters begin to diffuse out of the vesicular matrix (step b). Resulting from the ionic exchange (catecholamine cations vs. Na^+ or H_3O^+) and water entry between the intravesicular and extracellular media, local matrix swelling is induced, which triggers fusion pore expansion (step c).

neuronal cells, called the synaptic cleft. More generally, these messengers may be released by a similar mechanism into biological fluids (e.g., into the blood flow) or the extra-body environment (as for pheromones).

The different steps involved in the exocytotic process can be briefly described as follows (Fig. 1.1). First, secretory vesicles located in the emitting cell cytoplasm are initially filled with chemical messengers. Second, following appropriate cell stimulation, which, in all cases discussed hereafter, induces Ca^{2+} entry and/or concentration increase, the available vesicles dock to the cell membrane. This step requires overcoming of the natural electrostatic repulsions between the charged cell and bilipidic vesicular membranes and is assisted through the formation of soluble *N*-ethylmaleimide-sensitive fusion protein attachment receptors (SNAREs) complexes by multiple protein–protein interactions. Subsequent fusion between cells and vesicular membranes then occurs by mutual electroporation and local reorganization of phospholipidic bilayers. This allows for the formation of a nanometric fusion pore through which the release of the vesicular content into the extracellular medium begins. In most experimental cell models, "dense core" vesicles are involved, in which the cationic messengers are initially compacted into a matrix constituted of polyanionic proteins that can store high concentrations of secretory charged products. In that case, due to the fusion pore, ionic exchanges between this matrix and the extracellular medium spontaneously occur and allow release. This most possibly induces matrix destructuration/swelling. Following this brief stage (generally $<$1–2 ms), the initial fusion pore generally expands, thus leading to the release of a larger flux of the chemical messengers into the extracellular medium. This second phase is generally termed "full fusion" instead of the first one (fusion pore), though nowadays a debate exists about the fact that "full fusion" proceeds up to complete fusion, that is, complete merging of the vesicle membrane into that of the cell.

1.1.1.2 Analytical requirements: necessity of real-time monitoring for understanding exocytosis

The factors that allow the tight control and regulation of exocytosis are still a source of debate and are related to the presynaptic plasticity, that is, the ability of a neuron to change/adapt its release. The nature of the factors (physicochemical vs. biological) that govern the exocytotic activity, and more particularly their respective part/influence, remains to be fully discovered and understood. For example, the biological goal of exocytotic release can potentially be reached through a sequence of successive fully emptying vesicles or conversely through controlling of the amount delivered by a single vesicle. In that sense, the possible unvariance of the vesicular "quanta" or the nature of fusion states (from full release to "kiss and run" events) is a relevant question. However, developing analytics adapted to precisely monitor the released fluxes represents a challenge. Indeed, whatever the goal of exocytotic release is, an infinitely minute number of molecules (a few thousands to a few millions) is delivered during a brief fraction of time (i.e., milliseconds to seconds) by a living cell in its extracellular medium. However, as evidenced above, exocytosis cannot be viewed only as a simple act of vesicular release but has to be considered in its global complexity, that is, involving a succession of events leading to locally "pulsed" releases featuring a population of vesicles that is transported into the cytoplasm to eventually release their contents at different times and different sites of the cell membrane. The present challenges are therefore to design analytical techniques able to record in real time the whole sequence of each individual event leading to secretion, particularly the vesicle motion into the cytoplasm, the fusion of the membranes, and the number of neurotransmitters released. This needs to be performed within a very short time window, while monitoring the cell activity long enough over comparatively large volumes encompassing the release point.

1.1.2 *Analytical Techniques for Real-Time Detection of Exocytosis*

Nowadays, three major analytical approaches enable gathering of complementary information on temporal, quantitative or spatial monitoring of exocytotic events on single living cells (Fig. 1.2). Optical observations are mainly based on fluorescent microscopies by means of the imaging of secretory vesicles with tagged membranes or tagged content while they undergo fusion. This can be performed in real time though with an accuracy limited by cameras (pixels, time frames). Electrophysiological techniques based on patch-clamp measurements offer the possibility to monitor the surface area of the cell membrane around the very place of the fusion pore through recording with high accuracy and fast temporal resolution the conductivity between intra and extracellular media. Since both parameters vary during exocytosis due to the global increase of the cell membrane surface, the fusion pore and its transition toward fusion are thus fully analyzable. Amperometry at a carbon fiber ultramicroelectrode (UME) relies on detection, at a UME positioned on the top of the cell (semiartificial synapse configuration), of temporal fluxes of released messengers during single exocytotic events through oxidation of the released molecules.

1.1.2.1 Optical techniques

Optical techniques are based on fluorescence microscopy and rely on fluorescent markers whose fluorescence signal depends

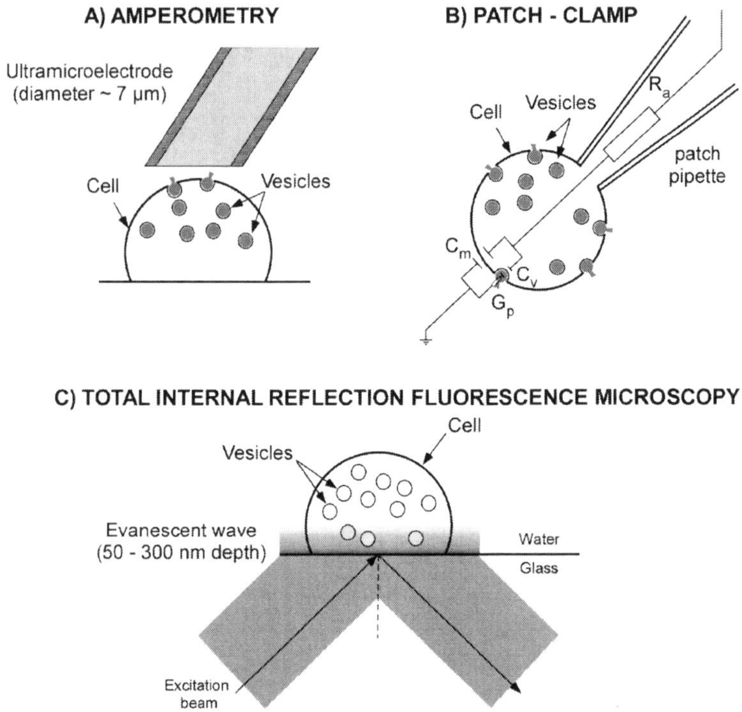

Figure 1.2 The three main techniques for investigating exocytosis at the single-cell level in real time. (a) Amperometry at a carbon fiber UME: the vesicular content (mostly catecholamines) is oxidized at the electrode surface. The measured current witnesses from the dynamics of the release. (b) Patch-clamp: a patch of the cell membrane is sealed at the top of a glass micropipette (whole-cell configuration in the figure). This impedance technique, through assuming an appropriate equivalent circuit, allows monitoring of the cell surface (capacitances C_m, C_v) during exocytosis and the conductance of fusion pores (G_p). (c) Total internal reflection fluorescence (TIRF) microscopy: fluorescent probed vesicles are visualized in the close vicinity of the cell membrane through an evanescent wave resulting from the total reflection of the excitation beam. Each exocytotic event can be located and detected as a flash or an extinction of fluorescence.

on the physicochemical time status of the cell compartment in which they are accumulated. Specific fluorescent probes can thus be targeted to different cell areas (membranes, cytoplasm, and vesicles), depending on their specific binding properties, thus offering through pixel color coding a view of the event in which this particular component is involved. Hence, the different steps of exocytosis may be visualized: vesicle displacement, docking, and fusion. In a conventional fluorescence microscope (epifluorescence), most fluorescent vesicles are excited by the beam traveling through the cell, so the emitted signal from the vesicles at the coverslip interface is blurred by signals from the numerous amount of other vesicles not effectively involved in exocytotic events [12]. Conversely, imaging single vesicles using confocal detection (in which not all the planes are visible but only the focus plane) has been adapted for the study of release by neuroendocrine cells and hippocampal neurons [13]. With this technique, out-of-focus signals are eliminated, but several handicaps still limit its use for complete vesicle motion studies. Indeed the thickness of the confocal is not adapted to tracking the approach of individual vesicles (e.g., 300 nm diameter in mean for chromaffin cells) toward their docking sites on the cell membrane. Also, the issue of photobleaching prevents acquisition of images during long time periods unless through using signals at low excitation intensity and low frame rates in order to collect a satisfactory number of photons per pixel. Two- or multiphoton excitation of vesicular fluorescence reduces the drawback of photodamage and restricts the fluorescence excitation to a small focal area, but this then demands scanning and point-by-point signal integration through the sample to obtain three-dimensional resolution, thus preventing dynamic imaging of a single exocytotic event [14]. Conversely, the performance of imaging vesicles undergoing exocytosis in real time is highly improved by total internal reflection fluorescence (TIRF) microscopy or evanescent wave/field microscopy [15, 16]. In this optical configuration, the excitation beam enters the coverslip with an incidence angle whose value is above a critical angle determined by the refractive indexes of the sample and the coverslip. The beam is then reflected but in the process generates an evanescent field that penetrates beyond the glass/cell interface, that is, over a few hundred nanometers inside the cell. The incident energy penetrating through the sample in the evanescent field range may excite fluorophores present within this range. More precisely, since the intensity of the field decays exponentionally orthogonally to the interface, the effective penetration depth depends on the refractive properties of the interface and the wavelength of the excitation light, thus ranging between 50 and 300 nm. Thanks to a minimization of the out-of-focus intracellular fluorescence, images of vesicles near the plasma membrane are acquired with a very good signal-to-noise (S/N) ratio [17]. Because of low background noise and high contrast, the vertical resolution in the cell may be as small as a few nanometers allowing the measuring of small vesicle displacements [18]. Various fluorescent probes are compatible with the method—chemical probes like acridine orange, fluorescent styryl compounds known as FM dyes, genetically encoded proteins introduced by gene transfer techniques, and pH-sensitive fluorescent proteins (green fluorescent protein [GFP] pHluorin, bright at pH 7.4 but not at acidic pH in vesicles, that is, around 5.5)—and their specific use depends generally on the type of cell and goal of the measurement.

Optical techniques like TIRF microscopy report in real time the vesicles status before and during fusion (displacement, docking) but not on the released fluxes themselves. Information, though rich, is visual and semiquantitative at most. Besides, constitutively fluorescent or beforehand-marked vesicles are required, while individual resolution of small vesicles in a population remains hampered by diffraction laws. Finally, the time scale of typical exocytotic release is in the range of a few milliseconds, which makes it below the limit of the present temporal resolution of TIRF imaging systems (of tens of milliseconds). Furthermore, though reduced photobleaching and cell damage are of high concern for long experiments, all these facts limit the gathering of statistically significant series of events from each single cell

1.1.2.2 Electrophysiological techniques

Electrophysiological techniques rely on impedance electric measurements of the cell membrane surface area with particular focus on tracking its capacitance variations [19]. Vesicle exocytosis and endocytosis induce changes in the cell membrane area and thus in its capacitance. A patch-clamp measurement attempts to extract these capacitance changes on the basis of the effect of the three-component equivalent electrical circuit. Four configurations (whole-cell, cell-attached, inside-out, and outside-out configurations) are generally used to gather different information about cell secretion. For instance, in whole-cell recordings, a glass pipette (diameter 1 μm) is gently sealed to the cell membrane after aspirating by sucking a portion of the cell membrane it approaches. This gives access to the global secretion activity of the patched cell (Fig. 1.2b). The equivalent circuit is assumed to involve the whole cell membrane resistance and capacitance as well as resistance through the patch-clamp pipette. On the basis of such an RC-series model, application of a sinusoidal voltage to the glass micropipette and recording of the phase and intensity of the resulting sinusoidal measured current give rise to instant capacitance values. Patch-clamp techniques allow single-cell analyses with a temporal resolution of around a millisecond, and because they are based on electrical measurements they provide an instant description without filtering by molecular motions. Hence, they provide access to conductance and expansion kinetics of the fusion pore within the framework of the rudimentary RC-series model adopted. The weaknesses of patch-clamp techniques are as follows:

(1) There are difficulties in discriminating signals from exocytosis and endocytosis (underestimation of the amount of secretion).
(2) Cell membrane capacitance increases could arise from other membrane insertion rather than neurotransmitter vesicles fusion.
(3) Capacitance changes may find their origin in events other than changes in membrane cell surfaces.
(4) The techniques do not provide individual kinetics information on single-vesicle exocytosis.
(5) Capacitance measurements are limited to cell types that are assumed to be correctly modeled by the most simple RC-series equivalent circuit in which the cell resistance is predictable due to the fact that the impedance is analyzed at a single-frequency circuit based on spherical-cell models [13, 20, 21].

1.1.2.3 Electrochemistry at ultramicroelectrodes: the semiartificial synapse configuration

The initial challenge of unraveling the dynamics of release by vesicular exocytosis with the same temporal and spatial accuracy had been taken up by mimicking the analytical performances of the synaptic cleft that has a volume around a zeptoliter in which no more than a thousand molecules are emitted within a few milliseconds, though leading to optimal detection. This optimal configuration was transposed in the beginning of the nineties by using a UME to mimic the postsynaptic neuron, and a single living cell in culture in a petri dish plays the part of the emitting neuron [22, 23]. Effectively, faradic signals are proportional to the concentration of the detected species and not to the quantities of them, contrary to spectroscopic techniques. The semiartificial synapse built on the UME/cell assembly is achieved by positioning, thanks to micromanipulators, a UME (of micrometric size comparable to the cell size) above the cell on the stage of an inverted microscope. By doing so, minimization of the distance between cell and the UME (~100 nm) allows restricting the volume where the electroactive species are emitted, resulting in transient concentrations of around 1 mM. Nevertheless, this technique requires that the detected species be electroactive so that they may be oxidized or reduced at the electrode surface faster than their emission by the cell. In that case, their release leads to a faradic signal, that is, a current of oxidation or reduction at the UME surface that is proportional to the released flux of the detected species. Numerous cell models may fit this constraint by releasing dopamine, epinephrine, norepinephrine, histamine, serotonin, or some neuropeptides [24]. The fact that the UME has the same dimensions as the cell optimizes the collection efficiency, while preventing the detection of nondesired signals like capacitive currents that are a source of noise, and depends on the electrode surface area.

Each exocytotic event representing one fusing vesicle thus appears as an amperometric spike, whose time course reflects the kinetics of release from an individual granule. Actually, the time of flight of the species in the semiartificial synaptic cleft of ca. 100 nm thickness and the temporal response of the UME avoid any diffusional filtering of the detected current and permit a quantitative detection with a precision of ca. 1000 molecules per millisecond (Fig. 1.3) [23]. In brief, electrochemical methods offer real-time monitoring of single exocytotic events and afford very fine quantitative information about kinetics and the number of released

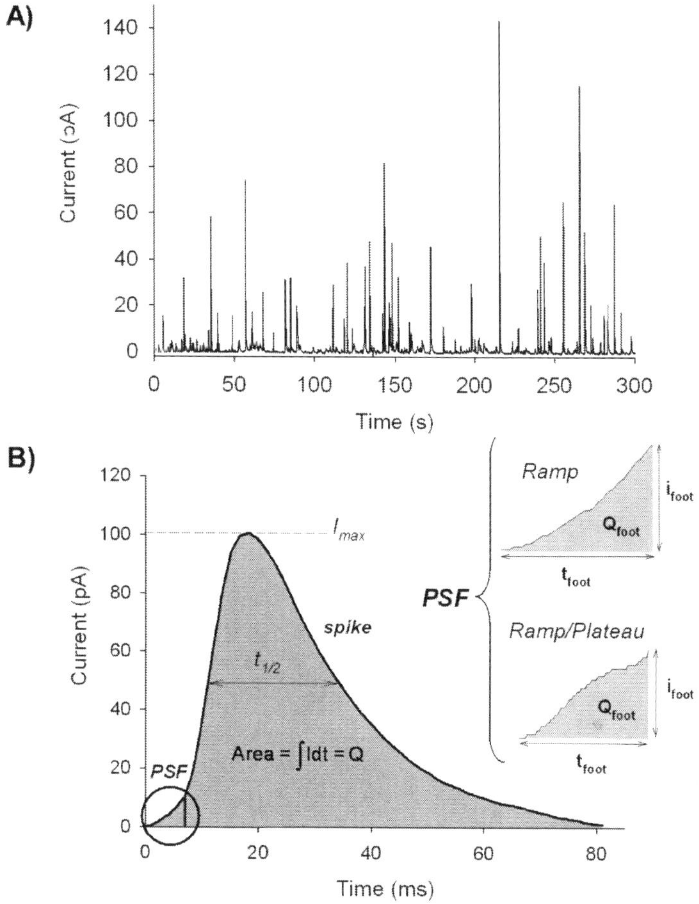

Figure 1.3 (a) Representative amperometric trace obtained during exocytosis of a single chromaffin cell. The potential of the carbon fiber microelectrode is set at 650 mV vs. Ag/AgCl, thus allowing oxidation of all the catecholamines released during an exocytotic event. (b) An amperometric spike selected from the amperogram and the quantitative/kinetic parameters that can be extracted. The considered spike is represented with a prespike feature (PSF). The different shapes of a PSF ("ramp" or "ramp/plateau") are thus depicted, and the usual quantitative foot parameters (I_{foot}, t_{foot}, and Q_{foot}) are given for both PSFs.

molecules. However, its advantage is also its main drawback, since it cannot offer any information relative to the vesicle geometry prior or during fusion pore formation or about vesicle motion, and electrochemical methods require electroactivity of emitted species.

1.2 Electrochemical Detection of Exocytosis: from Basic Carbon Fiber Ultramicroelectrodes to Combinatory Systems

1.2.1 *Properties of Ultramicroelectrodes for Biological Measurements*

Although no official definition prevails, a UME can be defined as an electrode whose diffusion layer can reach a comparable or even larger size than the dimension of the electrode at the time scale of an electrochemical experiment [25, 26]. However, the intrinsic properties of diffusion layers and of the convection-free domain directly set an upper boundary for UME dimensions that respect the definition introduced above. As a consequence, a UME is an electrode whose at least one dimension (radius for a disk, width for a band, etc.) is below a couple of tens of micrometers. From a practical point view, in classical electrochemical solvents and water, the UME critical dimension is considered smaller than 25 μm. This makes a UME *perfectly suited for biological investigations at single cells* that have similar dimensions.

A UME presents specific analytical properties. To describe hereafter the main UME properties, we will consider a solution containing a species named "R" (bulk concentration C_{R^*}.) and possibly oxidized at a UME surface ($R = O + ne^-$). The electrode surface area is assumed to be small by comparison with the volume of solution so that the passage of current will not alter the bulk concentrations of the electroactive species.

For conventional millimetric electrodes, due to the micrometric thickness of the diffusion layer, diffusion essentially proceeds along a direction orthogonal to the electrode surface (1D, see Fig. 1.4). If the working electrode (disk of radius r_0) potential is maintained constant (vs. a reference electrode) at the appropriate value to

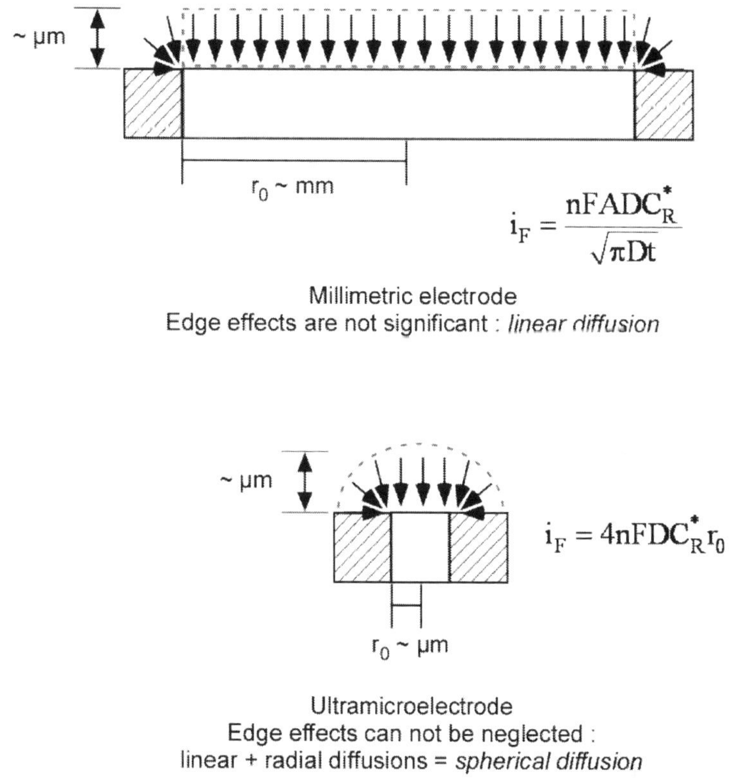

Figure 1.4 Schemes of diffusion layer shapes for conventional millimetric electrodes (top) and UMEs (bottom). The corresponding faradic current equation is added for each case (disk electrode of radius r_0). The arrows display the electroactive species' diffusive flux at the electrode surface.

oxidize R (amperometry at constant potential), the ensuing current follows Cottrell's law, while the diffusion layer remains contained within the convection-free domain. Then, the diffusion layer—of dimension $\sim(Dt)^{1/2}$, where D is the diffusion coefficient of the electroactive species—increases, thus leading to the time faradic current i_F decrease ($\propto t^{-1/2}$) with time and is proportional to the electrode surface, that is, $r_0{}^2$. Conversely, edge effects cannot be neglected for UMEs. The radial diffusion must be taken into account, and after a transient period, the 1D diffusion field is rapidly ($t > r_0{}^2/D \approx$ ms) replaced by an essentially radial one (2D). At disk or spherical UMEs this convergent regime leads to a steady state, that is, to a constant diffusion layer (a few r_0) and a constant faradic current recorded at the electrode, the latter only depending on r_0 (Fig. 1.4). Therefore, *UMEs can detect concentration changes occurring within a volume whose radius is at most a few* r_0. *They also respond within a millisecond time scale to any change in this microscopic domain* without signal distortion *and are thus perfectly able to resolve fast kinetics of cell secretion.*

The first electroanalytical difficulty is related to the control of the electrode potential. Briefly, the real working electrode potential could be smaller than the applied value due to an ohmic drop, *iR*, resulting in the current passing through the solution resistance R. Such an ohmic drop, partially compensated through the use of a counterelectrode, possibly alters the potential value of the working electrode versus that of the reference electrode. Under the linear diffusion transient regime, the ohmic drop directly scales with the electrode radius and hence decreases when the electrode size is reduced. For UMEs, once diffusion has switched to the spherical diffusion regime, the ohmic drop value reaches its minimum, which is generally negligible in electrolytes compatible with biological experiments, so *the need for a counterelectrode is removed* and a two-electrode configuration may be used as soon as the reference electrode has a millimetric size at least.

From an electroanalytical point of view, electrochemical measurements can also be distorted by capacitive phenomena providing from the double layer formed at the electrode/solution interface. The charging of this electrical capacitance C (proportional to the electrode area $\propto r_0^2$ induces the capacitive current i_C, whose exponential decrease is governed by a time constant $\tau = RC$ (R is the resistance of the solution). At disk electrodes, $\tau \propto r_0$, irrespective of the value of the electrode radius, and hence decreases with r_0. *UMEs thus allow distortion-free investigations of short time scales* by comparison with conventional millimetric electrodes. Finally, and most importantly for our purpose here, it must be noted that any analytical measurement is dependent on the S/N ratio. In electrochemistry, intrinsic noise arises from the electrochemical capacitance and is thus proportional to r_0^2. Under the spherical diffusion regime, the signal, that is, the faradic current, is proportional to the electrode radius. It follows that $S/N \propto 1/r_0$; therefore, everything else being the same, *the S/N ratio increases upon decreasing the electrode size.*

1.2.2 *Typical Single-Cell Models*

Electrochemical monitoring of exocytosis by single living cells requires that they are maintained under aerobic conditions. This prevents the exploring of cathodic ranges where reduction of dioxygen (0.23 mmol L^{-1}) dominates. Hence, the technique is limited (at least in its routine version) to cells secreting neurotransmitters that can be oxidized at the electrode surface. Yet, this is not a severe limitation for electroactive neurotransmitters that are most released in their reduced redox form. For instance, catecholamines (dopamine, noradrenaline, and adrenaline) and serotonin (5-hydroxytryptamine) are nowadays the most popular electrochemical probes and can be easily oxidized (E = 650 mV vs. Ag/AgCl) at the carbon surface (Scheme 1.1). Conversely, nonelectroactive species like glutamate or acetylcholine are not directly detectable by amperometry.

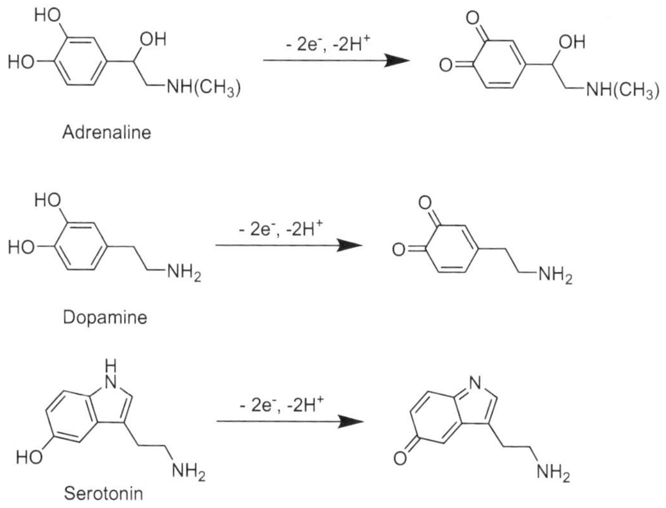

Scheme 1.1 Catecholamines and serotonin can be easily oxidized at the carbon surface.

Nonsynaptic cell models are preferentially investigated by neurobiologists since they possess rather large-sized vesicles (0.25–1 μm in diameter) packed at an extremely high neurotransmitter concentration (until 0.6 mol L^{-1}) that allows enhancing of the detection sensitivity (on average ca. 500,000 molecules are released per vesicle during ~10–100 ms). The cell models investigated are numerous and well documented in recent reviews [2, 3]. They mainly include adrenal (bovine, calf, rat, etc.) chromaffin cells (that release a mixture of catecholamines: adrenaline, noradrenaline, dopamine), pheochromocytoma (PC12) cells (that release dopamine and are derived from cancerous rat adrenal glands), and mast cells from normal or beige mice (corelease of 5-hydroxytryptamine [serotonin] and histamine stored in large vesicles). Furthermore, some other models that do not release endogenous electroactive species can be amperometrically investigated if their vesicles are artificially preloaded with serotonin or dopamine. One can mention RBL-2H3 mucosal mast cells (serotonin loaded), human or mouse pancreatic β cells (insulin release, preloading procedures with serotonin), and dog pancreatic duct epithelial cells (dopamine loaded). In certain cases, the nonelectroactive species released during exocytosis can still be detected through modifications of the electrode surface. This has been done for monitoring pancreatic β cells (insulin release) with the carbon fiber microelectrodes modified with a polynuclear ruthenium oxide/cyanoruthenate film, which allows fast catalysis of insulin oxidation. As such, insulin can be electrochemically monitored, though being nonelectroactive at the carbon surface [27, 28].

Works dealing with neuronal systems have been reported on invertebrate systems, although not having been performed in the real synapse configuration (i.e., electrochemical detection on the cell body, no intrusion of the electrochemical probe into the synaptic cleft). One can mention investigations dealing with the leech *Hirudo medicinalis* (serotonin), the pond snail *Planorbis corneus* (dopamine), mammalian neuronal systems, such as superior cervical ganglion neurons from neonatal rats (catecholamines) and cultured ventral midbrain neurons (dopamine) from the rat, or dissociated neurons of the sea pansy *Renilla koellikeri* (dopamine, noradrenaline).

Although less spatially controlled than studies at the single-cell level, the monitoring of neurotransmitters through electrochemical measurements can also be performed in brain slices or even in the rat living brain [29, 30].

1.2.3 *1990–2000: Carbon Fibers' Years for Apex Detection*

1.2.3.1 Types of ultramicroelectrodes

The most used UMEs to perform electrochemical measurements targeted at monitoring exocytosis on a single living cell are obtained from a carbon fiber with diameters ranging from $\sim$1 μm to 10 μm. The fiber is sealed into a glass capillary and before or after insulated through different existing procedures [31]. Such sealing reduces capacitive currents (hence noise) so that the faradic information comes from the tip of the UME with a controlled and knowledgeable active surface area. Insulation of the outer cylindrical surface may be performed by electrodeposition of a polymer (e.g., polyoxyphenylene, polypropylene, and polyethylene) on the carbon fiber shaft [32–34]. Insulating the UME shaft may also be realized by means of cathodic electrophoretic painting. The fibers are then usually cut to the sought dimension and polished to obtain a flat, reproducible, and controlled conductive area at the tip of the UME. The UME electroactive surface may also be modified, for example, by deposition of a thin film of Nafion (perfluorinated cation-exchange membrane), which has been shown to enhance selectivity toward cations like catecholamines that are released in their protonated form in chromaffin cells and can easily cross over the Nafion film. Though, the resulting drawback of this procedure is a loss of temporal resolution due to the diffusion–migration inside the film [31]. Lately, modified platinum electrodes involving a thin layer of conducting diamond have been shown to be useful for monitoring secretion of serotonin by enterochromaffin cells isolated from the mucosal layer of the intestine. They are obtained by pulling a Pt wire originally with a 76 μm diameter up to a sharp point followed by deposition of the doped diamond layer [35]. However, most of the studies on secretion involve simple carbon fiber UMEs, since most electroactive species released by exocytosis are oxidizable at bare carbon surfaces that are very stable, robust, low in cost, and easy to handle, giving rise to reasonably low capacitive background currents.

1.2.3.2 Voltammetry vs. amperometry: advantages and drawbacks

Electrochemical measurements performed in oxygen-containing buffers to mimic as closely as possible physiological conditions prohibit working in reduction to avoid any artefacts due to O_2 present at a much larger concentration ($\sim$0.24 mmol L^{-1}) than the detected species. Using the semiartificial synapse configuration, recorded currents directly report about the nature, quantities, and kinetics of the release of emitted molecules. Two main electroanalytical techniques are frequently adopted for real-time monitoring of exocytotic release, amperometry and fast-scan cyclic voltammetry. Amperometry consists of recording the current as a function of time at a fixed potential. Amperometry provides priceless time resolution (submillisecond event can be observed) that offers dynamic information about exocytotic events, as well as precise quantification of release. This double essential feature explains its large popularity. However, it does not allow discrimination between different species oxidizable at the selected working electrode potential. In fast-scan cyclic voltammetry, the current is measured at time intervals as a function of the triangular pulsed potential waveform applied at the UME. Resting time periods are necessary in between scans, so the time resolution is poorer than in amperometry. Moreover capacitive currents are significant and blur the faradic information unless they are sufficiently stable to be subtracted [36, 37]. For example, successive cyclic voltammograms performed on chromaffin cells at a reasonable scan rate (10 V/s) during exocytosis showed that at least three classes of vesicles may be involved in adrenal glands: one that releases adrenaline; the second one, noradrenaline; and the last one, a mixture of both [38, 39].

Finally, cell releases are commonly elicited by secretagogues delivered with a glass micropipette (diameter of a few microns) positioned close (at about tens of microns) of the emitting cell.

Stimulating agents (nicotine, digitonine coupled with Ca^{2+}, Ba^{2+}, a mixture of K^{+} and Ca^{2+}, etc.) are then ejected via pressure through the glass microcapillary. The ensuing release is signaled by a series of current spikes named "amperometric trace" (Fig. 1.3a) on which each spike (Fig. 1.3b) corresponds to one exocytotic event [22].

1.2.3.3 Applications of amperometry to the investigation of biological or/and physicochemical parameters

The use of UMEs in electrochemical investigations of real-time monitoring of exocytosis from single cells has been intensive in the nineties. Owing to the wide variety of amperometric investigations, this section cannot be exhaustive but will illustrate some representative examples of such works. Amperometric traces obtained from a single cell display a sequence of amperometric spikes from which four main parameters can be extracted. The frequency of release or the number of detected events per cell informs about the cell status. Each individual spike electrical charge $(Q)(fC$ or $pC)$ represents the quantity of species detected per vesicle by virtue of Farada's law ($Q = nNF$, where n is the number of exchanged electrons in the oxidation process, F is Farada's constant, and N is the species quantity in moles). Kinetic parameters are also obtained from each spike (Fig. 1.3b): the maximum spike oxidation current I_{max} (pA) linked to the maximal flux of neurotransmitters released during the event and time parameters such as the half-width time $t_{1/2}$ (ms) measured at 50% of I_m, the rising time (generally between 20% and 90% of I_{max}), or the decreasing time (measured from the maximum and the end of the amperometric peak) that characterize different phases of a releasing event [40, 41]. Besides, fine amperometric kinetic resolution (around a millisecond) provides information about the initial fusion pore that is identified as a small current prespike feature (PSF) apparent in ca. 30% of cases for chromaffin cells, though this percentage depends on the experimental conditions and the analysis [42, 43]. This pore signature is also called an "amperometric foot." Its duration, t_{foot}, and maximum current, I_{foot}, provide an estimation of the dynamics and size of the fusion pore, while its charge, Q_{foot}, gives the number of neurotransmitters diffusing through the pore structure (Fig. 1.3b) [33, 44].

1.2.3.3.1 Detection of the initial fusion pore

Since all vesicular release events should be initiated by the formation of a nanometric fusion pore, the amperometric foot should always be observed. Their absence in 70% of the events is probably due to the fact that their duration is too fast compared to amperometric resolution. When observable, the fusion pore may be described through four parameters (percentage, t_{foot}, Q_{foo}, and I_{foot}) that allow precious information on its dynamic and stability. Though this contains precious information, PSFs are seldom analyzed, may be because their precise measurement requires very high resolution and an excellent S/N ratio.

When observable, the amount of catecholamines released during a PSF (Q_{foot}) and the probability of foot observation do not vary when modifying the cell membrane with exogenous lipids, whereas the kinetics of the fusion pore are greatly affected [40, 45–48]. The same trend is observed when altering the cell membrane tension and viscosity, that is, by submitting the cell to brief hypotonic (200 mOsm) or hypertonic (750 mOsm) shocks (note that normal control conditions correspond to 315 mOsm) [47, 49, 50] or to reduced temperature (15°C and 22°C compared to 37°C) [46]. The foot is thus representative of a catecholamine flux whose observation may be due to a specific secretion granule composition and not due to the stability of the initial fusion pore that otherwise seems to be a fleeting feature. All these considerations show that the physical properties of the cell membrane play a dynamic role in foot features but have no role in the probability of its statistical observation. Alternatively, the modification of the vesicular internal composition (by L-DOPA and reserpine treatments) appears to have strong effects on the dynamics of the initial fusion pore and on the percentage of exocytotic events providing an amperometric foot at least in chromaffin cells [51, 52]. All this data has raised the question of whether the amperometric foot represents a special state in which the vesicular matrix is predissociate [42, 52, 53] or features special vesicles equipped with a liquid halo between the vesicle membrane and the vesicular matrix, as evidenced in PC12 cells [54, 55]. In fact, when observed by electron microscopy some vesicles show an electron-lucent halo between the vesicle membrane and the dense-core matrix [56], which was suggested to be the signature of a medium in which the diffusion coefficient of neurotransmitters is higher than in the compacted matrix. From such a perspective, the foot frequency and kinetics should not be linked to dynamics of the fusion pore but rather to the presence of halo-equipped vesicles. Alternatively, the stability of the fusion pore may be an intrinsic property of the fusing membrane vesicle, and docking site arrangement, independent of the vesicle's inner structure [46]. At this stage, the debate remains open.

1.2.3.3.2 Effects of biological parameters: SNARE proteins

The SNARE protein complex formed by vesicle membrane proteins (v-SNAREs) linked to cell membrane ones (t-SNAREs) plays a crucial role in allowing the last step of the vesicle approach near the cell membrane to occur by progressively tightening their association and thus counteracting the electrostatic repulsion between phospholipidic head groups [57, 58]. This is directly observable by amperometry: for example, when the SNARE amino acid sequence is modified or cleaved by special neurotoxins, the frequency of amperometric spikes falls dramatically [59–63]. SNARE assembling might also play a role in modifying the local membrane curvature or/and its viscosity. For example, some target proteins implied in the regulation of the SNARE complex formation have been reported to alter the normal dynamics of fusion, presumably by modifying the speed at which the cell and vesicle membranes merge during fusion, and consequently affect the kinetics of the exocytotic event itself through energetic modifications of viscous dissipation of energy released by the fusing membranes. This happens depending on which stage the aforementioned regulatory-specific proteins act, like α-SNAP, protein kinase C, or munc 18 [64–68].

1.2.3.3.3 Multiple vesicular populations

Statistical analyses of amperometric charges in different cell types suggest that at least two, maybe three, vesicular populations (i.e., size or diameter) coexist [69]. Interestingly, the released charge, Q, per amperometric spike is proportional to the third power of the vesicle radius, assuming that the intravesicular concentration of neurotransmitters is a constant for a given cell model [22]. Distributions of the vesicular radius, obtained by optical microscopy and by amperometry (as from $Q^{1/3}$) in chromaffin cells, are coherent in suggesting that this cell type offers at least two types of vesicle populations with different properties of secretion [70]. Playing on extracellular environment properties like osmolarity, for example, by bathing the chromaffin cells in hypotonic conditions, results in increasing the proportion of large- versus small-diameter-population vesicles compared to normal isotonic conditions [49]. Convincing and identical figures, namely, the presence of different vesicle populations acting in concurrence in exocytosis, were obtained on PC12 cells [71], Retzius cells of leec [72–74], and neurons of *Planorbis corneus*, whose body contains dopamine [75]. Finally, the number of populations is perhaps three, as described by Tang et al., showing that three Gaussian are required to adequately fit the radius distribution, the relative proportion of each population as well as the distribution of Q being dependent on the delay of cell culture before amperometric recordings [76].

1.2.3.3.4 pH effects on the exocytotic process

The dependence of exocytotic release on pH was investigated along two different axes, as a function of modified extracellular pH and through modifications of intravesicular vesicle pH.

In the first case, the results obtained in different cell types were difficult to rationalize and, most of all, to generalize because of intrinsic variations in the vesicle' granule composition. For example, in chromaffin cells, catecholamines are positively charged and the rate of release should respond to the intravesicle–extracellular pH gradient (7.4 against 5.5 in the vesicle). The external pH may also affect the swelling of the vesicular matrix since the catecholamine cations' departure must be compensated. Accordingly, at different extracellular values (5.5, 7.4, and 8.2), changes were observed by amperometry bearing on spike size and shape and on exocytosis frequency. These results were qualitatively consistent with the reported interactions between neurotransmitters and other matrix components, even if it is noted that the results observed depended on the secretagogue [53, 77]. Other amperometric measurements performed on pancreatic β cells releasing insulin and 5-HT showed that external acid pH plays a role only in insulin release, which then does not occur at pH 6.4 [78]. Increasing the extracellular pH by steps from 6.9 up to 7.9 induced faster insulin release but without modifying the average released quantities and with no effect on 5-HT release [79]. On PC12 cells, decreasing the external pH down to 6.8 was reported to elicit exocytosis by probable intracellular acidification [80]. However, despite interesting effects being observed for these cells types, any definitive interpretation is hampered by the fact that the biological ways through which a cell responds to changes in the extracellular pH remain unknown.

In the second case, in chromaffin cells, decreasing the intravesicular pH enhances the uptake of secretory granules [56], and doing so by using drugs blocking vesicle proton pumps induces a deceleration in the dynamics and a fall in the number of neurotransmitters released [81].

Another interesting problem was described recently about the effect of pH in semiartificial synapse configuration used in conventional UME amperometry [82]. Indeed the electrochemical oxidation of catechols in chromaffin cells produces quinone derivatives as well as protons. As a consequence, the local pH between the cell membrane and the electrode should drop within the UM–cell cleft volume. This point is generally ignored, but simulations of this pH drop have shown that depending on fusion frequencies and release quantities the cleft pH may drop as low as inside the granules after the detection of a few events. Furthermore, the corresponding acidification was shown to depend on the microelectrode radius, since this controls the rate of buffering by diffusion from the extracleft medium. However, no change in kinetics are observed as a function of the UME radius, suggesting that cells may be equipped for regulating pH in the nanometric vicinity of their membranes.

1.2.3.3.6 Role of cell membrane curvature

Exocytosis was studied amperometrically at adrenal chromaffin cells to evaluate the effect due to transinsertion of exogenous lipids (lysophosphatidylcholine [LPC] or arachidonic acid [AA]) on the kinetics of exocytotic events. Results showed that for short exposition times and micromolar concentrations of exogenous lipid solutions, LPC favored catecholamines release in terms of frequency and charge released, whereas AA disfavored frequency without altering quantities. Such kinetic trends were rationalized quantitatively by considering the physical constraints applied to the lipidic fusion pore by the presence of small fractions of LPC and AA diluted in its external leaflet (transinsertion). Thus alterations of the membrane curvature through changes of the lipid cone angle seem to play a role in the exocytosis course [45]. As a consequence, such experiments on PC12 cells were carried on with other phospholipids but using higher concentrations and longer incubation times. These experiments also led to the conclusion that differences in membrane composition significantly alter the fusion process [83].

1.2.3.3.7 Scanning electrochemical microscopy for studying exocytosis

Scanning electrochemical microscopy (SECM) is a scanning probe technique allowing imaging of single living cells topographically but also reporting cell activities through electrochemical measurements. It consists of bringing a UME tip in the close vicinity of the studied surface (cell in our application), thanks to piezo-positioner/computer-controlled stepper motors with fine precision (around 10 nm per step). The UME amperometric current is recorded as a function of (i) the UME–surface distance (z) in the case of the approach curve, (ii) the (x and y) position for topographic imaging (iii), or time at a fixed (x, y, z) position in chronoamperometry (already described above in classical semi-artificial synapse configuration). The use of the first mode is irrelevant for monitoring exocytosis since the SECM tip movement is necessarily slow compared to the duration of one secretion event; furthermore, the optimal tip–cell distance can be readily controlled by micromanipulators with submicrometric resolution (sufficient for amperometric analysis). The second mode would be of interest to map the position of favored release sites on a cell membrane, but again this is still not feasible because of the lack of temporal resolution inherent to the technique (the UME speed can reach a few micrometers per second when the exocytotic event lasts tens of milliseconds). However, the method is improving and may meet the required characteristics in the future. For example, Schuhmann et al. designed a new bio-SECM in which shear force–based distance control allows us to properly and accurately position the tip of the UME with nanometric resolution above the outer membrane of PC12 or chromaffin cells [84, 85]. Amperometric detection could then be performed after subsequent application of high K^+ solutions using carbon fiber UMEs (one unetched to 8 μm and one conically etched down to 2 μm) poised at 700 mV versus Ag/AgCl.

1.2.4 *Beginning of the 2000s: Development of Coupled Analytical Approaches*

1.2.4.1 Limitations of basic single UMEs with apex detection

Amperometry is nowadays a mature and routine method used in numerous laboratories over the world for investigating secretion phenomena of electroactive species from isolated cells. This is based on its performance, ease of application, and, most of all, the face that it gives access to statistically significant quantitative analytical information. However, despite these crucial advantages this technique has some limitations. First, as already mentioned above, it requires that the released species be electroactive to get a direct faradic signal corresponding to oxidation or reduction of the biomolecules. This prevents detection of numerous species implied in neurotransmission, like glutamate and acetylcholine [86, 87]. This may be compensated by modifying the UME electroactive surface by enzymes or modified layers, thus offering an indirect detection of secretion, but in return, the time resolution is highly degraded. Moreover, employing the above-described configuration of the semiartificial synapse generates, obviously, detection at the apical pole of the cell; thus traditional UME electrochemistry is sightless to exocytotic events happening on side or basal cell poles. A third disadvantage is that conventional UMEs are blind to any mechanism happening inside the cell or just beneath the cell membrane. Thus, the vesicles' traffic and 3D motion, last steps of vesicle/cell membrane approaches, cannot be investigated because amperometry may detect and analyze exocytotic events only after the initial fusion pore between vesicles/cells has been formed. Conversely, optical techniques like TIRF microscopy, though not quantitative, are informative about such prefusion mechanisms. Hence, ideally, coupling such optical methods with amperometry should afford the comprehensive total life of a fusing vesicle from its trail in the cytoplasm up to fusion. Although both methodologies operate on single cells, experiments have to be repeated on many cells to diminish the biases related to cellular variability (corresponding to different cell physiologies). In that sense, the combinatorial possibilities offered by lab-on-a-chip microsystems provide an emerging solution to this limitation. This short analysis of advantages and drawbacks evidences the need for coupled methods and improvements. Namely, the following objectives are highly desirable: (i) nanometer-sized electrochemical networks for the simultaneous monitoring of a cell surface at the nanoscale level, (ii) harmless penetration inside a living cell to investigate directly at the source the central contribution of chemical messengers in communication processes, (iii) integrated microsystems including microfluidic cell manipulations and electrochemical detection to overcome cellular variability and the ensuing time-consuming experiments imposed by single-cell studies, and (iv) coupling of amperometry, electrophysiological, optical, and microfluidic techniques usually employed to individually monitor exocytosis in order to gather complementary inputs. Not surprisingly, all these routes are currently explored.

1.2.4.2 Coupling amperometry with other techniques

1.2.4.2.1 Combining patch clamp and amperometr

Combinations of amperometric and patch-clamp techniques have been intensively used since the beginning of the nineties (Fig. 1.5) [33, 88, 89]. In the seminal works, electrophysiological measurements were achieved in the whole-cell configuration and the carbon fiber UME was present outside the patch-glass micropipette in the usual semiartificial synapse configuration on chromaffin and beige mast cells. Such simultaneous recordings allow verifying

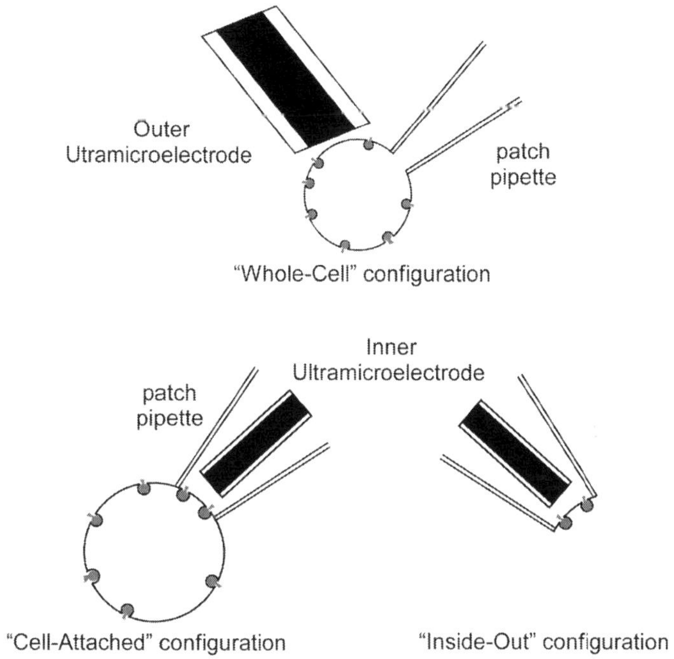

Figure 1.5 Three experimental configurations for studying exocytosis with patch amperometry.

that the PSF preceding amperometric spike (in around 30% of cases) effectively corresponded to the release of biomolecules through the initial narrow fusion pore. Numerous studies followed showing that monitoring amperometric spikes and cell membrane capacitance changes simultaneously increased our understanding of the release processes: its occurrence, dynamics, fusion pore formation, and evolution, estimation of the exocytosis/endocytosis relative shares, evidence of different exocytotic modes, etc. [90–95]. A significant advancement was made when the coupling involved inside-out or cell-attached electrophysiological modes (Fig. 1.5) [89, 96, 97]. Indeed in the whole cell configuration, the capacitance measurements reflect the whole cell surface variations, whereas amperometry is confined spatially to area in contact with the UME that covers an upper fraction of cell surface. This limitation is overcome when the UME is inserted inside the patch pipette

(Fig. 1.5). These two above-cited configurations thus brought very refined capacitance measurements resolution, consecutively enabling demonstration of "kiss and run" events presence [98], evidence of links between modifications in vesicle membrane area with alterations in their material content [99] and improvement in understanding of the initial fusion pore role [100]. The only major drawback resides in distortion of recorded amperometric spikes because of diffusional broadening while neurotransmitters cover the distance between the patched cell membrane and the UME tip, which is hard to control and minimize accurately [89, 96, 97].

1.2.4.2.2 Coupling amperometry and fluorescence techniques

Coupled amperometry and intracellular calcium concentration recordings by standard fluorimetry have been used over the past 20 years to establish the causality between calcium entry or increase in cytoplasm compartment with exocytotic onset [77, 101–106]. Fluorescence calcium imaging and amperometry were also performed, for example, in chromaffin cells, by using etched electrodes (1 μm at the tip) to demonstrate the colocalization between calcium "hotspots" (fluorescence) with zones of catecholamine release (amperometry) on cell membranes [107]. A dual microsensor was also built to detect at the same time calcium (with calcium green-1 dextran as the fluorescent dye linked to the sensor tip by interaction with glutaraldehyde) and catecholamines (by standard carbon fiber UMEs) from chromaffin cells [108]. On the other hand, confocal microscopy and UME amperometry have been associated for demonstrating zinc and insulin corelease in pancreatic β cells [109]. However, simultaneous recording of exocytosis events by amperometry and TIRF microscopy appeared only recently. Indeed, direct and simple combination of these two highly powerful and complementary techniques is precluded a priori when thinking in terms of carbon fiber electrodes since TIRF microscopy and amperometry since then each detection must occur on a different cell side (bottom and apex, respectively). Conversely, as evidenced in the late 2000s, this can be solved through elaboration of adequate microsystems (see Section 1.2.4.4 below).

1.2.4.3 Decreasing the electrode size

From an electroanalytical point of view, decreasing the electrode size should improve its analytical properties and notability increases the S/N ratio, reducing the noise level and the time constant. However, the present UMEs already meet these requirements. Therefore, decreasing the electrode area is mostly interesting to enhance the spatial resolution of the detection and thus identify active exocytotic zones on the cell membrane (see above for the intrinsic limitation). More interestingly, the simultaneous recording of cell secretion with two etched electrodes revealed that different release sites coexist on endocrine cells, the correlation between these active zones and a high local intracellular calcium concentration (hotspots) being also clearly evidenced [107]. Using etched electrodes ($\sim$1 μm radius) evidenced the existence of hotspots on neurite-emitting chromaffin cells [110] or pancreatic β cells [109, 111]. Reducing the electrode size also offers an additional advantage for data treatment. Hence, because the size-reduced sensor does not collect cell release from a large surface area, the probability of overlapping spikes decreases with the electrode size. This has been evidenced at chromaffin cells by showing that the frequency of detected events drastically dropped when the electrode area diminished from 90 μm^2 to 25 μm^2 [112]. Nevertheless, comparisons between data recorded with usual 7 μm diameter carbon fiber or etched electrodes evidenced that the decrease in the potential spikes' superimposition did not alter the extracted information (PSFs, spikes' intensity and duration, etc.), establishing that even at 7 μm UMEs the probability of the spikes' superimposition cannot be considered, provided a sufficient number of cells is analyzed [112].

This raises the question of the potential benefit of nanoelectrodes. For example, amperometric detection of exocytosis at PC12 cells was reported using a 100 nm diameter carbon fiber nanoelectrode whose size corresponds to the vesicle diameter range ($\sim$200 nm). Ideally, this should allow detection at the single vesicle level. However, as expected with such low dimensions, exocytosis is scarcely detected (no amperometric events in 70% of the recordings) with only a few spikes per cell [113], thus decreasing the statistical significance of the data. A more comprehensive study

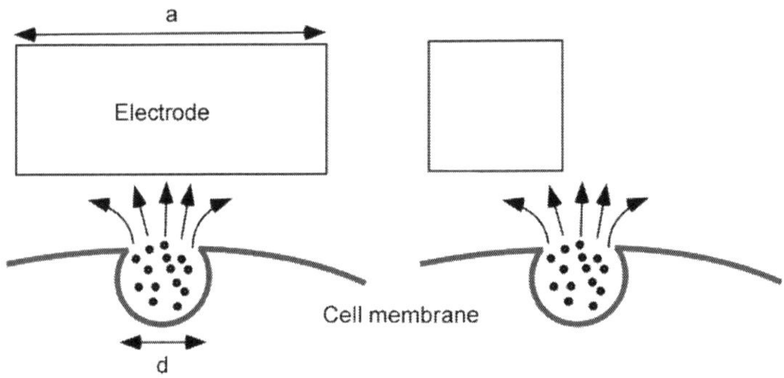

Figure 1.6 Representation of the electrode-size effect for electrochemical detection of exocytosis. A vesicle (diameter d) releases its content beneath the electrode (diameter a). Left: When $a >> d$, the intravesicular release is ensured to reach the electrode active surface. Right: When $a \approx d$, the probability that the vesicle releases its content directly in the middle of the electrode is lower. A significant portion of the release can be done in a nonelectroactive zone.

at PC12 cells has been performed with carbon fiber nanoelectrodes of 500 ±100 nm diameter. Exocytosis was thus detected for 60% of the cells investigated. However, contrary to the results obtained with 1 μm size electrodes, the spikes' features (area, duration, and maximum intensity) seemed slightly affected by comparison to those monitored at conventional 7 μm diameter UMEs [114]. However, this may be an artefact due to the relative vesicle/electrode sizes. Indeed, provided that the electrode dimensions are sufficiently large versus those of a given vesicle, the probability for the sensor to entirely collect the vesicular release is reasonably high. Under such conditions, the features of the corresponding amperometric event should not depend on the electrode area. Conversely, when the electrode diameter is similar to (or less than) the vesicle one, the total coverage of the single vesicular release is not ensured. This may lead to a partial detection of the vesicular content, a non-negligible part of the intravesicular medium being released outside of the electroactive surface (Fig. 1.6). A 1 μm diameter thus appears as an optimal value for this type of cells.

All these attempts to reduce electrode dimensions suggest that, except for applications aimed at cell surface mapping, the analytical treatment and reliability of the extracted data do not necessarily benefit from decreasing the electrode size below a certain threshold. In that sense, using a single nanoelectrode for amperometrically monitoring exocytosis at isolated cells could be limited by the distortion of the amperometric spikes induced by a partial collection of the vesicular content described above. Conversely, its excellent S/N ratio would obviously allow detection of some very small important events (PSs, stand-alone feet, etc.). However, several potential difficulties could be also considered. As an example, stray capacitances must be considered so that the S/N ratio and the time constant do not benefit so highly as expected [115]. Additionally, theoretical and experimental behaviors remain to be established when the nanometric electrode dimension becomes comparable to the double-layer thickness or even to the molecular size. Yet, nanoelectrodes may represent an ideal electrochemical tool for studying real synapses, that is, allowing their placement in the synaptic cleft between the nerve terminals. In that case, the electrochemical probe could be inserted into a synapse (nanometer sized) without inducing its complete destructuration. This is probably the most valuable and future application of nanoelectrodes for exocytosis.

1.2.4.4 Integrating methods in microsystems

It is nowadays established that exocytotic activity is not homogenous and that secretion is concentrated at specific zones named hotspots. This suggests the importance of electrochemically mapping the cell surface for pinpointing exocytotic sites and evaluating, if any, correlations between their localization and the kinetic and quantitative features of the amperometric events. Decreasing the electrode size has thus been attempted (see below). Results confirmed the hotspots' existence. But, because a nanometric electrode cannot be scanned fast enough to locate a releasing site while an event occurs (~10–100 ms), it has to be positioned randomly on the cell surface. Hence, electrochemical mapping onto the whole surface cannot be envisioned with the present technologies. This constraint can be overpassed through the simultaneous operation of carbon microelectrode arrays (MEAs) [116, 117]. Presently, MEAs are composed of 7 individually addressed 5 μm diameter carbon microdisks arranged in a hexagonal array and embedded in glass, thus covering a zone of ca. 20–40 μm diameter of the cell apex. Amperometric and voltammetric measurements with MEAs have been achieved on PC12 cells, thus providing important information about the spatial heterogeneity of the exocytotic release sites.

If one leaves the classical paradigm of UME detection, the recent advances in microfabrication and more particularly soft photolithography techniques offer new ways for solving this important issue. Yet, because electrodes are generally delimited on a glass support, *exocytosis at the single-cell level must be investigated at the bottom of cells* (called basal pole) and not at the apex (called apical pole), as usually achieved with carbon fibers or MEAs. For example, an electrochemical detector array composed of four platinum microelectrodes deposited on a glass coverslip (Fig. 1.7a) has been used to record exocytotic events simultaneously at four sections of a single chromaffin cell basal pole [118]. Because the resolution of such an array is intrinsically limited, fluorescence imaging from individual vesicles was simultaneously performed for assigning more precisely the location of each event [118]. Another similar array has been recently reported [119]. Then four boron-doped nanocrystalline diamond electrodes were microfabricated and isolated in an 18 μm diameter well delimited by an epoxy-based photoresist (Fig. 1.7b). As with the platinum array described above, detection of exocytosis could then be achieved simultaneously over four large sections of a single chromaffin cell basal pole.

Simultaneous optical and amperometric detections require that cells adhere to a material with excellent optical (transparency) and electrochemical (electroactivity and conductivity) properties. Indium tin oxide (ITO) perfectly matches these two constraints and is biologically compatible. As a consequence, several ITO microsystems have been reported in the past few years by the group of Kevin D. Gillis. For instance, a microchip device using a transparent ITO electrode crossing a microfluidic channel has been developed to measure exocytosis from a population of chromaffin cells (Fig. 1.8) [120]. A similar configuration with 24 cell-sized ITO electrodes has enabled electrochemical detection of catecholamine secretion from cell arrays and individual cells with concomitant

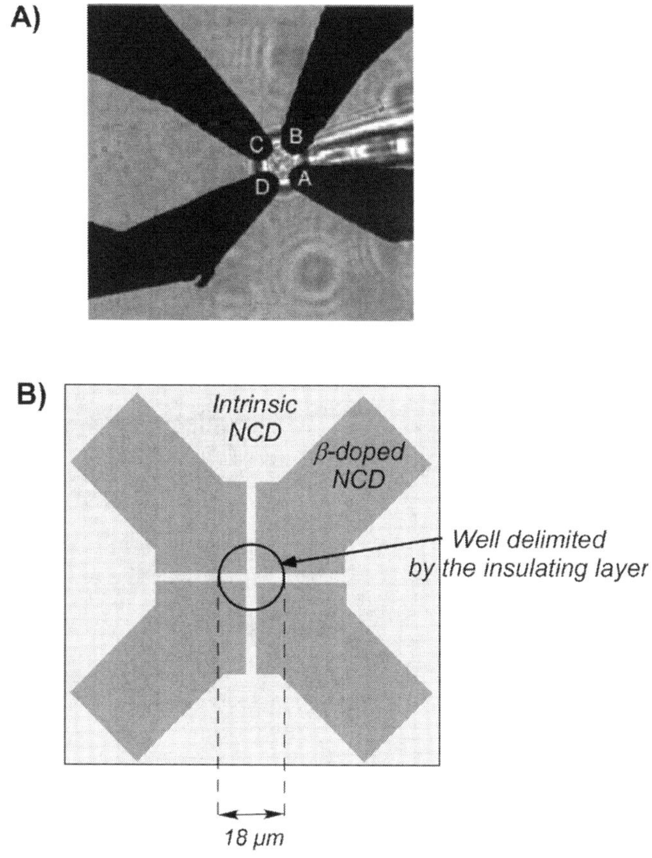

Figure 1.7 Two microelectrode arrays devoted to single-cell mapping. (a) Four platinum microelectrodes (3 μm wide) on a glass coverslip. Reprinted with permission from Ref. [118]. Copyright 2005 National Academy of Sciences. (b) An 18 μm diameter well in which 4 boron-doped nanocrystalline diamond electrodes are inserted. Adapted from Ref. [119].

fluorescent measurements of intracellular Ca^{2+} [121]. A different strategy was implemented in our group. Thus, a micrometric ITO disk surface was delimited by a 40–100 μm diameter well with a photoresist (Fig. 1.9a) to record amperometrically exocytosis at a single chromaffin cell, while fluorescence observation was feasible [122]. Comparing secretion events monitored with this device at the basal pole with those monitored with a classical carbon UME

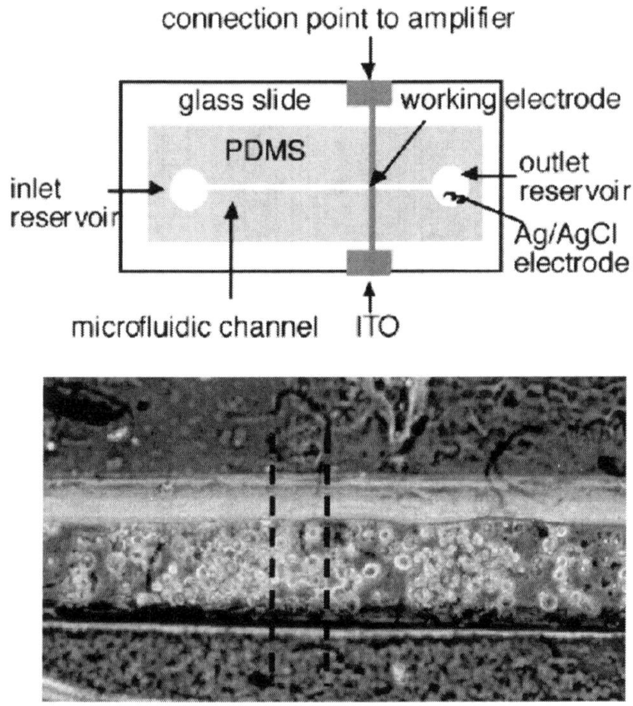

Figure 1.8 Microsystem containing a single ITO electrode crossing a microfluidic channel (vertical dashed lines depict the position of the ITO transparent electrode. The horizontal microfluidic channel is 100 μm wide). Reprinted with permission from Ref. [120]. Copyright 2006 American Chemical Society.

at the apical pole indicated that the processes occurring at each pole differed [123]. This device has been extended to test different configurations that preserve a limited electrode surface (for the minimization of electrical noise and capacitive currents) and a high S/N ratio of detection without compromising the probability of finding observable cells on the electrode. Such devices consisted of four ITO bands (200 μm width; 750 μm length) whose active surfaces could be delimited by a spiraling channel (Fig. 1.9b) [124, 125]. Simultaneous amperometrical and optical TIRF microscopy recordings of enterochromaffin BON cells (containing constitutively a neuropeptide Y tagged with a green fluorescent protein [GFP] and

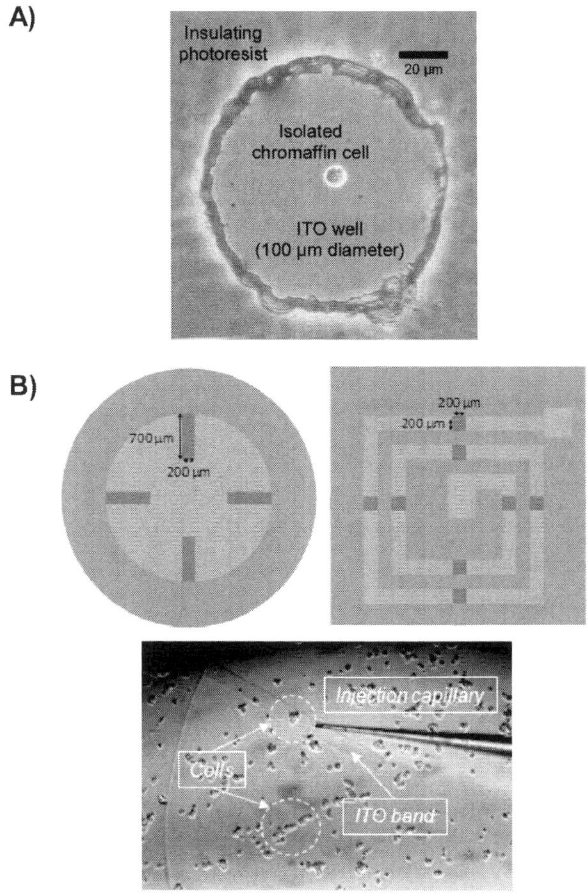

Figure 1.9 ITO devices for the combination between optical and electrochemical measurements. (a) ITO micrometric well containing a single chromaffin cell [122]. (b) Top: ITO four-band device, with each band being 200 μm wide and 750 μm long (left) [125]. ITO four bands delimited by a spiral well (right) [124]. Bottom: photograph of an ITO band with BON cells [125].

releasing serotonin) were obtained, thanks to the low frequency of exocytosis of these cells (0.1 Hz). In these experiments, the precise location of releasing points recorded by TIRF microscopy and amperometry was achieved by time correlations, thus eliminating the need for the use of dense microelectrode arrays.

Another drawback of classical amperometry stems from the single-cell-level necessity. Indeed, cellular variability prevents obtaining of statistically significant experimental results from the analysis of only one or two cells. Carbon fiber amperometry is thus a low-throughput technique that leads to a huge and time-consuming number of experiments. An obvious issue is performing electrochemical measurements from a cell population without losing single-cell analysis specificities. In the last few years, recent developments of microsystems have thus been elaborated to perform parallel recordings from multiple single cells.

The most popular solution consisted of microfabricating arrays of UMEs for the simultaneous recording of several cells. From an analytical point of view, using UME arrays drastically enhances the S/N ratio for an electrochemical experiment performed at constant potential [25, 126, 127]. Briefly, for appropriate time scales and interelectrode distances, the diffusion layers of each electrode overlap. The array thus behaves like a global electrode, and the faradic current is proportional to the entire array surface area, that is, to the sum of the electrodes and insulating surface areas. Conversely, the capacitive current (viz., the noise) only depends on the electroactive array surface. Hence, the S/N ratio is increased, typically by 2–3 orders of magnitude, in comparison to a conventional electrode of a surface area identical to that of the array. However, this argument is valid only for homogenous solutions and does not make any sense with regard to electrochemical measurements of exocytosis when the cells adhere to the array. Then, detection occurs only from cell surfaces that overlap with one exposed electroactive part of the array.

However, examples of UME arrays for exocytosis are numerous. A biochip involving a 5×5 array of gold disk UMEs (from 10 to 90 μm diameter) disposed over ca. 1 mm^2 has been used to detect exocytosis of PC12 or MN9D cells (Fig. 1.10a) [128, 129]. Arrays of disk UMEs (4 nanocrystalline diamond wells; 20 μm in diameter) were used for investigating chromaffin cells (Fig. 1.10b) [119, 130]. The UME array can be localized into a microfluidic channel, as recently reported with 16 UMEs (nitrogen-doped diamond-like carbon [DLC:N], 30×40 μm) used for investigating chromaffin cells (Fig. 1.10c) [131]. Finally, one can mention a recent array of

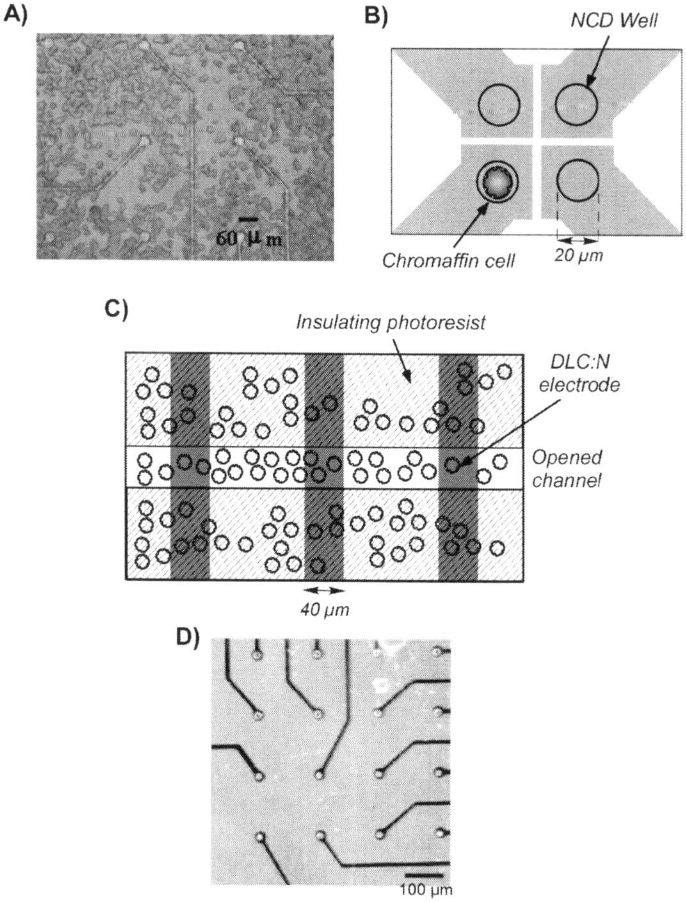

Figure 1.10 Photographs or schemes of different configurations of UME arrays [129–132]. (a) Reprinted from Ref. [129]. Copyright 2006 American Chemical Society. (b) and (c) Adapted from Refs. [130, 131]. (d) Reprinted from Ref. [132]. Copyright 2011 American Chemical Society.

platinum UMEs (6 × 6; 25 μm in diameter) combined with a cell culture chamber (0.9 mm diameter) and perfusion system, thus allowing a proper cell culture to be achieved for a long time. This constitutes a versatile platform in which cell behaviors at different cell growth stages and development phases can be electrochemically investigated in situ and in real time (Fig. 1.10d) [132].

Cell seeding onto UME arrays is generally achieved through random settling out of a cell suspension. As a consequence, a significant portion of the whole electrode surface cannot be covered by cells when working with a diluted cell suspension. In that case, all free electrode surfaces generate electrical noise and capacitive current without contributing to faradic signals, thus altering the global S/N ratio. Alternatively, covering all the UMEs by "carpeting" the surface of the chip with a high density of cells is not adequate. Such a configuration must indeed be avoided if single-cell addressing per electrode is desired, since a given electrode then detects the responses of several cells. Elegant ways to enforce one cell to one electrode in UME arrays have been recently reported by Gillis et al. This was achieved in a microchip device equipped with microfluidic traps that automatically address individual cells (or at least small groups of cells) to platinum UMEs [133]. This device has been successfully tested with adrenal bovine chromaffin cells. Furthermore, some spikes' parameters (charge and intensity) have been recorded. These results confirmed that different dynamics of secretion occur at the top and bottom poles of cells. However, the question is raised about whether or not mechanical stress applied to cells may have significant effects on exocytotic release characteristics.

Another method for bringing cells only onto electrode surfaces relies on the surface chemistry approach. For example, a UME array device composed of cytophilic UMEs has been recently reported [134]. The array is made from 40 wells containing UMEs (20 μm in diameter). On the one hand, the electrode material is DLC:N for which cell affinity promotes adhesion. On the other hand, the insulating surface is made from Teflon AF cytophobic film that inhibits cell attachment and thus entices the cells to preferentially adhere to the UMEs. In a different version, the Teflon film was replaced by a SU-8 photoresist whose surface was grafted with cytophobic poly(ethylene glycol) (PEG) using a polydopamine adhesion layer, while the electrode surfaces were patterned with poly(L-lysine) to enhance cell adhesion (Fig. 1.11) [135]. Both devices allowed electrochemical recording of exocytosis from chromaffin cells, while offering the possibility of analyzing the cells' responses separately or simultaneously.

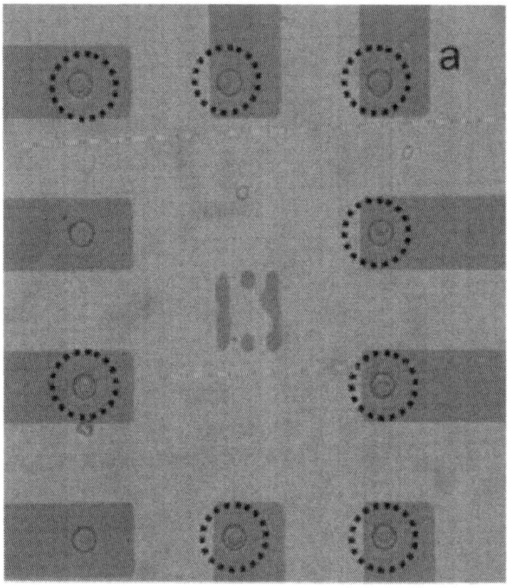

Figure 1.11 Representation of an array devoted to multicell analysis able to bring one cell per electrode by chemical treatment of the surfaces: cytophilic electrodes and cytophobic insulating part of the array. Reprinted with permission from Ref. [135]. Copyright 2011 American Chemical Society.

It must be emphasized that the design and use of microsystems just described resulted from arbitrary choices. Indeed, such devices are versatile and have therefore many potential applications. As examples, devices containing many transparent UMEs can be used not only for coupling electrochemical and optical recordings but also for combinatorial analyses. For instance, a UME array containing 16 picoliter-sized wells equipped with gold electrodes (15 μm opening for each well) has been applied for the catecholamine release of chromaffin cells (Fig. 1.12) [136]. Each well electrode was designed to conform to the shape of one cell. Each UME thus captures a larger fraction of released catecholamines in comparison to a carbon fiber electrode, though the temporal resolution is not identical for each event. Additionally, the combination of amperometry with patch-clamp recordings of membrane capacitance (whole-cell configuration) can be performed at the single-cell level. Recently, devices for bottom electrochemical detection, based on carbon nanotubes, have been reported [137, 138].

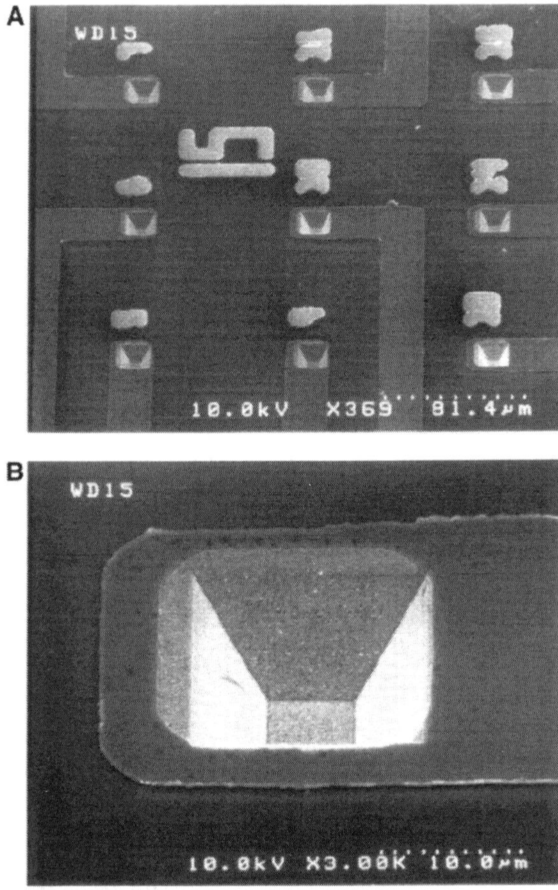

Figure 1.12 Photographs of a well gold electrode array described in Ref. [136]. The insulating film is a SiO_2 layer. (a) Nine of sixteen wells on the chip are depicted. The well openings are 15 μm, and the distance between wells is 100 μm. (b) An individual well. Reprinted with permission from Ref. [136]. Copyright 2003 American Chemical Society.

1.3 Conclusion

Several important aspects of electrochemistry at UMEs aimed to provide a fine analytical and physicochemical understanding of exocytosis mechanisms have been summarized, with particular emphasis on carbon fiber microelectrodes used in a semiartificial synapse. Indeed, they have proven to be spectacular tools for addressing quantitatively central issues at the single-cell level that could not be examined by any other techniques. Owing to this extremely important advantage, their great simplicity, and large versatility, single-carbon-fiber microelectrodes, including arrays of them, will most certainly remain a choice configuration for routine measurements.

However, over the past decade, the principle of the semiartificial synapse configuration has known several technical implementations, thanks to the development and availability of microfabrication. On the one hand, integration of microelectrodes into microfluidic-driven chips provided an effective way to overcome cellular variability by gathering quantitative information directly from populations of cells, though this is generally accomplished at the expenses of kinetic resolution. On the other hand, microfabrication techniques also enabled coupling of amperometric measurements at UMEs with different spectroscopic methods with microscopic resolution. These two present successes perfectly demonstrate that amperometry at microelectrodes is an extremely lively area of research that will certainly lead to many new applications in the future.

References

1. Huang, Y. X., Cai, D., Chen, P. (2011) Micro- and nanotechnologies for study of cell secretion, *Anal. Chem.*, **83**, 4393–4406.
2. Wang, W., Zhang, S.-H., Li, L.-M., Wang, Z.-L., Cheng, J.-K., Huang, W.-H. (2009) Monitoring of vesicular exocytosis from single cells using micrometer and nanometer-sized electrochemical sensors, *Anal. Bioanal. Chem.*, **394**, 17–32.
3. Amatore, C., Arbault, S., Guille, M., Lemaître, F. (2008) Electrochemical monitoring of single cell secretion: vesicular exocytosis and oxidative stress, *Chem. Rev.*, **108**, 2585–2621.
4. Cans, A. S., Ewing, A. G. (2011) Highlights of 20 years of electrochemical measurements of exocytosis at cells and artificial cells, *J. Solid State Electrochem.*, **15**, 1437–1450.
5. Lin, Y., Trouillon, R., Safina, G., Ewing, A. G. (2011) Chemical analysis of single cells, *Anal. Chem.*, **83**, 4369–4392.

6. Mellander, L., Cans, A. S., Ewing, A. G. (2010) Electrochemical probes for detection and analysis of exocytosis and vesicles, *ChemPhysChem*, **11**, 2756–2763.
7. Omiatek, D. M., Cans, A. S., Heien, M. L., Ewing, A. G. (2010) Analytical approaches to investigate transmitter content and release from single secretory vesicles, *Anal. Bioanal. Chem.*, **397**, 3269–3279.
8. Spegel, C., Heiskanen, A., Skjolding, L. H. D., Emneus, J. (2008) Chip based electroanalytical systems for cell analysis, *Electroanalysis*, **20**, 680–702.
9. Burgoyne, R. D., Morgan, A. (2003) Secretory granule exocytosis, *Physiol. Rev.*, **83**, 581–632 and references therein.
10. Alberts, B., Johnson, A., Lewis, J., Raff, M., Roberts, K., Walter, P. (2002) *Molecular Biology of the Cell* (Garland Science, New York).
11. Siegel, G. J., Albers, R. W., Brady, S. T., Price, D. L. (2006) *Basic Neurochemistry* (Academic Press, Elsevier).
12. Ravier, M. A., Tsuboi, T., Rutter, G. A. (2008) Imaging a target of Ca(2+) signalling: dense core granule exocytosis viewed by total internal reflection fluorescence microscopy, *Methods*, **46**, 233–238.
13. Oheim, M. (2001) Imaging transmitter release. I. Peeking at the steps preceding membrane fusion, *Lasers Med. Sci.*, **16**, 149–158.
14. Takahashi, N., Kishimoto, T., Nemoto, T., Kadowaki, T., Kasai, H. (2002) Fusion pore dynamics and insulin granule exocytosis in the pancreatic islet, *Science*, **297**, 1349–1352.
15. Oheim, M., Loerke, D., Stuhmer, W., Chow, R. H. (1998) The last few milliseconds in the life of a secretory granule: docking, dynamics and fusion visualized by total internal reflection fluorescence microscopy (TIRFM), *Eur. Biophys. J.*, **27**, 83–98.
16. Steyer, J. A., Horstmann, H., Almers, W. (1997) Transport, docking and exocytosis of single secretory granules in live chromaffin cells, *Nature*, **388**, 474–478.
17. Mattheyses, A. L., Simon, S. M., Rappoport, J. Z. (2010) Imaging with total internal reflection fluorescence microscopy for the cell biologist, *J. Cell Sci.*, **123**, 3621–3628.
18. Tran, V. S., Huet, S., Fanget, I., Cribier, S., Henry, J. P., Karatekin, E. (2007) Characterization of sequential exocytosis in a human neuroendocrine cell line using evanescent wave microscopy and "virtual trajectory" analysis, *Eur. Biophys. J.*, **37**, 55–69.
19. Neher, E., Marty, A. (1982) Discrete changes of cell-membrane capacitance observed under conditions of enhanced secretion in

bovine adrenal chromaffin cells, *Proc. Natl. Acad. Sci. U. S. A.*, **79**, 6712–6716.

20. Borges, R., Camacho, M., Gillis, K. D. (2008) Measuring secretion in chromaffin cells using electrophysiological and electrochemical methods, *Acta Physiol.*, **192**, 173–184.
21. Zhang, H., Qu, A. L., Luo, J., Luo, J. (2010) Error analysis of C(m) measurement under the whole-cell patch-clamp recording, *J. Neurosci. Methods*, **185**, 307–314.
22. Wightman, R. M., Jankowski, J. A., Kennedy, R. T., et al. (1991) Temporally resolved catecholamine spikes correspond to single vesicle release from individual chromaffin cells, *Proc. Natl. Acad. Sci. U. S. A.*, **88**, 10754–10758.
23. Schroeder, T. J., Jankowski, J. A., Kawagoe, K. T., Wightman, R. M., Lefrou, C., Amatore, C. (1992) Analysis of diffusional broadening of vesicular packets of catecholamines released from biological cells during exocytosis, *Anal. Chem.*, **64**, 3077–3083.
24. Mosharov, E. V., Sulzer, D. (2005) Analysis of exocytotic events recorded by amperometry, *Nat. Methods*, **2**, 651–658.
25. Amatore, C. (1995) Electrochemistry at ultramicroelectrodes, in *Physical Electrochemistry: Principles, Methods and Applications.*, ed. I. Rubinstein (M. Dekker, New York), 131–208.
26. Bard, A. J., Faulkner, L. R. (2001) *Electrochemical Methods: Fundamentals and Applications* (John Wiley and Sons, New York).
27. Kennedy, R. T., Huang, L., Atkinson, M. A., Dush, P. (1993) Amperometric monitoring of chemical secretions from individual pancreatic beta-cells, *Anal. Chem.*, **65**, 1882–1887.
28. Huang, L., Shen, H., Atkinson, M. A., Kennedy, R. T. (1995) Detection of exocytosis at individual pancreatic beta-cells by amperometry at a chemically-modified microelectrode, *Proc. Natl. Acad. Sci. U. S. A.*, **92**, 9608–9612.
29. Kita, J. M., Wightman, R. M. (2008) Microelectrodes for studying neurobiology, *Curr. Opin. Chem. Biol.*, **12**, 491–496.
30. Robinson, D. L., Hermans, A., Seipel, A. T., Wightman, R. M. (2008) Monitoring rapid chemical communication in the brain, *Chem. Rev.*, **108**, 2554–2584.
31. Leszczyszyn, D. J., Jankowski, J. A., Viveros, O. H., Diliberto, E. J., Near, J. A., Wightman, R. M. (1991) Secretion of catecholamines from individual adrenal-medullary chromaffin cells, *J. Neurochem.*, **56**, 1855–1863.

32. Kawagoe, K. T., Jankowski, J. A., Wightman, R. M. (1991) Etched carbon-fiber electrodes as amperometric detectors of catecholamine secretion from isolated biological cells, *Anal. Chem.*, **63**, 1589–1594.
33. Chow, R. H., Vonruden, L., Neher, E. (1992) Delay in vesicle fusion revealed by electrochemical monitoring of single secretory events in adrenal chromaffin cells, *Nature*, **356**, 60–63.
34. Zhou, Z., Misler, S. (1996) Amperometric detection of quantal secretion from patch-clamped rat pancreatic beta-cells, *J. Biol. Chem.*, **271**, 270–277.
35. Patel, B. A., Bian, X., Quaiserova-Mocko, V., Galligan, J. J., Swain, G. M. (2007) In vitro continuous amperometric monitoring of 5-hydroxytryptamine release from enterochromaffin cells of the guinea pig ileum, *Analyst*, **132**, 41–47.
36. Walsh, P. L., Petrovic, J., Wightman, R. M. (2011) Distinguishing splanchnic nerve and chromaffin cell stimulation in mouse adrenal slices with fast-scan cyclic voltammetry, *Am. J. Physiol.: Cell Ph.*, **300**, C49–C57.
37. Ge, S. C., Woo, E., Haynes, C. L. (2011) Quantal regulation and exocytosis of platelet dense-body granules, *Biophys. J.*, **101**, 2351–2359.
38. Hawley, M. D., Tatawawa, S. V., Piekarsk, S., Adams, R. N. (1967) Electrochemical studies of oxidation pathways of catecholamines, *J. Am. Chem. Soc.*, **89**, 447–450.
39. Ciolkowski, E. L., Cooper, B. R., Jankowski, J. A., Jorgenson, J. W., Wightman, R. M. (1992) Direct observation of epinephrine and norepinephrine cosecretion from individual adrenal-medullary chromaffin cells, *J. Am. Chem. Soc.*, **114**, 2815–2821.
40. Amatore, C., Bouret, Y., Travis, E. R., Wightman, R. M. (2000) Interplay between membrane dynamics, diffusion and swelling pressure governs individual vesicular exocytotic events during release of adrenaline by chromaffin cells, *Biochimie*, **82**, 481–496.
41. Amatore, C., Bouret, Y., Travis, E. R., Wightman, R. M. (2000) Adrenaline release by chromaffin cells: constrained swelling of the vesicle matrix leads to full fusion, *Angew. Chem., Int. Ed.*, **39**, 1952–1955.
42. Wightman, R. M., Schroeder, T. J., Finnegan, J. M., Ciolkowski, E. L., Pihel, K. (1995) Time-course of release of catecholamines from individual vesicles during exocytosis at adrenal-medullary cells, *Biophys. J.*, **68**, 383–390.
43. Amatore, C., Arbault, S., Bonifas, I., Guille, M. (2009) Quantitative investigations of amperometric spike feet suggest different controlling

factors of the fusion pore in exocytosis at chromaffin cells, *Biophys. Chem.*, **143**, 124–131.

44. DeToledo, G. A., Fernandezchacon, R., Fernandez, J. M. (1993) Release of secretory products during transient vesicle fusion, *Nature*, **363**, 554–558.
45. Amatore, C., Arbault, S., Bouret, Y., Guille, M., Lemaître, F., Verchier, Y. (2006) Regulation of exocytosis in chromaffin cells by trans-insertion of lysophosphatidylcholine and arachidonic acid into the outer leaflet of the cell membrane, *ChemBioChem*, **7**, 1998–2003.
46. Amatore, C., Arbault, S., Bonifas, I., Guille, M., Lemaître, F., Verchier, Y. (2007) Relationship between amperometric pre-spike feet and secretion granule composition in chromaffin cells: an overview, *Biophys. Chem.*, **129**, 181–189.
47. Amatore, C., Arbault, S., Bonifas, I., Bouret, Y., Erard, M., Guille, M. (2003) Dynamics of full fusion during vesicular exocytotic events: release of adrenaline by chromaffin cells *ChemPhysChem*, **4**, 147–154.
48. Schroeder, T. J., Borges, R., Finnegan, J. M., Pihel, K., Amatore, C., Wightman, R. M. (1996) Temporally resolved, independent stages of individual exocytotic secretion events, *Biophys. J.*, **70**, 1061–1068.
49. Amatore, C., Arbault, S., Bonifas, I., Lemaître, F., Verchier, Y. (2007) Vesicular exocytosis under hypotonic conditions evidences two distinct populations of dense core vesicles in bovine chromaffin cells, *ChemPhysChem*, **8**, 578–585.
50. Borges, R., Travis, E. R., Hochstetler, S. E., Wightman, R. M. (1997) Effects of external osmotic pressure on vesicular secretion from bovine adrenal medullary cells, *J. Biol. Chem.*, **272**, 8325–8331.
51. Colliver, T. L., Pyott, S. J., Achalabun, M., Ewing, A. G. (2000) VMAT-mediated changes in quantal size and vesicular volume, *J. Neurosci.*, **20**, 5276–5282.
52. Amatore, C., Bouret, Y., Midrier, L. (1999) Time-resolved dynamics of the vesicle membrane during individual exocytotic secretion events, as extracted from amperometric monitoring of adrenaline exocytosis from chromaffin cells, *Chem. Eur. J.*, **5**, 2151–2162.
53. Jankowski, J. A., Schroeder, T. J., Ciolkowski, E. L., Wightman, R. M. (1993) Temporal characteristics of quantal secretion of catecholamines from adrenal-medullary cells, *J. Biol. Chem.*, **268**, 14694–14700.
54. Amatore, C., Arbault, S., Bonifas, I., et al. (2005) Correlation between vesicle quantal size and fusion pore release in chromaffin cell exocytosis, *Biophys. J.*, **88**, 4411–4420.

55. Sombers, L. A., Hanchar, H. J., Colliver, T. L., et al. (2004) The effects of vesicular volume on secretion through the fusion pore in exocytotic release from PC12 cells, *J. Neurosci.*, **24**, 303–309.
56. Pothos, E. N., Mosharov, E., Liu, K. P., et al. (2002) Stimulation-dependent regulation of the pH, volume and quantal size of bovine and rodent secretory vesicles, *J. Physiol. (London)*, **542**, 453–476.
57. Almers, W. (2001) Fusion needs more than SNAREs, *Nature*, **409**, 567–568.
58. An, S. J., Almers, W. (2004) Tracking SNARE complex formation in live endocrine cells, *Science*, **306**, 1042–1046.
59. Criado, M., Gil, A., Viniegra, S., Gutierrez, L. M. (1999) A single amino acid near the C terminus of the synaptosome-associated protein of 25 kDa (SNAP-25) is essential for exocytosis in chromaffin cells, *Proc. Natl. Acad. Sci. U. S. A.*, **96**, 7256–7261.
60. Fisher, R. J., Burgoyne, R. D. (1999) The effect of transfection with Botulinum neurotoxin C1 light chain on exocytosis measured in cell populations and by single-cell amperometry in PC12 cells, *Pflügers Arch.*, **437**, 754–762.
61. Gil, A., Viniegra, S., Gutierrez, L. M. (1998) Dual effects of botulinum neurotoxin A on the secretory stages of chromaffin cells, *Eur. J. Neurosci.*, **10**, 3369–3378.
62. Graham, M. E., Fisher, R. J., Burgoyne, R. D. (2000) Measurement of exocytosis by amperometry in adrenal chromaffin cells: effects of clostridial neurotoxins and activation of protein kinase C on fusion pore kinetics, *Biochimie*, **82**, 469–479.
63. Quetglas, S., Iborra, C., Sasakawa, N., et al. (2002) Calmodulin and lipid binding to synaptobrevin regulates calcium-dependent exocytosis, *EMBO J.*, **21**, 3970–3979.
64. Burgoyne, R. D., Fisher, R. J., Graham, M. E., Haynes, L. P., Morgan, A. (2001) Control of membrane fusion dynamics during regulated exocytosis, *Biochem. Soc. Trans.*, **29**, 467–472.
65. Fisher, R. J., Pevsner, J., Burgoyne, R. D. (2001) Control of fusion pore dynamics during exocytosis by Munc18, *Science*, **291**, 875–878.
66. Graham, M. E., Barclay, J. W., Burgoyne, R. D. (2004) Syntaxin/Munc18 interactions in the late events during vesicle fusion and release in exocytosis, *J. Biol. Chem.*, **279**, 32751–32760.
67. Graham, M. E., Burgoyne, R. D. (2000) Comparison of cysteine string protein (Csp) and mutant alpha-SNAP overexpression reveals a role for Csp in late steps of membrane fusion in dense-core granule exocytosis in adrenal chromaffin cells, *J. Neurosci.*, **20**, 1281–1289.

68. Schutz, D., Zilly, F., Lang, T., Jahn, R., Bruns, D. (2005) A dual function for Munc-18 in exocytosis of PC12 cells, *Eur. J. Neurosci.*, **21**, 2419–2432.

69. Anderson, B. B., Zerby, S. E., Ewing, A. G. (1999) Calculation of transmitter concentration in individual PC12 cell vesicles with electrochemical data and a distribution of vesicle size obtained by electron microscopy, *J. Neurosci. Methods*, **88**, 163–170.

70. Grabner, C. P., Price, S. D., Lysakowski, A., Fox, A. P. (2005) Mouse chromaffin cells have two populations of dense core vesicles, *J. Neurophysiol.*, **94**, 2093–2104.

71. Westerink, R. H. S., de Groot, A., Vijverberg, H. P. M. (2000) Heterogeneity of catecholamine-containing vesicles in PC12 cells, *Biochem. Biophys. Res. Commun.*, **270**, 625–630.

72. Bruns, D. (2004) Detection of transmitter release with carbon fiber electrodes, *Methods*, **33**, 312–321.

73. Bruns, D., Jahn, R. (1995) Real-time measurement of transmitter release from single synaptic vesicles, *Nature*, **377**, 62–65.

74. Bruns, D., Riedel, D., Klingauf, J., Jahn, R. (2000) Quantal release of serotonin, *Neuron*, **28**, 205–220.

75. Anderson, B. B., Chen, G. Y., Gutman, D. A., Ewing, A. G. (1998) Dopamine levels of two classes of vesicles are differentially depleted by amphetamine, *Brain Res.*, **788**, 294–301.

76. Tang, K. S., Tse, A., Tse, F. W. (2005) Differential regulation of multiple populations of granules in rat adrenal chromaffin cells by culture duration and cyclic AMP, *J. Neurochem.*, **92**, 1126–1139.

77. Jankowski, J. A., Finnegan, J. M., Wightman, R. M. (1994) Extracellular ionic composition alters kinetics of vesicular release of catecholamines and quantal size during exocytosis at adrenal-medullary cells, *J. Neurochem.*, **63**, 1739–1747.

78. Kennedy, R. T., Lan, H. A., Aspinwall, C. A. (1996) Extracellular pH is required for rapid release of insulin from Zn-insulin precipitates in beta-cell secretory vesicles during exocytosis, *J. Am. Chem. Soc.*, **118**, 1795–1796.

79. Aspinwall, C. A., Brooks, S. A., Kennedy, R. T., Lakey, J. R. T. (1997) Effects of intravesicular H+ and extracellular H+ and Zn2+ on insulin secretion in pancreatic beta cells, *J. Biol. Chem.*, **272**, 31308–31314.

80. Taylor, S. C., Roberts, M. L., Peers, C. (1999) Acid-evoked quantal catecholamine secretion from rat phaeochromocytoma cells and its interaction with hypoxia-evoked secretion, *J. Physiol. (London)*, **519**, 765–774.

81. Camacho, M., Machado, J. D., Montesinos, M. S., Criado, M., Borges, R. (2006) Intragranular pH rapidly modulates exocytosis in adrenal chromaffin cells, *J. Neurochem.*, **96**, 324–334.

82. Amatore, C., Arbault, S., Bouret, Y., Guille, M., Lemaître, F. (2010) Prediction of Llocal pH variations during amperometric monitoring of vesicular exocytotic events at chromaffin cells, *ChemPhysChem*, **11**, 2931–2941.

83. Uchiyama, Y., Maxson, M. M., Sawada, T., Nakano, A., Ewing, A. G. (2007) Phospholipid mediated plasticity in exocytosis observed in PC12 cells, *Brain Res.*, **1151**, 46–54.

84. Bauermann, L. P., Schuhmann, W., Schulte, A. (2004) An advanced biological scanning electrochemical microscope (Bio-SECM) for studying individual living cells, *Phys. Chem. Chem. Phys.*, **6**, 4003–4008.

85. Hengstenberg, A., Blochl, A., Dietzel, I. D., Schuhmann, W. (2001) Spatially resolved detection of neurotransmitter secretion from individual cells by means of scanning electrochemical microscopy, *Angew. Chem., Int. Ed.*, **40**, 905–908.

86. Xin, Q., Wightman, R. M. (1997) Enzyme modified amperometric sensors for choline and acetylcholine with tetrathiafulvalene tetracyanoquinodimethane as the electron-transfer mediator, *Anal. Chim. Acta*, **341**, 43–51.

87. Chen, G. Y., Ewing, A. G. (1997) Chemical analysis of single cells and exocytosis, *Crit. Rev. Neurobiol.*, **11**, 59–90.

88. De Toledo, G. A., Fernandez-Chacon, R., Fernandez, J. M. (1993) Release of secretory products during transient vesicle fusion, *Nature*, **363**, 554–558.

89. Dernick, G., Gong, L. W., Tabares, L., de Toledo, G. A., Lindau, M. (2005) Patch amperometry: high-resolution measurements of single-vesicle fusion and release, *Nat. Methods*, **2**, 699–708.

90. Fernandez-Peruchena, C., Navas, S., Montes, M. A., De Toledo, G. A. (2005) Fusion pore regulation of transmitter release, *Brain Res. Rev.*, **49**, 406–415.

91. Zhou, Z., Misler, S., Chow, R. H. (1996) Rapid fluctuations in transmitter release from single vesicles in bovine adrenal chromaffin cells, *Biophys. J.*, **70**, 1543–1552.

92. Engisch, K. L., Chernevskaya, N. I., Nowycky, M. C. (1997) Short-term changes in the Ca2+-exocytosis relationship during repetitive pulse protocols in bovine adrenal chromaffin cells, *J. Neurosci.*, **17**, 9010–9025.

93. Di, A., Krupa, B., Bindokas, V. P., et al. (2002) Quantal release of free radicals during exocytosis of phagosomes, *Nat. Cell Biol.*, **4**, 279–285.

94. Oberhauser, A. F., Robinson, I. M., Fernandez, J. M. (1995) Do Caged-Ca2+ Compounds mimic the physiological stimulus for secretion, *J. Physiol. (Paris)*, **89**, 71–75.

95. Oberhauser, A. F., Robinson, I. M., Fernandez, J. M. (1996) Simultaneous capacitance and amperometric measurements of exocytosis: a comparison, *Biophys. J.*, **71**, 1131–1139.

96. Albillos, A., Dernick, G., Horstmann, H., Almers, W., deToledo, G. A., Lindau, M. (1997) The exocytotic event in chromaffin cells revealed by patch amperometry, *Nature*, **389**, 509–512.

97. Dernick, G., de Toledo, G. A., Lindau, M. (2003) Exocytosis of single chromaffin granules in cell-free inside-out membrane patches, *Nat. Cell Biol.*, **5**, 358–362.

98. Ales, E., Tabares, L., Poyato, J. M., Valero, V., Lindau, M., De Toledo, G. A. (1999) High calcium concentrations shift the mode of exocytosis to the kiss-and-run mechanism, *Nat. Cell Biol.*, **1**, 40–44.

99. Gong, L. W., Hafez, I., de Toledo, G. A., Lindau, M. (2003) Secretory vesicles membrane area is regulated in tandem with quantal size in chromaffin cells, *J. Neurosci.*, **23**, 7917–7921.

100. Tabares, L., Lindau, M., De Toledo, G. A. (2003) Relationship between fusion pore opening and release during mast cell exocytosis studied with patch amperometry, *Biochem. Soc. Trans.*, **31**, 837–841.

101. Anderova, M., Duchene, A. D., Barbara, J. G., Takeda, K. (1998) Vasoactive intestinal peptide potentiates and directly stimulates catecholamine secretion from rat adrenal chromaffin cells, *Brain Res.*, **809**, 97–106.

102. Chow, R. H., Klingauf, J., Neher, E. (1994) Time-course of Ca2+ concentration triggering exocytosis in neuroendocrine cells, *Proc. Natl. Acad. Sci. U. S. A.*, **91**, 12765–12769.

103. Finnegan, J. M., Borges, R., Wightman, R. M. (1996) Comparison of cytosolic Ca2+ and exocytosis responses from single rat and bovine chromaffin cells, *Neuroscience*, **71**, 833–843.

104. Haynes, C. L., Buhler, L. A., Wightman, R. M. (2006) Vesicular Ca2+-induced secretion promoted by intracellular pH-gradient disruption, *Biophys. Chem.*, **123**, 20–24.

105. Kang, G. X., Holz, G. G. (2003) Amplification of exocytosis by Ca (2+)-induced Ca2+ release in INS-1 pancreatic beta cells, *J. Physiol. (London)*, **546**, 175–189.

106. Kim, T. D., Eddlestone, G. T., Mahmoud, S. F., Kuchtey, J., Fewtrell, C. (1997) Correlating Ca2+ responses and secretion in individual RBL-2H3 mucosal mast cells, *J. Biol. Chem.*, **272**, 31225–31229.

107. Robinson, I. M., Finnegan, J. M., Monck, J. R., Wightman, R. M., Fernandez, J. M. (1995) Colocalization of calcium-entry and exocytotic release sites in adrenal chromaffin cells, *Proc. Natl. Acad. Sci. U. S. A.*, **92**, 2474–2478.

108. Xin, Q., Wightman, R. M. (1998) Simultaneous detection of catecholamine exocytosis and Ca2+ release from single bovine chromaffin cells using a dual microsensor, *Anal. Chem.*, **70**, 1677–1681.

109. Qian, W. J., Aspinwall, C. A., Battiste, M. A., Kennedy, R. T. (2000) Detection of secretion from single pancreatic beta-cells using extracellular fluorogenic reactions and confocal fluorescence microscopy, *Anal. Chem.*, **72**, 711–717.

110. Gutierrez, L. M., Gil, A., Viniegra, S. (1998) Preferential localization of exocytotic active zones in the terminals of neurite-emitting chromaffin cells, *Eur. J. Cell Biol.*, **76**, 274–278.

111. Paras, C. D., Qian, W. J., Lakey, J. R., Tan, W. H., Kennedy, R. T. (2000) Localized exocytosis detected by spatially resolved amperometry in single pancreatic beta-cells, *Cell Biochem. Biophys.*, **33**, 227–240.

112. Amatore, C., Arbault, S., Bouret, Y., Guille, M., Lemaître, F., Verchier, Y. (2009) Invariance of exocytotic events detected by amperometry as a function of the carbon fiber microelectrode diameter, *Anal. Chem.*, **81**, 3087–3093.

113. Wu, W. Z., Huang, W. H., Wang, W., et al. (2005) Monitoring dopamine release from single living vesicles with nanoelectrodes, *J. Am. Chem. Soc.*, **127**, 8914–8915.

114. Li, Z. Y., Zhou, W., Wu, Z. X., Zhang, R. Y., Xu, T. (2009) Fabrication of size-controllable ultrasmall-disk electrode: monitoring single vesicle release kinetics at tiny structures with high spatio-temporal resolution, *Biosens. Bioelectron.*, **24**, 1358–1364.

115. Amatore, C., Maisonhaute, E., Schollhorn, B. (2008) Molecular electrochemistry pushed to its limits: from nanosecond kinetics to the dynamic study of nanometric objects, *Actualité Chim.*, 69–74.

116. Zhang, B., Adams, K. L., Luber, S. J., Eves, D. J., Heien, M. L., Ewing, A. G. (2008) Spatially and temporally resolved single-cell exocytosis utilizing individually addressable carbon microelectrode arrays, *Anal. Chem.*, **80**, 1394–1400.

117. Zhang, B., Heien, M., Santillo, M. F., Mellander, L., Ewing, A. G. (2011) Temporal resolution in electrochemical imaging on single pc12 cells using amperometry and voltammetry at microelectrode arrays, *Anal. Chem.*, **83**, 571–577.

118. Hafez, I., Kisler, K., Berberian, K., et al. (2005) Electrochemical imaging of fusion pore openings by electrochemical detector arrays, *Proc. Natl. Acad. Sci. U. S. A.*, **102**, 13879–13884.

119. Pasquarelli, A., Carabelli, V., Xu, Y. L., et al. (2011) Diamond microelectrodes arrays for the detection of secretory cell activity, *Int. J. Environ. Anal. Chem.*, **91**, 150–160.

120. Sun, X. H., Gillis, K. D. (2006) On-chip amperometric measurement of quantal catecholamine release using transparent indium tin oxide electrodes, *Anal. Chem.*, **78**, 2521–2525.

121. Chen, X. H., Gao, Y. F., Hossain, M., Gangopadhyay, S., Gillis, K. D. (2008) Controlled on-chip stimulation of quantal catecholamine release from chromaffin cells using photolysis of caged Ca2+ on transparent indium-tin-oxide microchip electrodes, *Lab Chip*, **8**, 161–169.

122. Amatore, C., Arbault, S., Chen, Y., Crozatier, C., Lemaître, F., Verchier, Y. (2006) Coupling of electrochemistry and fluorescence microscopy at indium tin oxide microelectrodes for the analysis of single exocytotic events, *Angew. Chem., Int. Ed.*, **45**, 4000–4003.

123. Amatore, C., Arbault, S., Lemaître, F., Verchier, Y. (2007) Comparison of apex and bottom secretion efficiency at chromaffin cells as measured by amperometry, *Biophys. Chem.*, **127**, 165–171.

124. Meunier, A., Fulcrand, R., Darchen, F., et al. (2012) Indium tin oxide devices for amperometric detection of vesicular release by single cells, *Biophys. Chem.*, **162**, 14–21.

125. Meunier, A., Jouannot, O., Fulcrand, R., et al. (2011) Coupling amperometry and total internal reflection fluorescence microscopy at ITO surfaces for monitoring exocytosis of single vesicles, *Angew. Chem., Int. Ed.*, **50**, 5081–5084.

126. Amatore, C., Saveant, J. M., Tessier, D. (1983) Charge-transfer at partially blocked surfaces: a model for the case of microscopic active an inactive sites, *J. Electroanal. Chem.*, **147**, 39–51.

127. Yeh, J. I., Shi, H. (2010) Nanoelectrodes for biological measurements, *WIRE: Nanomed. Nanobiotechnol.*, **2**, 176–188.

128. Cui, H. F., Ye, J. S., Chen, Y., et al. (2006) In situ temporal detection of dopamine exocytosis from L-dopa-incubated MN9D cells using microelectrode array-integrated biochip, *Sens. Actuators, B*, **115**, 634–641.

129. Cui, H. F., Ye, J. S., Chen, Y., Chong, S. C., Sheu, F. S. (2006) Microelectrode array biochip: tool for in vitro drug screening based on the detection of a drug effect on dopamine release from PC12 cells, *Anal. Chem.*, **78**, 6347–6355.
130. Carabelli, V., Gosso, S., Marcantoni, A., et al. (2010) Nanocrystalline diamond microelectrode arrays fabricated on sapphire technology for high-time resolution of quantal catecholamine secretion from chromaffin cells, *Biosens. Bioelectron.*, **26**, 92–98.
131. Gao, Y. F., Chen, X. H., Gupta, S., Gillis, K. D., Gangopadhyay, S. (2008) Magnetron sputtered diamond-like carbon microelectrodes for on-chip measurement of quantal catecholamine release from cells, *Biomed. Microdevices*, **10**, 623–629.
132. Li, L. M., Wang, W., Zhang, S. H., et al. (2011) Integrated microdevice for long-term automated perfusion culture without shear stress and real-time electrochemical monitoring of cells, *Anal. Chem.*, **83**, 9524–9530.
133. Gao, Y. F., Bhattacharya, S., Chen, X. H., Barizuddin, S., Gangopadhyay, S., Gillis, K. D. (2009) A microfluidic cell trap device for automated measurement of quantal catecholamine release from cells, *Lab Chip*, **9**, 3442–3446.
134. Barizuddin, S., Liu, X., Mathai, J. C., Hossain, M., Gillis, K. D., Gangopadhyay, S. (2010) Automated targeting of cells to electrochemical electrodes using a surface chemistry approach for the measurement of quantal exocytosis, *ACS Chem. Neurosci.*, **1**, 590–597.
135. Liu, X., Barizuddin, S., Shin, W., Mathai, C. J., Gangopadhyay, S., Gillis, K. D. (2011) Microwell device for targeting single cells to electrochemical microelectrodes for high-throughput amperometric detection of quantal exocytosis, *Anal. Chem.*, **83**, 2445–2451.
136. Chen, P., Xu, B., Tokranova, N., Feng, X. J., Castracane, J., Gillis, K. D. (2003) Amperometric detection of quantal catecholamine secretion from individual cells on micromachined silicon chips, *Anal. Chem.*, **75**, 518–524.
137. Shi, B. X., Wang, Y., Zhang, K., Lam, T. L., Chan, H. L. W. (2011) Monitoring of dopamine release in single cell using ultrasensitive ITO microsensors modified with carbon nanotubes, *Biosens. Bioelectron.*, **26**, 2917–2921.
138. Sudibya, H. G., Ma, J. M., Dong, X. C., et al. (2009) Interfacing glycosylated carbon-nanotube-network devices with living cells to detect dynamic secretion of biomolecules, *Angew. Chem., Int. Ed.*, **48**, 2723–2726.

Chapter 2

Adsorptive Stripping Voltammetric Determination of Metabolites

Xiaoquan Lu, Hui Li, and Yaya Li

Key Laboratory of Bioelectrochemistry and Environmental Analysis of Gansu Province, Chemical Engineering, Northwest Normal University, Lanzhou 730070, China

luxq@nwnu.edu.cn

2.1 Introduction

Adsorptive stripping voltammetry (AdSV), developed in the late 1970s, is an analytical method. Many researchers have paid attention to this method, due to it can determine extensive objects. The combination of an effective preconcentration step with advanced electrochemical measurements of the accumulated analytes makes AdSV the most sensitive electroanalytical technique [1, 2]. Its remarkable sensitivity is attributed to the built-in preconcentration step in which the target molecules are accumulated onto the working electrode. Proper choice of the working electrode is crucial for the success of the adsorptive stripping operation. The ideal working electrode should offer effective preconcentration, a favorable redox reaction of the target molecules, a reproducible and renewable surface, and a low background current over a wide potential

Electrochemical Biosensors
Edited by Serge Cosnier

ISBN 978-981-4411-46-2 (Hardcover), 978-981-4411-47-9 (eBook)
www.panstanford.com

range. Mercury has been the electrode material of choice for many AdSV applications. Two basic electrode systems, the mercury film electrode (MFE) and the hanging mercury drop electrode (HMDE), have gained wide acceptance in the development of AdSV [1–4]. While these small-volume mercury electrodes offer an attractive stripping performance, new alternative electrode materials with similar performance are urgently desired for addressing growing concerns regarding the toxicity, handling, volatility, and disposal of mercury. Future regulations and occupational health considerations may severely limit (and even ban) the use of mercury as an electrode material. Such concerns regarding the toxicity of mercury electrodes are particularly important in view of the growing demands for on-site environmental analysis and decentralized clinical material testing. Different bare carbon, gold, silver, or iridium electrodes have been used as possible alternatives to mercury [5, 6]. While offering useful stripping signals for several materials, the overall performance of these "nonmercury" electrodes has not approached that of mercury ones due to a low cathodic potential limit, multiple distorted (multiple/broad) peaks, large background contributions, or poor precision and resolution. Hence, the development of reliable nonmercury electrodes is considered a major challenge for AdSV in the early stage of the 21st century.

The Joseph Wang group has done a detailed discussion of the theory and technology of AdSV. In 2000, the group introduced a bismuth film electrode (on glassy carbon and carbon fiber substrates) for AdSV measurements of heavy metals [7, 8]. Further studies by the group and Ogorevc demonstrated that bismuth-coated glassy carbon electrodes (GCEs) offer an extremely attractive stripping voltammetric performance compared favorably with that of mercury electrodes [9–14]. Parallel investigations in the Czech Republic focused on stripping voltammetry and potentiometry at bismuth-coated carbon paste electrodes (CPEs) [15, 16]. Subsequently, Bobrowski [17, 18] and Economou [19, 20] extended the scope of bismuth electrodes to adsorptive stripping measurements for tracing metals that cannot be plated electrolytically.

A new electrically heated CPE has been developed by the Joseph Wang group for performing adsorptive stripping measurements of trace nucleic acids. Such coupling of electrochemistry at electrically heated electrodes with adsorptive constant-current stripping chronopotentiometry offers distinct advantages for trace measurements of nucleic acids [21, 22]. The application of increased temperatures during the deposition step results in dramatic enhancement of the stripping signal. Such improvement is attributed to the accumulation step at the heated electrode. Forced thermal convection near the electrode surface facilitates the use of quiescent solutions, and hence ultrasmall volumes were used. Using an electrode temperature of 32°C and a quiescent solution during the 1 min accumulation, the response is linear over the 1–8 mg/L range tested, with a detection limit of 0.5 mg/L. Such electrode heating technology offers great promise for various applications involving thermal manipulations of nucleic acids [23].

In 2003, Michal Masarik and his group reported that the proteins streptavidin and avidin were electrochemically detected in solution by adsorptive transfer stripping square-wave voltammetry (AdTS SWV) at a CPE. AdTS SWV was used to quantify biotinylated oligonucleotides, DNA hybridizations, and avidin in extracts of transgenic avidin maize. The detection limits of denatured and native streptavidin were 6 pM and 120 nM, respectively. The results demonstrated that streptavidin/avidin AdTS SWV is a sensitive and specific method for quantifying DNA and proteins in biological samples such as foods and tissue extracts, including genetically modified crops (avidin maize) and other plants in neighboring fields [24].

In 2007, Samo and his group reported an antimony film electrode (SbFE) for electrochemical stripping analysis. In this work, an SbFE is reported for the first time as a possible alternative for electrochemical stripping analysis of trace heavy metals. The SbFE was prepared in situ on a glassy carbon substrate electrode and employed in combination with either anodic stripping voltammetry or stripping chronopotentiometry in nondeaerated solutions of 0.01 M hydrochloric acid (pH 2) [25]. Several key operational parameters influencing the electroanalytical response of SbFE were examined and optimized, such as deposition potential, deposition time, and composition of the measurement solution. The SbFE exhibited well-defined and separated stripping signals for both model metal ions, Cd(II) and Pb(II), surrounded with low background contribution and a relatively large negative potential range. Comparing the SbFE with the commonly used MFE and the recently introduced bismuth film electrode, the newly proposed electrode offers a remarkable performance in more acidic solutions (pH $\leq$ 2), which can be advantageous in electrochemical analysis of trace heavy metals. Thus, it contributes to the wider applicability of electrochemical stripping techniques in connection with “mercury free” electrodes [26].

2.2 Classification, Mechanism, and Theory of Adsorptive Stripping Voltammetry

Determination of objects by AdSV could be divided into two categories. One category of measured objects is some metal ions or valence ions, and enrichments are complexes that are formed by a test element and an organic complexing agent solution on the electrode. So the enrichment process involves a chemical reaction, then the adsorption process, or the organic complexing agent adsorbed to the electrode surface first, then complexes formed with the measured elements, and then the stripping electrode reaction proceeds finally. The other category of measured objects includes surfactants, alkaloids, dyes, drugs, etc., adsorbed to the electrode surface spontaneously. These measured objects are enrichments; meanwhile, the stripping process is desorption, which accompanies electrode reaction. In addition, adsorption of organic enrichment can also be a product of the electrochemical reaction. In other words, adsorption, electrode reaction, and stripping process involve in oxidation or reduction of adsorbates. The category is much more complex than the first one.

Organic complexes of certain metals, first of all, exclude water molecules on the electrode surface and then are adsorbed to the electrode surface. The water molecule is a dipole, which has a certain dipole moment, and is strongly affected by electric fields. When the electrode surface charge is very positive or negative and the electric field in the liquid layer near the electrode surface is large, water molecules will be on priority absorbed. Large organic molecules whose dipole moment is much smaller than that of water molecules, when the electrode surface is not charged (i.e., near zero charge) and the electric field near the electrode surface is very weak, will be adsorbed to the electrode surface. So the accumulation potential for adsorption in the AdSV is requested not to be too positive or too negative, or it cannot achieve the purpose of adsorption enrichment. In particular, some metal–organic complexes have a common feature, that is, adsorption enrichment proceeds in the case of an open circuit, which may be related to the above reasons.

Q represents the power required (coulomb) of the adsorbed reactants to the electrode reaction according to the following relationship:

$$Q = nFA\Gamma \tag{2.1}$$

where Γ represents the amount (mol/cm^2) of reactants adsorbed to the electrode surface, n is the number of electrons in the electrode reaction, F is the Faraday constant, and A is the electrode area.

Equation 2.1 shows that for the concentration of reactants in a certain range, the stripping peak current is proportional to the concentration of reactants and can be quantitatively measured, which is the AdSV quantitative basis. When Γ reaches Γ_m (reactants on the electrode surface reach the maximum adsorption capacity), the entire electrode surface was covered by a tight monolayer of the adsorbate, showing adsorption isotherms of the process of reactants on the electrode surface, and then the stripping current peak has no linear relationship with the concentration of reactants. That is the measurable boundary.

2.2.1 *Mechanism and Characteristics of the First Adsorptive Stripping Voltammetry*

The mechanism and characteristics of the first AdSV are discussed below with examples.

If there is a solution to form complexes with the measured ion surfactant A (for convenience, eliminating the charge of A), then it is possible with the measured ionic surfactant adsorption to the electrode surface, which is the enrichment process, and the stripping process is the oxidation of the adsorbate (which is the electrical activity of the ligand) or restore (the central ion or ligand is electrical

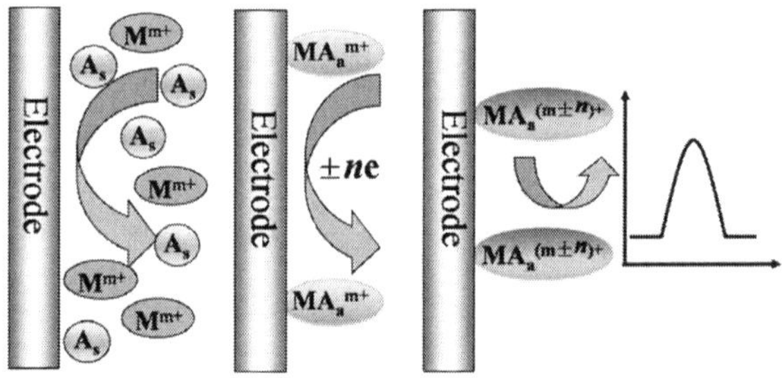

Scheme 2.1 Mechanism of M^{m+} adsorption and stripping. M^{m+}: measured ion; A_s: A_{solution}; MA_a^{m+}: $MA_{\text{adsorption}}^{m+}$.

activity) (see Scheme 2.1) the phase equation can be expressed as follows:

$$A_{\text{solution}} \rightarrow A_{\text{adsorption}} \quad (2.2)$$

$$M^{m+} + A_{\text{adsorption}} \rightarrow MA_{\text{adsorption}}^{m+} \quad (2.3)$$

$$MA_{\text{adsorption}}^{m+} \pm ne \rightarrow MA^{(m\pm n)+} \quad (2.4)$$

or

$$M^{m+} + A_{\text{solution}} \rightarrow MA_{\text{solution}}^{m+} \quad (2.5)$$

$$MA_{\text{solution}}^{\text{m}+} \rightarrow MA_{\text{adsorption}} \quad (2.6)$$

$$MA_{\text{adsorption}}^{\text{m}+} \pm ne \rightarrow MA^{(\text{m}\pm\text{n})+} \quad (2.7)$$

In Reactions (2.2)–(2.4), the adsorption process ahead of the formation of complexes can be written as the adsorption–chemical–electrochemical (ACE) reaction, and in Reactions (2.5)–(2.7), the complex formation ahead of the adsorption process is recorded as the chemical–adsorption–electrochemical (CAE) reaction.

In fact, ACE and CAE are just two extreme situations; in many cases, the adsorption process and the complex formation process often occur at the same time. We have used two categories just for the convenience of presentation. Reactions (2.2)–(2.4) and (2.5)–(2.7) have been used for many inorganic depolarizer enrichments and determinations. Reaction (2.4) or (2.7) indicates the generation of stripping current.

In this type of AdSV, the original oxidation state of the measured elements and the oxidation state of complexes are the same. The latter on the electrode surface has the adsorption isotherm, and the relationship of the adsorption to the electrode with enrichment time, accumulation potential is less. Enrichment can also be conducted in "open circuit" conditions. The stripping current depends on the amount of adsorption, and the amount of adsorption depends on the ion concentration measured in solution. Under appropriate conditions, the measured ion concentration is linear with the stripping current. But when adsorption of the reactant to the electrode surface reaches saturation, the stripping peak has no linear relationship with the concentration. That is to say the quantitative analysis can only be carried out lower than the concentration limits.

The approach requires electrode reaction products $MA^{(m+n)}$ deposited on the electrode surface effectively. This method requires the organic complexing agent (ligand) having some adsorption capacity on the electrode surface. Their structural features are that they contain unsaturated bonds; in particular, the organic ligands contain benzene and a conjugated double-bond structure, in which π electrons can be overlapped, shared, and strongly adsorbed to the electrode surface. Many organic ligands such as 2,2′-bipyridine, diethyl carbamate (DDTC), dithizone, 1,10-phenanthroline, 8-hydroxyquinoline, α-sub-nitro-β-naphthol, dimethylglyoximeand, and other metal complexes can be adsorbed to the electrode surface. For example, voltammograms of nortriptyline hydrochloride in Tween 20 exhibit very well-defined cathodic peaks by square-wave cathodic adsorptive stripping voltammetry (SWC AdSV). The current is mainly adsorption controlled and proportional to the concentration over a convenient range [27] (Fig. 2.1).

2.2.2 *Mechanism and Characteristics of the Second Adsorptive Stripping Voltammetry*

The mechanism and characteristics of the second AdSV are discussed with examples as follows.

Organic compounds containing hydrophobic groups adsorb to the electrode easily, and the adsorption characteristics will sometimes cause trouble with inorganic polarographic analysis.

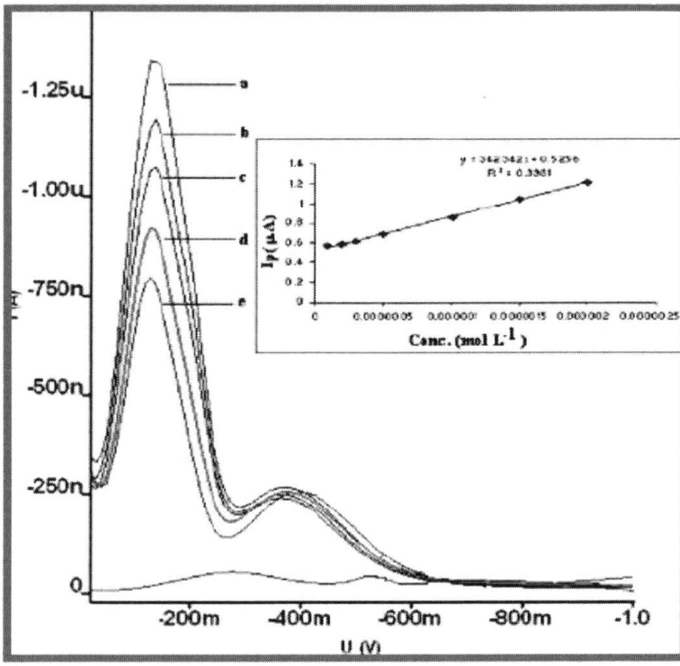

Figure 2.1 The dependence of the SWC AdSV current for nortriptyline hydrochloride at different concentrations in Tween 20, $E_{acc} = -1.0$ V, $t_{acc} = 200$ s, frequency $f = 140$ Hz, pulse amplitude $\Delta E_{sw} = 50$ mV, and scan increment $\Delta s = 10$ mV. (a) 2×10^{-6} mol L^{-1}, (b) 1.21×10^{-4} mol L^{-1}, (c) 1.5×10^{-6} mol L^{-1}, (d) 1×10^{-6} mol L^{-1}, and (e) 5×10^{-7} mol L^{-1}. From Ref. [27].

But this feature is also available for the determination of the adsorption concentration of organics by AdSV. This type of organic concentration can be a pure adsorption process and can also be accompanied by electrochemical changes in the adsorption process; in other words, it is the adsorption process of the electrochemical reaction product of organics. Therefore, this AdSV can be broadly grouped into the adsorption–electrochemical (AE) reaction or the electrochemical–adsorption–electrochemical (EAE) reaction, but the actual situation is even more complicated (see Scheme 2.2).

Robert Kalvada has reported that the adsorptive stripping method generally was applied to surface-active organic compounds

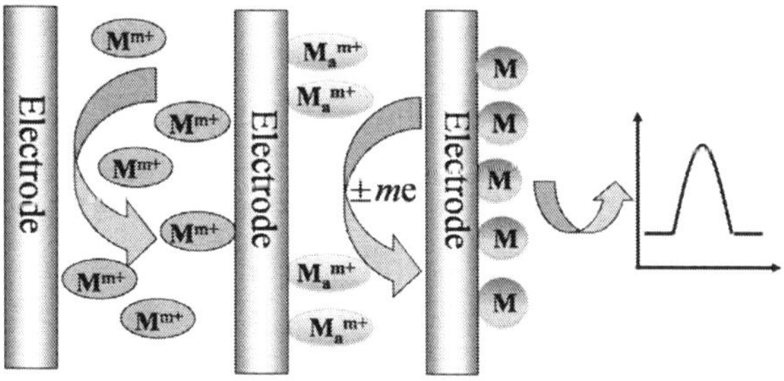

Scheme 2.2 Mechanism of the adsorption–electrochemical reaction. M^{m+}: measured object; M_a^{m+}: $M_{adsorption}^{m+}$.

within the range of 10^{-6}–10^{-5} M concentration with a well-relaxed peak. The absorption coefficient of these compounds is typically 1×10^{-9} M^{-1}; this value should be higher in an alkaline solution. Most of the time, their dripping cannot achieve equilibrium at the dropping mercury electrode (DME). But within the accumulation time, a material concentration less than 1×10^{-6} M in the electrode achieves a balance and a relaxation balance significantly at the static mercury drop electrode as (SMDE). As a rule of the peak potential, the stripping peak potential moves toward more negative values with increasing amounts of adsorbed compounds at the electrode surface. The stripping peak height depends on the accumulation time and reaches a limit. The linear range of the peak height and concentration varies with different compounds and also depends on the experimental conditions. In short, the adsorption of organic compounds and the stripping process are more complicated, and the theory is being researched and developed continually.

2.3 Experimental Techniques of Adsorptive Stripping Voltammetry

AdSV experimental procedures are as the same as the usual stripping voltammetry: After removing oxygen, a constant voltage was applied to the SMDE or other solid electrodes in the solution (relative to the reference electrode) in a specified time interval. Accumulation (enrichment) can be carried out under mixing and without mixing conditions. In the latter case, a shorter accumulation time is sufficient. Over the accumulation time, rest for a few seconds, the electrode potential moves to a more negative (or positive) potential scan and the polarization curve is recorded. The above experimental steps (especially the dissolution process) can be carried out by AC polarography, pulse and differential pulse polarography, sweep polarography, and Kalsousek polarography. For actual sample analysis at low concentrations, it is very important to strictly control the purity of the base solution. The best way is to check the dissolution step by using a blank supporting electrolyte solution. The impurities in solution can adsorb to electrode surface in a long enrichment process, which can affect adsorption of studied compounds. The measurement error is 5%–10%.

If an SMDE is used, under normal circumstances, each experiment should be switched to a new mercury drop.

There are two experimental measurement techniques especially suitable for AdSV combined with differential pulse polarography, *subtraction* and *media exchange*.

2.3.1 *Subtraction*

Subtraction stripping voltammetry is used to determine organics that adsorb (or extract) and accumulate on the carbon electrode, by means of different preconcentration times and recording of the difference between the stripping currents. The background current is independent of the preconcentration time, so the background current can be eliminated. The combination of the current peak enhanced in the preconcentration stage and the background current corrected by subtractive methods is a major advantage of the method. Equation 2.8 expresses the difference between dissolution current Δi measured by different preconcentration times T_1 and T_2:

$$\Delta i = (i_{p(T_1)} + i_p) - (i_{p(T_2)} + i_b) = i_{p(T_1)} - i_{p(T_2)} \qquad (2.8)$$

where $i_{p(T)}$ is Faraday current generated by oxidation–reduction of boundary chemical species on the electrode surface and i_b is the background current. Because the background current is independent of the differential pulse enrichment time, the background current can be offset when the total current subtracted from another. In most cases (the same as conventional AdSV), to maintain the maximum speed of the analysis, set $T_2 = 0$. However, a very short period of time (usually 15 s) is required using medium exchange. Owing these features and benefits, this method has been applied to the determination of epinephrine and dopamine. It is known that these compounds adsorb and accumulate on the modified gold electrode and the GCE. Figure 2.2a shows differential pulse voltammograms of 2×10^{-7} epinephrine. It is seen that conventional differential pulse voltammetry (DPV) after preconcentration does not have sufficient sensitivity. Because the decomposition and surface oxidation–reduction reaction of solvent, the desorption/stripping peak (at $+0.2$ V) and high differential pulse background are obtained, as showed in Fig. 2.2b. Therefore, a 5 min preconcentration is clearly not enough to the level of 1×10^{-7} M for common measurement. However, background subtraction is able to distinguish components so that the measure meets the requirements, as the curve in Fig. 2.2c (indicated by a flat baseline) shows. In view of the good limit of determination, its pre-enrichment time can be shortened. Figure 2.2b also has this advantage, which is obtained by preconcentration of 5×10^{-7} M dopamine in phosphate buffer solution. And the "background curve" and "analysis curve" of Fig. 2.2b were recorded through preconcentration of 15 s and 120 s, respectively. The determination was completed in 0.05 M phosphate buffer (through medium exchange). The subtractive curve in Fig. 2.2B(c) shows a good description peak of dopamine, and it is not interfered by decomposition of solvent.

2.3.2 *Media Exchange*

An important feature of stripping voltammetry is "the steps of stripping." It can be completed in a better solution that has a different composition compared to the original solution. This method is the so-called media exchange. Some benefits have been found in heavy-metal complexes by stripping voltammetry. For example, interference from the original solution, which has a

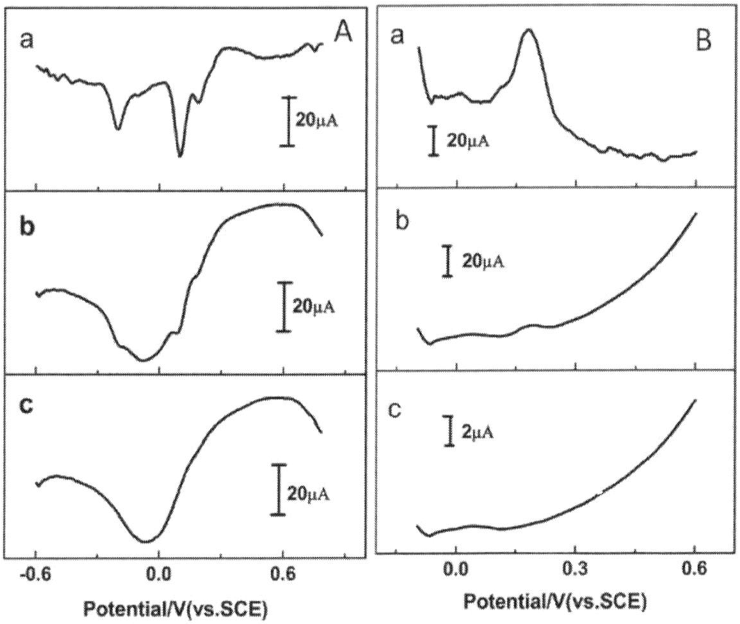

Figure 2.2 (a) 2×10^{-8} M adrenaline and (b) 5×10^{-7} M dopamine differential pulse voltammograms. (a) The background without a preconcentration curve. (b) The analysis of conventional methods to get the curve; it is the result of a 5 min preconcentration. (c) The subtraction curve (b)–(a).

large number of components, low resolution, and high slope of the background current can reduce stripping. Joseph et al. have shown that using medium exchange combined with DPV analysis makes the determination of materials of electrical activity that strongly adsorbe to the electrode to achieve higher sensitivity and good selectivity. The general step of this method is that the adsorption enrichment is carried out under strictly controlled conditions and then the working electrode is transferred to a more suitable electrolyte solution for dissolution before determination. Its features and advantages can be seen from determination of submicromolar concentrations objects in the presence of a large number (10^{-8}–10^{-4} times) of other components (even with the same redox potential).

Of course, media exchange and subtraction were used at the same time.

2.4 Electrodes

Proper choice of the working electrode is crucial for the success of the stripping operation. As mentioned earlier, the ideal working electrode offers an effective preconcentration, a favorable redox reaction of the target metal, a reproducible and renewable surface, and a low background current over a wide potential range.

2.4.1 *Hanging Mercury Drop Electrode*

The above requirement of a reproducible surface is best fulfilled by the HMDE [28–30]. The best results are obtained with an SMDE in which the flow rate of mercury in the capillary and thus the formation of a mercury drop with the required, precisely reproducible volume are ensured by a needle valve or other type of valve controlled by an electronic circuit.

2.4.2 *Mercury Film Electrode*

An MFE may be preplated (prior to the measurement) or codeposited with the target metal analytes. Glassy carbon, carbon fiber, iridium microdisks, and screen-printed carbon are the most common supports for the mercury film. MFEs possess a higher surface-to-volume ratio (compared to the HMDE) and consequently offer a more efficient preconcentration [31]. Alternative nonmercury electrodes, including gold, carbon, silver, or bismuth ones, have been developed for addressing concerns regarding the toxicity and handling of mercury. Particularly attractive are bismuth film electrodes, which offer an attractive stripping performance similar to that of mercury electrodes, while eliminating toxicity problems [32]. Gold surfaces are characterized by a wider potential window (compared to mercury ones) to allow the convenient detection of important metals, such as arsenic, selenium, and mercury [33].

2.4.3 *Ultramicroelectrodes*

Ultramicroelectrodes have been shown to be useful for eliminating the need for forced convection during the deposition step and for obviating the need for a supporting electrolyte. The latter is particularly attractive for speciation work in low-ionic-strength natural waters and for minimization of the risk of contamination. Arrays of microelectrodes have shown to be useful in measuring metal concentration profiles in high spatial resolution [34, 36]. Microfabricated (mass-produced) electrodes, particularly screen-printed electrode strips [37] and gold ultramicroelectrodes, have been developed as low-cost disposable ("one shot") metal sensors [38, 39]. The deliberate modification of conventional electrodes can benefit various stripping applications by imparting higher selectivity, sensitivity, and stability. Various surface coatings have thus been used for protection against surface-active foulants or for preconcentration of target molecules.

2.4.4 *Carbon Paste Electrode and Platinum Electrode*

A CPE, made by mixing carbon powder or graphite with a binder (various types of mineral oils, etc.) and pressing the mixture into a glass tube, has been widely used. Procedures for its manufacture have been given [40–45]. Similarly, a platinum disk electrode can also be employed. Both these types of electrodes are especially suitable for studying adsorbable substances that can be oxidized at the electrode, as they can be polarized to much more positive potentials (e.g., +1.0 V vs. a saturated calomel electrode [SCE]) than mercury electrodes, which, on the other hand, can be used in a wider negative potential range. Thus, mercury electrodes are preferable for studying reducible substances. However, it is somewhat more difficult to work with nonmercury electrodes. The electrode must be conditioned prior to use, for example, by periodic polarization from negative to positive potentials and back again for a certain period of time in the base electrolyte. After recording the curve, the electrode surface must be renewed, for example, by removing the surface layer of the paste. The difficulties connected with producing a good-quality CPE are reflected in the lower precision in analytical determinations than that obtained for HMDEs. The choice of binder is also important for paste electrodes: suitable selection can sometimes lead to more specific adsorption. It should be noted when using paste electrodes that the substance can also be accumulated as a result of dissolution in the binder during the accumulation period. This is then a combined adsorption–extraction effect or can consist purely of extraction [46, 48]. Whether extraction is involved can be determined by removing the surface layer of the paste after the accumulation period and recording the curve. As quinoid groups in either the reduced or the oxidized state (of the quinone–hydroquinone type) may be present on the carbon particles, chemisorption of the substance in solution can play an important role in some cases. This type of electrode can be considered to be modified by the presence of these groups in a certain sense.

2.5 Representative Examples

Numerous applications of AdSV for many environmental, clinical, or industrial problems have been reported. The technique has been widely used for the measurement of trace metals in natural waters, including pristine and polluted inland and marine waters [46]. The technique is particularly suitable for the study of chemical speciation, that is, the determination of different physical–chemical forms of the element [47]. Measurements of oxidation states, complexation capacity, stability constants, or "labile metal" concentration represent typical stripping speciation experiments. Because of its fundamentally different detection principles, the adsorptive stripping approach provides different speciation information (based on ligand competition) compared to conventional AdSV procedures. Recent advances in miniaturization and remote sensing have facilitated the application of AdSV for on-site monitoring of toxic metals, including in situ (real-time) profiling [48]. Other environmental applications of AdSV include assays of sediments, fly ash, rain, or airborne particulate matters.

With regard to direct voltammetry, adsorptive stripping techniques exhibit some additional advantages. Since a great number

Table 2.1 Representative applications of stripping analysis

Metal	Sample matrix	Stripping mode	Working electrode	References
Co	Soil	AdSV	Bi	[49, 50]
Cr	River water	AdSV	Bi	[50]
Pb	Drinking water	SW AdSV	Bi	[51]
Cr	Soil and tobacco	SW AdSV	Bi	[52]
Ni	Aerosols	AdSV	HMDE	[53]
Ti	Seawater	AdSV	HMDE	[54]
Pt	Plants	AdSV	HMDE	[55]
U	Groundwater	AdSV	MFE	[29]

SW AdSV: square-wave adsorptive stripping voltammetry

of organic compounds, including pharmaceutical and biological substances, exhibit surface-active properties, they can be determined at very low levels, generally ranging from 10^{-6} M to 10^{-1} M. Moreover, several metal ions that can be reduced at mercury electrodes cannot be amalgamated or exhibit an irreversible metal–metal ion couple, extreme redox potentials, or the formation of intermetallic compounds. They can be analyzed after their complexation by some surface-active organic ligand, followed by their adsorptive collection at the electrode surface and the subsequent reduction of the adsorbed layer. As the analytical concentrations are very low, aqueous solutions may be used, increasing the adsorption capability of most organic compounds (e.g., Tables 2.1 and 2.2).

Table 2.2 Representative applications of stripping analysis

Compound	Sample matrix	Stripping mode	Working electrode	References
Drotaverine hydrochlorie	Human urine	AdSV	HMDE	[56]
Lamotrigine	Pharmaceutical preparations	AdSV	CSPE	[57]
Paracetamol	VitaminC	AdSV	CPE	[58]
Rosiglitazone	Human urine, plasma samples	SW AdSV	HMDE	[59]
Folk acid		AdSV	HMDE	[60]
trans-Resveratrol	Red wine	SW AdSV	GCE	[61]
Nicardipine	Blood, urine	AdSV	HMDE	[62]

CSPE: carbon screen-printed electrode

AdSV has been widely used for measuring trace metals in biological fluids. The use of AdSV for decentralized testing of blood lead in children has been particularly successful. Measurements of mercury in fish, of antimony in gunshot residues, or of cadmium impurities in zinc plant electrolytes represent typical food, forensic, and industrial applications, respectively, of AdSV. These, and other applications, are illustrated in Tables 2.1 and 2.2.

References

1. Wang, J. (1989) *Stripping Analysis: Principles, Instrumentation and Applications* (VCH, Deerfield Beach Press, USA).
2. Batley, G. (1983) Electroanalytical techniques for the determination of heavy metals in seawater, *Mar. Chem.*, **12**, 107–117.
3. Florence, T. M. (1984) Recent advances in stripping analysis, *J. Electroanal. Chem.*, **168**, 207–218.
4. Economou, A., Fielden, P. R. (2003) Mercury film electrodes: developments, trends and potentialities for electroanalysis, *Analyst*, **128**, 205–213.
5. Bonfil, Y., Kirowa-Eisener, E. (2002) Determination of nanomolar concentrations of lead and cadmium by anodic-stripping voltammetry at the silver electrode, *Anal. Chim. Acta*, **457**, 285–296.
6. Wang, J., Tian, B. (1993) Mercury-free disposable lead sensors based on potentiometric stripping analysis of gold-coated screen-printed electrodes, *Anal. Chem.*, **65**, 1529–1532.
7. Wang, J., Lu, J., Hocevar, S., Farias, P., Ogorevc, B. (2000) Bismuth-coated carbon electrodes for anodic stripping voltammetry, *Anal. Chem.*, **72**, 3218–3222.
8. Yong, L., Armstrong, Kristie, C., et al. (2006) Quantitative analysis of trace chromium in blood samples. combination of the advanced oxidation process with catalytic adsorptive stripping voltammetry, *Anal. Chem.*, **78**, 7582–7587.
9. Wang, J., Lu, J., Hocevar, S., Farias, P., Ogorevc, B. (2001) Insights into the anodic stripping voltammetric behavior of bismuth film electrodes, *Anal. Chim. Acta*, **434**, 29–34.
10. Wang, J., Kirgçz, R. A., Lu, J. M. (2001) Stripping voltammetry with the electrode material acting as a 'built-in' internal standard, *Electrochem. Commun.*, **3**, 703–706.

11. Hoeevar, S. B., Wang, J., Deo, R. P., Ogorevc, B. (2002) Potentiometric stripping analysis at bismuth-film electrode, *Electroanalysis*, **14**, 112–115.
12. Wang, J., Lu, J. M., Hoeèvar, S. B., Ogorevc, B. (2001) Bismuth-coated screen-printed electrodes for stripping voltammetric measurements of trace lead, *Electroanalysis*, **13**, 13–16.
13. Wang, J., Deo, R. P., Thongngamdee, S., Ogorevc, B. (2001) Effect of surface-active compounds on the stripping voltammetric response of bismuth film electrodes, *Electroanalysis*, **13**, 1153–1156.
14. Wang, J., Lu, J. M. (2000) Bismuth film electrodes for adsorptive stripping voltammetry of trace nickel, *Electrochem. Commun.*, **2**, 390–393.
15. Królicka, A., Pauliukaite, R., Svancara, I., et al. (2002) Bismuth-film-plated carbon paste electrodes, *Electrochem. Commun.*, **4**, 193–196.
16. Vytras, K., Svancara, I., Metelka, R. (2002) A novelty in potentiometric stripping analysis: total replacement of mercury by bismuth, *Electroanalysis*, **14**, 1359–1364.
17. Krolicka, A., Bobrowski, A., Kalcher, K., Mocak, J., Svancara, I., Vytras, K. (2003) Study on catalytic adsorptive stripping voltammetry of trace cobalt at bismuth film electrodes, *Electroanalysis*, **15**, 1859–1863.
18. Królicka, A., Bobrowski, A. (2004) Bismuth film electrode for adsorptive stripping voltammetry-electrochemical and microscopic study, *Electrochem. Commun*, **6**, 99–104.
19. Morfobos, M., Economou, A., Voulgaropoulos, A. (2004) Simultaneous determination of nickel(II) and cobalt(II) by square wave adsorptive stripping voltammetry on a rotating-disc bismuth-film electrode, *Anal. Chim. Acta*, **519**, 57–64.
20. Sun, H. Z., Li, H. Y., Sadler, P. J. (1999) Interactions of bismuth complexes with metallothionein(II), *J. Biol. Chem.*, **274**, 29094–29101.
21. Palecek, E., Fojta, M. (1994) Differential pulsed voltammetric determination of RNA at the picomole level in the presence of DNA and nucleic acid components, *Anal. Chem.*, **66**, 1566–1571.
22. Anderson, J. L., Coury, L. A., Leddy, J. (1998) Dynamic electrochemistry: methodology and application, *Anal. Chem.*, **70**, 519–589.
23. Michal, M., Rene, K. Kramer, K. J., et al. (2003) Application of avidin–biotin technology and adsorptive transfer stripping square-wave voltammetry for detection of DNA hybridization and avidin in transgenic avidin maize, *Anal. Chem.*, **75**, 2663–2669.

24. Hocevar, S. B., Svancara, I. Ogorevc, B., Vytras, K. (2007) Antimony film electrode for electrochemical stripping analysis, *Anal. Chem*, **79**, 8639–8643.

25. Jain, R., Dwivedi, A., Mishra, R. (2009) Adsorptive stripping voltammetric behavior of nortriptyline hydrochloride and its determination in surfactant media, *Langmuir*, **25**, 10364–10369.

26. Wang, J. (2007) *Stripping Analysis* (Wiley-VCH Press, USA).

27. Economou, A., Fielden, P. R. (2003) Mercury film electrodes: developments, trends and potentialities for electroanalysis, *Analyst*, **128**, 205–213.

28. Moreira, M. M., Carvalho, A. M., Valente, I. M., et al. (2011) Novel application of square-wave adsorptive-stripping voltammetry for the determination of xanthohumol in spent hops, *J. Agric Food Chem.*, **59**, 7654–7658.

29. Wang, J., Wang, J. Y., Tian, B. M., Jiang, M. (1997) Ultrasound-enhanced anodic stripping voltammetry using perfluorosulfonated ionomer-coated mercury thin-film electrodes, *Anal. Chem*, **69**, 1657–1656.

30. Wang, J. (2005) Stripping analysis at bismuth electrodes, *Electroanalysis*, **17**, 1341–1346.

31. Vega, M., Constant, M. G., Berg, V. D. (1997) Determination of cobalt in seawater by catalytic adsorptive cathodic stripping voltammetry, *Anal. Chem.*, **69**, 874–881.

32. Abbasi, S., Khodarahmiyan, K., Abbasi, F. (2011) Simultaneous determination of ultra trace amounts of lead and cadmium in food samples by adsorptive stripping voltammetry, *Food Chem.*, **128**, 254–257.

33. Wang, J., Lu, J. M., Tian, B. M., Yarnitzky, C. (1993) Screen-printed ultramicroelectrode arrays for on-site stripping measurements of trace metals, *J. Electroanal Chem.*, **361**, 77–83.

34. Collins, C. J., Arrigan, A. W. M. (2009) Ion-transfer voltammetric determination of the β-blocker propranolol in a physiological matrix at silicon membrane-based liquid—liquid microinterface arrays, *Anal. Chem.*, **81**, 2344–2349.

35. Chou, C. H., Chang, J. L., Zen, J. M. (2010) Effective analysis of gaseous formaldehyde based on a platinum-deposited screen-printed edge band ultramicroelectrode coated with Nafion as solid polymer electrolyte, *Sens. Actuators, B*, **147**, 669–675.

36. McKnight, T. E., Melechko, A. V., Austin, D. W., et al. (2004) Microarrays of vertically-aligned carbon nanofiber electrodes in an open fluidic channel, *J. Phys. Chem. B*, **108**, 7115–7125.

37. Wang, J., Tian, B. M. (1993) Gold ultramicroelectrodes for on-site monitoring of trace lead, *Electroanalysis*, **5**, 809–814.
38. Vydra, F., Stulik, K., E. Julakova (1976) *Electrochemical Stripping Analysis* (E. Horwood, Chichester UK).
39. Wang, J., Freiha, B. A. (1983) Preconcentration and differential pulse voltammetry of butylated hydroxyanisole at a carbon paste electrode, *Anal. Chim. Acta*, **154**, 87–94.
40. Wang, J., Bollo, S., Lopez Paz, J. L., Sahlin, E., Mukherjee, B. (1999) Ultra-trace measurements of nucleic acids by baseline-corrected adsorptive stripping square-wave voltammetry, *Anal. Chem.*, **71**, 1910–1913.
41. Masarik, M., Kizek, R., Kramer, K. J., et al. (2003) Application of avidin–biotin technology and adsorptive transfer stripping square-wave voltammetry for detection of DNA hybridization and avidin in transgenic avidin maize, *Anal. Chem.*, **75**, 2663–2669.
42. Wang, J., Grundler, P., Flechsig, G. U., et al. (2000) Stripping analysis of nucleic acids at a heated carbon paste electrode, *Anal. Chem.*, **72**, 3752–3756.
43. Ortiz, M. C., Arcos, J., Juarros, J. V., Lopez-Palacios, J., Sarabiat, L. A. (1993) Robust procedure for calibration and calculation of the detection limit of trimipramine by adsorptive stripping voltammetry at a carbon paste electrode, *Anal. Chem.*, **65**, 678–682.
44. Wang, J., Freiha, B. A. (1984) Extractive preconcentration of organic compounds at carbon paste electrodes, *Anal. Chem.*, **56**, 849–852.
45. Wang, J., Luo, D. B. (1984) Competition studies in voltammetric measurements based on extractive accumulation into carbon paste electrodes, *J. Electroanal. Chem. Interface*, **179**, 251–261.
46. Armalis, S. Krikŝĉiuniene, I., Kubiliene, E., et al. (1999) Stripping analysis of trace metals at a flow-through reticulated vitreous carbon electrode after the preconcentration by supported liquid membrane technique, *Int. J. Environ. Anal. Chem.*, **74**, 233–242.
47. Jones, P. W., Taylor, D. M., Williams, D. R. (2000) Analysis and chemical speciation of copper and zinc in wound fluid, *J. Inorg. Biochem*, **81**, 1–10.
48. Brett, C. M. A. (1999) Electroanalytical techniques for the future-the challenges of miniaturization and of real-time measurements, *Electroanalysis*, **11**, 1013–1016.
49. Hutton, E. A., van Elteren, J. T., Ogorevc, B., Smyth, M. (2004) Validation of bismuth film electrode for determination of cobalt and cadmium in soil extracts using ICP-MS, *Talanta*, **63**, 849–855.

50. Lin, L., Lawrence, N. S., Thongngamdee, S., Wang, J., Lin, Y. (2005) Catalytic adsorptive stripping determination of trace chromium (VI) at the bismuth film electrode, *Talanta*, **65**, 144–148.

51. Wang, J., Kirgçz, R. A., Lu, J. M. (2001) Stripping voltammetry with the electrode material acting as a 'built-in' internal standard, *Electrochem. Commun.*, **3**, 703–706.

52. Chatzitheodorou, E., Economou, A., Voulgaropoulos, A. (2004) Trace determination of chromium by square-wave adsorptive stripping voltammetry on bismuth film electrodes, *Electroanalysis*, **16**, 1745–1754.

53. Brett, C. M. A., Oliveira Brett, A. M. C. F., Jorge, C. P. L. (1991) Adsorptive stripping voltammetry of cobalt and nickel in flow systems at wall-jet electrodes, *Electroanalysis*, **3**, 683–689.

54. Wang, J., Mahmoud, J. S. (1986) Stripping voltammetry with adsorptive accumulation for trace measurements of titanium, *J. Electroanal. Chem. Interface*, **208**, 383–394.

55. Huszal, S., Kowalska, J., Krzeminska, M., Golimowski, J. (2005) Determination of platinum with thiosemicarbazide by catalytic adsorptive stripping voltammetry (AdSV), *Electroanalysis*, **17**, 209–304.

56. Zayed, S. I. M., Issa, Y. M. (2009) Cathodic adsorptive stripping voltammetry of drotaverine hydrochloride and its determination in tablets and human urine by differential pulse voltammetry, *Bioelectrochemistry*, **75**, 9–12.

57. Domínguez-Renedo, O., Calvo, M., Arcos-Martínez, M. (2008) Determination of lamotrigine in pharmaceutical preparations by adsorptive stripping voltammetry using screen printed electrodes, *Sensors*, **8**, 4201–4212.

58. Zhang, Y., Luo, L. Q., Ding, Y. P., Liu, X., Qian, Z. Y. (2010) A highly sensitive method for determination of paracetamol by adsorptive stripping voltammetry using a carbon paste electrode modified with nanogold and glutamic acid, *Microchim Acta*, **171**, 133–138.

59. Al-Ghamdi, A. F, Hefnawy, M. M., Al-Majed, A. A., Belal, F. F. (2012) Development of square-wave adsorptive stripping voltammetric method for determination of acebutolol in pharmaceutical formulations and biological fluids, doi:10.1186/1752-153X-6-15.

60. Han, J. L., Chen, H. Y., Gao, H. (1991) Alternating current adsorptive stripping voltammetry in a flow system for the determination of ultratrace amounts of folic acid, *Anal. Chim. Acta*, **252**, 47–52.

61. Airado-Rodríguez, D., Galeano-Díaz, T., Durán-Merás, I. (2010) Determination of trans-resveratrol in red wine by adsorptive stripping square-wave voltammetry with medium exchange, *Food Chem.*, **122**, 1320–1326.
62. Obendorf, D., Stubauer, G., Pharm, B. J. (1995) Adsorptive stripping voltammetry of nicardipine at a HMDE; determination of trace levels nicardipine in blood and urine, *J. Pharm. Biomed. Anal.*, **13**, 1339–1348.

Chapter 3

Electrochemical Nucleic Acid Aptamer–Based Biosensors

Ilaria Palchetti and Marco Mascini
Dipartimento di Chimica, Università degli Studi di Firenze, Sesto Fiorentino (Fi), Italy
ilaria.palchetti@unifi.it

Aptamers are nucleic acid ligands that can be generated against amino acids, drugs, proteins, and other molecules. They are isolated from a large random library of synthetic nucleic acids by an iterative process of binding, separation, and amplification, called systematic evolution of ligands by exponential enrichment (SELEX). In this chapter, the selection procedure and peculiar characteristics of these innovative biomolecules are described. The advancements of a wide range of electrochemical aptamer-based biosensors are reviewed.

3.1 Introduction

In recent years, there has been increasing interest in finding new molecules that are able to mimic antibodies and replace them for diagnostic as well as for therapeutic purposes [1]. Antibodies encounter the challenges of animal immunization, the cost of production, a short shelf life, easy degradation, and pH- and

Electrochemical Biosensors
Edited by Serge Cosnier

ISBN 978-981-4411-46-2 (Hardcover), 978-981-4411-47-9 (eBook)
www.panstanford.com

temperature-dependent stability. Different strategies are currently studied to identify antibody alternatives. Among these, there are nucleic acid aptamers. This chapter deals with these molecules, briefly describing their production, highlighting their peculiarities and their drawbacks, and focusing on biosensing applications [2–6]. In particular, electrochemical aptamer-based biosensors, or aptasensors [7], will be reviewed.

3.1.1 *Nucleic Acid Aptamers*

Nucleic acid aptamers are short, single-stranded DNA or RNA oligonucleotides that adopt stable three-dimensional sequence-dependent structures. This intrinsic property makes them efficient binding molecules capable of binding to molecular targets ranging from small ions and organic molecules to large proteins, viruses, and even cells. Aptamers can vary in size between 25 and 90 bases and adopt complex secondary and tertiary structures (Fig. 3.1), which facilitate specific interactions with other molecules; their name "aptamer" is derived from the Latin word *aptus* (meaning "to fit") and the Greek word *meros* (meaning "part"). This term was first introduced in 1990 by Ellington and Szostak [8] to describe RNA molecules that bind to a small organic dye. Since that time nucleic acid aptamers have been generated against a wide spectrum of targets that differ in dimensions and chemical classes, and currently, nucleic acid aptamers have been selected for more than 150 targets [9]. The functionality of nucleic acid aptamers is based on their stable three-dimensional structure (similar to those observed on natural RNA and DNA), which is dependent on the primary sequence, the length of the nucleic acid molecule, and the environmental conditions [10]. In the presence of a target, the aptamers undergo adaptive conformational changes [10] and their three-dimensional folding creates a specific binding site for the target. The intermolecular interactions between aptamer and target are characterized by a combination of complementarity in shape, stacking interactions between aromatic compounds and the nucleobases of the aptamers, electrostatic interactions between charged groups, and hydrogen bonds [11].

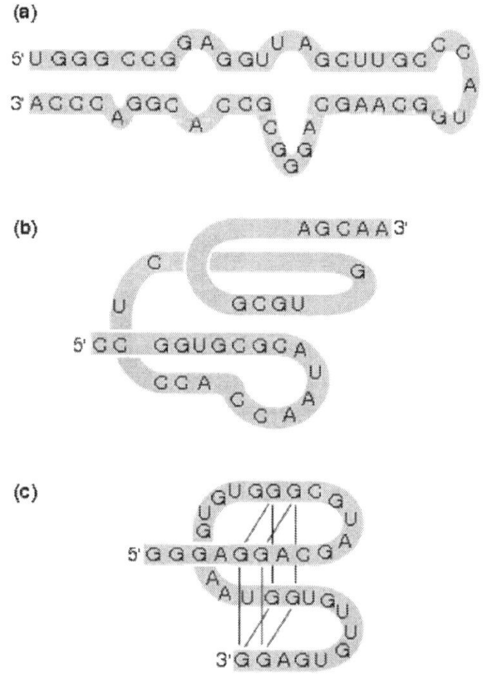

Figure 3.1 Selected examples of secondary structures of aptamers. (a) Internal loops and side bulges embedded within a long stem loop. (b) An aptamer that forms a pseudoknot. (c) An anti-IgE aptamer that contains a central G-quartet stack. Reprinted with permission from Ref. [24].

Aptamers show a big range of affinities for the target molecules, with a dissociation constant from the micromolar to the nanomolar range; they are able to distinguish minor structural differences between targets and their related molecules [12] and to recognize different enantiomers [13].

Aptamers can be obtained via isolation from synthetic combinatorial libraries of nucleic acids through an in vitro selection process called systematic evolution of ligands by exponential enrichment (SELEX) (Fig. 3.2). The SELEX process was first reported in 1990 quasi-simultaneously by two independent laboratories, Szostak [8] and Gold [14]. The technique essentially consists of the repeated binding, selection, and amplification of aptamers from the initial, synthetic combinatorial library of nucleic acids until one (or more)

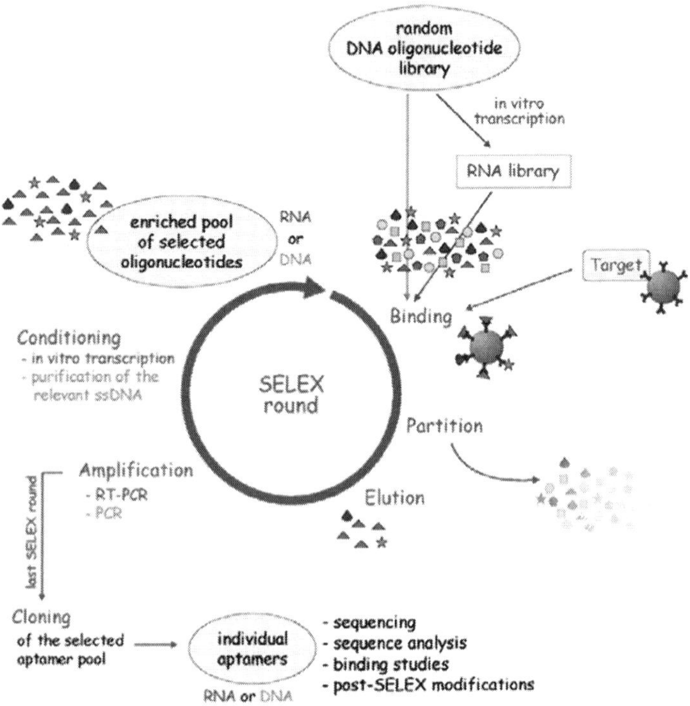

Figure 3.2 Scheme of the SELEX procedure. The starting point of each SELEX process is a synthetic random DNA oligonucleotide library consisting of a multitude of ssDNA fragments with different sequences (10^{15}). This library is used directly for the selection of DNA aptamers. For the selection of RNA aptamers, the library has to be transferred into an RNA library. The SELEX procedure is characterized by the repetition of successive steps consisting of selection (binding, partition, and elution), amplification, and conditioning. In the first SELEX round, the library and the target molecules are incubated for binding. Unbound oligonucleotides are removed by several stringent washing steps of the binding complexes. The target-bound oligonucleotides are eluted and subsequently amplified by PCR or RT-PCR. A new enriched pool of selected oligonucleotides is generated by preparation of the relevant ssDNA from the PCR products (DNA SELEX) or by in vitro transcription (RNA SELEX). This selected oligonucleotide pool is then used for the next selection round. In general, 8 to 15 SELEX cycles are needed for the selection of highly affine, target-specific aptamers. The last SELEX round is finished after the amplification step. The enriched aptamer pool is cloned, and several individual aptamers have to be characterized. Reproduced with permission from Ref. [19].

aptamer displaying the desired characteristics has been isolated. Excellent reviews, book chapters, and research papers dealing with detailed technical aspects of SELEX technology are available [15–26], and therefore, the methodology will be described here briefly and only in a general form, even if several modifications into the SELEX procedure have been introduced in the basic procedure described here [15–26]. The initial and very important step of the SELEX process is the choice and synthesis of the library. Part of the enormous potential of aptamers lies precisely in the fact that libraries with vast numbers of potential ligands can be created and screened rapidly. Typically, aptamer libraries consist of 10^{13}–10^{18} random oligonucleotide sequences, which are chemically synthesized and amplified with the polymerase chain reaction (PCR). In the next step, the target and the library are brought together under favorable binding conditions, where sequences with the highest affinity will bind the target. These sequences are then partitioned from those with lower affinity. This step can be performed by attaching the aptamers to a solid phase support, such as sepharose, and specifically eluting the desired aptamers after binding has taken place. Alternatively, the aptamer and the target could be allowed to interact freely in solution, after which the target–aptamer complex could be recovered by filtration through nitrocellulose. A round of negative selection or counterselection is of great importance to remove the nonspecific binders [27]. The high-affinity sequences are then amplified by reverse transcription polymerase chain reaction (RT-PCR) (for RNA aptamers) or by PCR for DNA aptamers in order to create a new aptamer library enriched with aptamers of high affinity. The entire process is then repeated, resulting in fewer and fewer unique sequences, with higher and higher affinity to the target, being retained. Note that during each round of selection, the binding conditions for the aptamer and the target are generally made more stringent to increase the selective pressure on the remaining aptamers. Generally, a complete SELEX process (between 8 and 15 cycles) will yield a final mixture of no more than 10 aptamers [14].

Characterization of the affinity of aptamer binding is required for estimating the sensitivity and selectivity of appropriate aptasensors. Both DNA and RNA aptamers bind their targets with dissociation constants (K_d) which in the case of proteins are in the low picomolar to low nanomolar range, discriminating between related proteins that share common sets of structural domains. Affinities in the micromolar range can be observed in the case of aptamers—smallmolecule complexes. Aptamers with affinity for a large variety of molecules, including virtually any class of proteins (enzymes, membrane proteins, viral proteins, etc., [27, 28]), drugs, toxins, small organic compounds [29–31], and ions [32] or even a complex target (i.e., living cells [33]) have been isolated. The folding of nucleic acids around the target provides numerous discriminatory intermolecular interactions. These interactions fall in the class of noncovalent bonding as already discussed. Multiple interactions contribute to the same aptamer–target complexes. These molecular interactions govern the specific recognition and discrimination between different target classes.

Nowadays, the most important class of affinity molecules, nonrelated to hybridization, includes antibodies that are produced in vivo by immunizing animals. As reported in Ref. [15], antibodies are generally competent in binding to mostly larger molecules; thus there is a potential niche for aptamers in diagnostics when small molecules are the target [15]. However, the following advantages of aptamers are mostly counted over antibodies: (i) universal approach to an in vitro selection procedure, which does not depend on a particular analyte (possibility to use toxins as well as molecules that do not elicit a good immune response) and that allows us to use nonphysiological conditions, including extremely high or low temperature and pH, and (ii) chemical synthesis, which does not require biological raw materials (bioethics requirements), has minimal batchtobatch differences in activity and is cost effective. Moreover aptamers possess high thermal stability and unlike antibodies can be denatured and renatured multiple times without significant loss of activity [34, 35]. Aptamers can be easily modified with a variety of functional groups (primary amine, thiol groups) or molecules (i.e., biotin) that provide the immobilization of aptamers onto a solid support or the introduction of a label, as well as with a molecular label itself (Fig. 3.3).

The size of aptamers is much smaller than that of antibodies and this feature helps to increase the binding yield on a solid substrate,

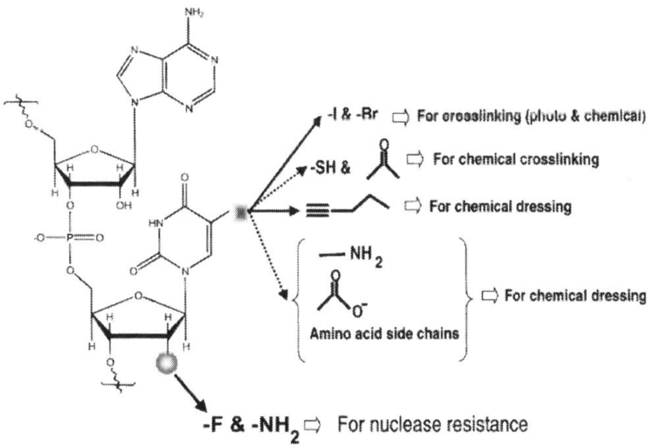

Figure 3.3 Possible modifications on an oligonucleotide strand to generate modified oligonucleotide libraries for the SELEX process. Modification at the 2' position of the sugar confers nuclease stability, whereas various modifications at the C5 position of the pyrimidines could be used either to attract certain classes of targets or to generate covalent crosslinks with targets. Modifications indicated by solid arrows have been used in aptamer identification; those indicated by broken arrows represent some examples of the potential substitutions that could be used for the SELEX process. Reprinted with permission from Ref. [36].

especially when dealing with micro- and nanodimensions of the solid support.

One of main important concerns is aptamer stability to nuclease degradation. Some procedures have been proposed to increase their stability to nuclease, such as the modification of aptamers with 2′-aminopyridine, 2′-fluoropyrimidine, or 2′-o-methyl nucleotides (Fig. 3.3) [36].

3.2 Electrochemical Aptasensors

The majority of electrochemical detection principles described for other affinity reactions [7] are applicable to electrochemical aptasensors. These can be classified basically, as label-free and label-based modes. Label-free modes are based upon the change in electrode surface behavior after the formation of the aptamer–

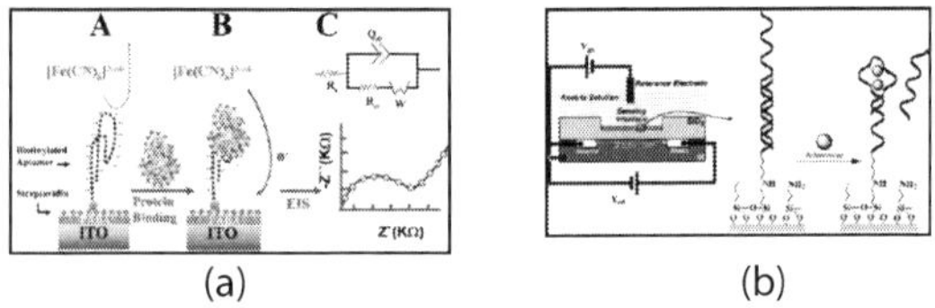

Figure 3.4 Examples of label-free aptasensors (a) Label-free EIS biosensing of proteins at an aptamer-functionalized indium tin oxide (ITO) electrode. Reprinted with permission from Ref. [37]. Aptamers are negatively charged due to the phosphate backbone of the nucleic acids. Without the target protein the negatively charged $[Fe(CN)_6]^{3-/4-}$ marker is repelled from the surface and its redox reaction is hindered (A). When the protein binds to the aptamer probe, the $[Fe(CN)_6]^{3-/4-}$ marker is attracted to the surface and the resistance to electron transfer is decreased (using pH below the pI) (B). Equivalent circuit $R_s(Q_{dl}[R_{et}W])$ used to fit the frequency scans along with an EIS response (C). R_s is the solution resistance, R_{et} is the electron transfer resistance, Q_{dl} is a constant phase element modeling the doublelayer capacitance and W is the Warburg impedance element. (b) Label-free, reagentless aptasensor for adenosine using an ion-sensitive field-effect transistor (ISFET) device. Reprinted with permission from Ref. [38].

target complex (generally monitored by electrochemical impedence spectroscopy [EIS] or a field-effect transistor [FET]) [3, 37–40] or upon the evaluation of the target properties, that is, intrinsic electrochemical responses of the protein (as in the direct electrochemical measurement of the enzymatic product of thrombin captured by the immobilized aptamer, reported in Refs. [4, 41]).

In Fig. 3.4 are reported two examples, reproduced from the literature, of label-free electrochemical aptasensors, based on EIS and FET detection respectively.

Different label modes are reported. Electroactive compounds that can be covalently tethered to an aptamer or bound to an aptamer complementary sequence (which modulates the electrochemical signal upon the formation of an aptamer–target complex) as well as enzymes or metal nanoparticles are some examples of electrochemical labels [7]. Electrochemical label-based aptasensors are here reviewed on the basis of the assay format, and three main classes are considered: 1) direct assay, 2) sandwich assay, and 3) competition-based assay.

The fact that some aptamers fold or make a conformational change upon associating with their molecular targets represents an interesting mechanism that can be exploited in the design of new aptamer-based assays in the direct format using an electroactive label. Various electrochemical assays, based on this approach, have been used for the detection of different targets such as theophylline [42], interferon [43], *Botulinum* neuorotoxin [44], adenosine [45–47], cocaine [48], and thrombin [49]. The interaction of a labeled aptamer with its target can modulate the distance of the electroactive labels from the sensor electrode, thereby altering the redox current. Two strategies are reported, *signal-on* and *signal-off*. In the case of the *signal-on* strategy, the interaction of the target with the aptamer leads to an increased current by shortening the distance from the redox moiety to the electrode surface. In the case of the *signal-off* strategy, a decrease in the current is observed. Examples of this conformational change-based strategy are reported in Fig. 3.5.

Well-folded aptamers that fail to undergo any significant conformational change upon target binding can be re-engineered to generate a binding-induced conformational change [50–54]. When the conformational change is absent or partial and it does not generate any signaling event, a change in the aptamer geometry is necessary through the introduction of an antisense oligonucleotide that hybridizes with the aptamer, thereby keeping it in unfolded form in the absence of the target, or through destabilization of the native aptamer fold by truncation or the introduction of point mutations (Fig. 3.6). An alternative possible alteration of the aptamer geometry exploits the splitting of the aptamers into two suitable segments [54].

The use of a sandwich format allows detection of the target analyte with very high sensitivity and selectivity. Secondary ligands labeled with nanomaterials or enzymes, such as horseradish peroxidase (HRP) or alkaline phosphatase (ALP), are generally employed (Fig. 3.7). Two conditions are required to perform a sandwich assay: 1) The analyte possesses two epitopes, which are so different that both receptors can bind to the analyte without the binding of one affecting the binding of the other, and 2) two aptamers are selected against such an analyte. The disadvantage

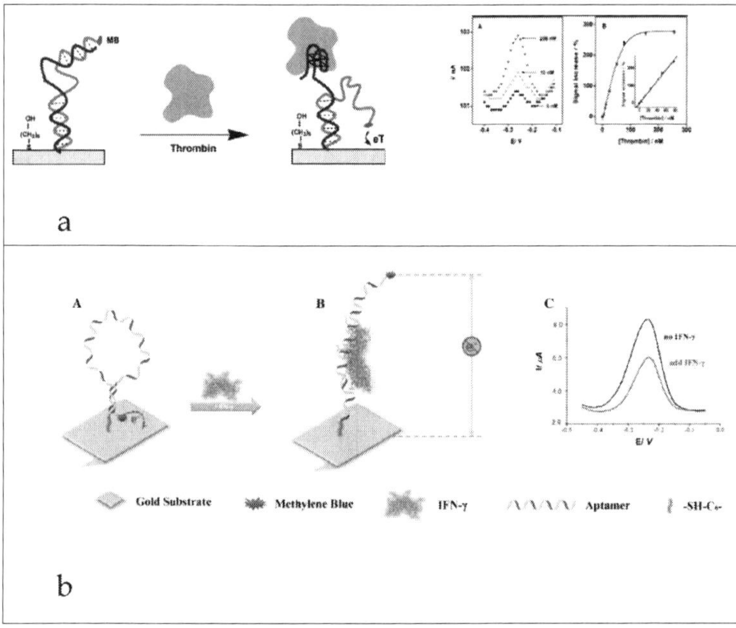

Figure 3.5 Examples of assays based on conformational changes (*signal-on* and *signal-off* detection schemes). (a) The *signal-on* detection scheme. Reprinted with permission from Ref. [49]. A short, methylene blue (MB)-tagged oligonucleotide that hybridizes with both the thrombin-binding portion of the aptamer and the DNA sequence linking the aptamer to the electrode. The formation of these regions of rigid, duplex DNA prevents the MB tag from approaching the electrode surface, suppressing faradic currents. Thrombin binding stabilizes the alternative G-quadruplex conformation of the aptamer, liberating the 5′ end of the tagged oligonucleotide to produce a flexible, single-stranded element. This, in turn, allows the MB tag to collide with the electrode surface, producing a readily detectable faradic current. The current values increase with the concentration of the target. (b) Example of the *signal-off* detection scheme applied to interferon (IFN-γ) detection. Reprinted with permission from Ref. [43]. (A) The aptamer hairpin was thiolated at the 5′ end, allowing self-assembly on gold electrodes. A redox label was attached at the 3′ end of the hairpin and was in close proximity to the electrode surface. (B) Upon addition of IFN-γ the aptamer hairpin changed confirmation and the redox label moved further away from the electrode, lowering the electron transfer efficiency. (C) The differences in the faradic current before and after addition of IFN-γ were quantified using square-wave voltammetry (SWV).

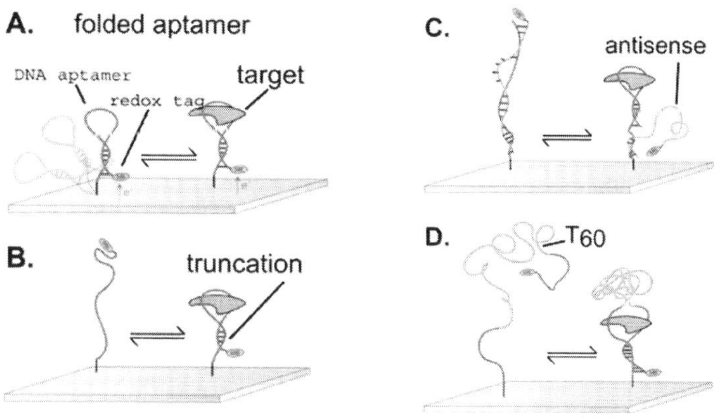

Figure 3.6 Some strategies for re-engineering aptamers to generate a binding conformational change. In (A) is reported a full-length, folded aptamer. Several approaches have been reported, however, by which aptamers can be engineered to undergo large-scale, binding-induced conformational changes that significantly improve the signal. These approaches include (B) the destabilization of the wild-type aptamer via introduction of sequence truncations or point mutations, (C) the introduction of antisense sequences, or (D) the introduction of long unstructured sequences internal to the aptamer. The overarching goal of each of these approaches is to create an alternative structure in equilibrium with the "native," target-binding fold such that the presence of the target pushes the equilibrium back to this native fold. Reprinted with permission from Ref. [52].

related to this format is the several incubation steps that make the assay time consuming. To overcome the fact that only very few molecules (thrombin and platelet-derived growth factor [PDGF]) possess two aptamers binding to two different sites, many authors have developed aptamer-based assays by using for the sandwich either the same aptamer as the primary and secondary ligands [55, 56] or an aptamer and an antibody as ligands for the sandwich [57].

The advantages of a competitive assay format are mainly related to the fact that only one aptamer is required (considering that two or more aptamers are not selected for many target analytes) and the time necessary for the assay is faster. An example of an electrochemical competitive assay is that reported in Ref. [58] for neomycin B. The gold electrode surface was modified with

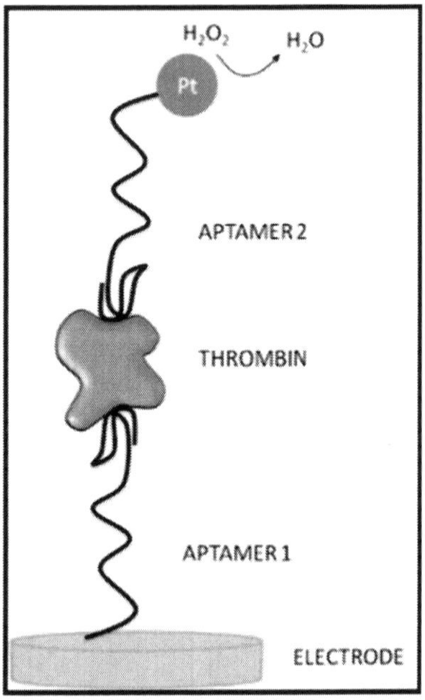

Figure 3.7 A sandwich assay for the detection of thrombin using aptamers and platinum nanoparticles (Pt-NPs) as labels. Au electrodes were modified with the thiolated thrombin aptamer. The Au electrodes were subjected to solutions with different concentrations of thrombin. The modified electrodes were then reacted with the Pt-NPs that were modified with the thiolated secondary aptamer. As thrombin includes two binding sites for the aptamer, the nucleic acid–functionalized Pt-NPs bind to the thrombin complex associated with the surface. The Pt-NP labels associated with thrombin were then used as sites for the electrocatalytic reduction of H_2O_2 monitored by voltammetry. Reprinted with permission from Ref. [63].

neomycin B, followed by exposition to varying concentrations of neomycin B and a fixed concentration of the aptamer, leading to competition to the aptamer. The authors preferred this design to the conventional electrode-bound aptamers to magnify the impedance change associated with aptamer displacement by the solution-phase analyte. A similar approach is reported in Ref. [59].

Other examples of electrochemical aptamer-based biosensors can be found in Refs. [5, 60–62].

3.3 Conclusion: Criticisms and Future Perspectives

This chapter did not attempt to provide a comprehensive survey of different electrochemical aptasensors but aimed to give some general overview on this field. One important comment that arose looking at the literature was the fact that despite the hundreds of aptamers available nowadays, more or less the same 10–15 aptamers are always used to develop electrochemical devices. Moreover, a great part of these studies report analysis on standard conditions only; major information regarding their behavior in real matrices is still needed in order to fully validate the aptasensor-based analytical methodologies. A few real samples or treated samples often fail to provide an adequate measure of capability for real-world samples, leading to failed technology transfer and investment.

Nevertheless, looking at the analytical parameters reported in the literature [64], we can conclude that nanomaterials are really interesting systems to obtain more and more sensitive devices. Nanomaterials stimulate the mind of researches to find new intriguing strategies for detection with really interesting analytical performances. These innovative materials will benefit from the peculiar features of aptamers, as affinity bioreceptors, and from electrochemical transduction to develop integrated nanoscale devices that will find applications in different areas from biosensing to nanomedicine.

References

1. Binz, H. K., Amstutz, P., Plückthun, A. (2005) Engineering novel binding proteins from nonimmunoglobulin domains, *Nat. Biotechnol.*, **23**, 1257.
2. Levy, M., Carter, S. F., Ellington, A. D. (2005) Quantum-dot aptamer beacons for the detection of proteins, *ChemBiochem*, **6**, 2163.
3. Willner, I., Zayats, M. (2007) Electronic aptamer-based sensors, *Angew. Chem., Int. Ed.*, **46**, 6408.
4. Centi, S., Messina, G., Tombelli, S., Palchetti, I., Mascini, M. (2008) Different approaches for the detection of thrombin by an electrochemical, *Biosens. Bioelectron.*, **23**, 1602.

5. Strehlitz, B., Nikolaus, N., Stoltenburg, R. (2008) Protein detection with aptamer biosensors, *Sensors*, **8**, 4296–4307.
6. Song, S., Wang, L., Li, J., Zhao, J., Fan, C. (2008) Aptamer-based biosensors, *TrAC*, 108–117.
7. Labuda, J., Oliveira Brett, A. M. Evtugyn, G. et al. (2010) Electrochemical nucleic acid-based biosensors: concepts, terms, and methodology, *Pure Appl. Chem.*, **82**, 1161–1187.
8. Ellington, A. D., Szostak, J. W. (1990) In vitro selection of RNA molecules that bind specific ligands, *Nature*, **346**, 818.
9. http://aptamerstream.blogspot.com
10. Ferre-D'Amare, A. R., Doudna, J. A. (1999) RNA folds: insights from recent crystal structures, *Annu. Rev. Biophys. Biomol. Struct.*, **28**, 57.
11. Hermann, T., Patel, D. J. (2000) Adaptive recognition by nucleic acid aptamers, *Science*, **287**, 820.
12. Jenison, R. D., Gill, S. C., Pardi, A., Polinsky, B. (1994) High resolution molecular discrimination by RNA, *Science*, **263**, 1425.
13. Geiger, A., Burgstaller, P., von der Eltz, H., Roeder, A., Famulok, M. (1996) In vitro selection analysis of neomycin binding, *Nucleic Acids Res.*, **24**, 1029.
14. Tuerk, C., Gold, L. (1990) Systematic evolution of ligands by exponential enrichment: RNA ligands to bacteriophage T4 DNA polymerase, *Science*, **249**, 505.
15. Lee, J. H., Yigit, M. V., Mazumdarc, D., Lu, Y. (2010) Molecular diagnostic and drug delivery agents based on aptamer-nanomaterial conjugates, *Adv. Drug Delivery Rev.*, **62**, 592.
16. James, W. (2000) *Encyclopedia of Analytical Chemistry*, Ed. R. A. Meyers (John Wiley and Sons, Chichester), 4848.
17. Mayer, G. (2009) The chemical biology of aptamers, *Angew. Chem., Int. Ed.*, **48**, 2672.
18. Mayer, G. (Ed.) (2009), *Nucleic Acid and Peptide Aptamers: Methods and Protocols*, Vol. 535 (Humana Press, New York).
19. Stoltenburg, R., Reinemann, C., Strehlitz, B. (2007) SELEX–a (r)evolutionary method to generate high-affinity nucleic acidligands, *Biomol. Eng.*, **24** 381–403.
20. Klussman, S. (Ed.) (2006) *The Aptamer Handbook* (Wiley-VCH, Weinheim).
21. Mascini, M., Palchetti, I., Tombelli, S. (2012) Nucleic acid and peptide aptamers: fundamentals and bioanalytical aspects, *Angew. Chem., Int. Ed.*, **51**, 1316.

22. Eulberg, D., Klussmann, S. (2003) Spiegelmers: biostable aptamers, *ChemBioChem*, **4**, 979.

23. Ellington, A. D., Szostak, J. W. (1992) Selection in vitro of single-stranded-DNA molecules that fold into specific ligand-binding structures, *Nature*, **355**, 850.

24. Bacher, J. M., Ellington, A. D. (1998) Nucleic acid selection as a tool for drug discovery, *Drug Discovery Today*, **3**, 265–273.

25. Kopylov, A. M., Spiridonova, V. A. (2000) Combinatorial chemistry of nucleic acids: SELEX, *Mol. Biol.*, **34**, 940.

26. Osborne, S. E., Ellington, A. D. (1997) Nucleic acid selection and the challenge of combinatorial chemistry, *Chem. Rev.*, **97**, 349–370.

27. Geiger, A., Burgstaller, P., von der Eltz, H., Roeder, A., Famulok, M. (1996) RNA aptamers that bind L-arginine with sub-micromolar dissociation constants and high enantioselectivity, *Nucleic Acids Res.*, **24**, 1029–1036.

28. Kraus, E., James, W., Barclay, A. N. (1998) Novel RNA ligands. Able to bind CD4 antigen and inhibit CD4+ T lymphocyte function, *J. Immunol.*, **160**, 5209.

29. Wilson, C., Szostak, J. W. (1998) Isolation of a fluorophore-specific DNA aptamer with weak redox activity, *Chem. Biol.*, **5**, 609.

30. Yang, Q., Goldstein, I. J., Mei, H. Y., Engelke, D. R. (1998) DNA ligands that bind tightly and selectively to cellobiose, *Proc. Natl. Acad. Sci. U. S. A.*, **95**, 5462.

31. Famulok, M., Huttenhofer, A. (1996) In vitro selection analysis of neomycin binding RNAs with a mutagenized pool of variants of the 16S rRNA decoding region, *Biochemistry*, **35**, 4265.

32. Ciesiolka, J., Gorski, J., Yarus, M. (1995) Selection of an RNA domain that binds Zn2+, *RNA*, **1**, 538.

33. Phillips, J. A., Lopez-Colon, D., Zhu, Z., Xu, Y., Tan, W. (2008) Applications of aptamers in cancer cell biology, *Anal. Chim. Acta*, **621**, 101.

34. Langer, R. (1998) Drug delivery and targeting, *Nature*, **392**, 5.

35. Liss, M., Petersen, B., Wolf, H., Prohaska, E. (2002) An aptamer-based quartz crystal protein biosensor, *Anal. Chem.*, **74**, 4488.

36. Jayasena, S. D. (1999) Aptamers: An emerging class of molecules that rival antibodies in diagnostics, *Clin. Chem.*, **45**, 1628–1650.

37. Rodriguez, M. C., Kawde, A.-N., Wang, J. (2005) Aptamer biosensor for label-free impedance spectroscopy detection of proteins based on recognition-induced switching of the surface charge, *Chem. Commun.*, **34**, 4267.

38. Zayats, M., Huang, Y., Gill, R., Ma, C. A., Willner, I. (2006) Label-free reagentless aptamer based sensors for small molecules, *J. Am. Chem. Soc.*, **128**, 13666.
39. Xu, D., Yu, X., Liu, Z., He, W., Ma, Z. (2005) Label-free electrochemical detection for aptamer-based array electrodes, *Anal. Chem.*, **77**, 5107.
40. Radi, A. E., Sanchez, J. L. A., Baldrich, E., O'Sullivan, C. K. (2005) Reusable impedimetric aptasensor, *Anal. Chem.*, **77**, 6320.
41. Mir, M., Vreeke, M., Katakis, I. (2006) Different strategies to develop an electrochemical thrombin aptasensor, *Electrochem. Commun.*, **8**, 506.
42. Ferapontova, E. E., Gothelf, K. V. (2009) Effect of serum on an RNA aptamer-based electrochemical sensor for theophylline, *Electroanalysis*, **21**, 261.
43. Liu, Y., Tuleouva, N., Ramanculov, E., Revzin, A. (2010) Aptamer-based electrochemical biosensor for interferon gamma detection, *Anal. Chem.*, **82**, 8131–8136.
44. Wei, F., Ho, C. M. (2009) Aptamer-based electrochemical biosensor for Botulinum neurotoxin, *Anal. Bioanal. Chem.*, **393**, 1943.
45. Wang, J., Wang, F., Dong, S. (2009) Methylene blue as an indicator for sensitive electrochemical detection of adenosine based on aptamer switch, *J. Electroanal. Chem.*, **626**, 1.
46. Zhou, X., Song, S., Zhang, J., Pan, D., Wang, L., Fan, C. (2007) A target-responsive electrochemical aptamer switch (TREAS) for reagentless detection of nanomolar ATP, *J. Am. Chem. Soc.*, **129**, 1042.
47. Ceretti, H., Ponce, B., Ramirez, S. A., Montserrat, J. M. (2010) Adenosine reagentless electrochemical aptasensor using a phosphorothioate immobilization strategy, *Electroanalysis*, **22**, 147.
48. Swensen, J. S., Xiao, Y., Ferguson, B. S. et al. (2009) Continuous, real-time monitoring of cocaine in undiluted, unmodified blood serum via a microfluidic, aptamer-based sensor, *J. Am. Chem. Soc.*, **131**, 4262.
49. Xiao, Y., Piorek, D., Plaxco, K. W., Heeger, A. J. (2005) A reagentless, signal-on design for electronic aptamer-based sensors via target-induced strand displacement, *J. Am. Chem. Soc.*, **127**, 17990.
50. Xiao, Y., Uzawa, T., White, R. J., DeMartini, D., Plaxco, K. W. (2009) On the signaling of electrochemical, aptamer-based sensors: collision- and folding-based mechanisms, *Electroanalysis*, **21**, 1267.
51. Radi, A.-E., Acero Sanchez, J. L., Baldrich, E., O'Sullivan, C. K. (2006) Reagentless, reusable, ultrasensitive electrochemical molecular beacon aptasensor, *J. Am. Chem. Soc.*, **128**, 117–124.

52. White, R. J., Rowe, A. A., Plaxco, K. W. (2010) Re-engineering aptamers to support reagentless, self-reporting electrochemical sensors, *Analyst*, **135**, 589.
53. Lai, R. Y., Plaxco, K. W., Heeger, A. J. (2007) Rapid, aptamer-based electrochemical detection of platelet-derived growth factor at picomolar concentrations directly in blood serum, *Anal. Chem.*, **79**, 229.
54. Golub, E., Pelossof, G., Freeman, R., Zhang, H., Willner, I. (2009) Detection of picomolar platelet-derived growth factor directly in blood serum, *Anal. Chem.*, **81**, 9291.
55. Wang, H., Liu, Y., Liu, C., Huang, J., Yang, P., Liu, B. (2010) Microfluidic chip-based aptasensor for amplified electrochemical detection of human thrombin, *Electrochem. Commun.*, **12**, 258–261.
56. Palchetti, I., Mascini, M. (2012) Electrochemical nanomaterial-based nucleic acid aptasensors, *Anal. Bioanal. Chem.*, **402**, 3103–3114.
57. Centi, S., Bonel Sanmartin, L., Tombelli, S., Palchetti, I., Mascini, M. (2009) Detection of C reactive protein (CRP) in serum by an electrochemical aptamer-based sandwich assay, *Electroanalysis*, **21**, 1309–1315.
58. de-los-Santos-Álvarez, N., Lobo-Castañón, M. J., Miranda-Ordieres, A. J., Tuñón-Blanco, P. (2007) Modified-RNA aptamer-based sensor for competitive impedimetric assay of neomycin B, *J. Am. Chem. Soc.*, **129**, 3808.
59. Papamichael, K. I., Kreuzer, M. P., Guilbault, G. G. (2007) Viability of allergy (IgE) detection using an alternative aptamer receptor and electrochemical mean, *Sens. Actuators, B*, **121**, 178.
60. Hianik, T., Wang, J. (2009) Electrochemical aptasensors: recent achievements and Perspective, *Electroanalysis*, **21**, 1223–1235.
61. Cheng, A. K. H., Sen, D., Yu, H. Z. (2009) Design and testing of aptamer-based electrochemical biosensors for small molecules and proteins, *Bioelectrochemistry*, **77**, 1.
62. Xu, Y., Cheng, G., He, P., Fang, Y. (2009) A review: electrochemical aptasensors with various detection strategies, *Electroanalysis*, **21**, 1251.
63. Polsky, R., Gill, R., Kaganovsky, L., Wilner, I. (2006) Nucleic acid-functionalized Pt nanoparticles: catalytic labels for the amplified electrochemical detection of biomolecules, *Anal. Chem.*, **78**, 2268.
64. Willner, I., Willner, B. (2010) Biomolecule-based nanomaterials and nanostructures, *Nano Lett.*, **10**, 3805–3815.

Chapter 4

Amperometric Enzyme Electrodes

Serge Cosnier[a] and Gareth P. Keeley[b]
[a]*UMR 5250, Département de Chimie Moléculaire, Université Joseph Fourier, 38041 Grenoble Cedex 9, France*
[b]*Department of Chemistry, Trinity College, Dublin 2, Ireland*
serge.cosnier@ujf-grenoble.fr

4.1 Introduction

Rapid detection and monitoring, in clinical and food diagnostics and in environmental and biodefense, have paved the way for the development of alternative analytical devices based on the molecular recognition properties of macromolecular biomolecules and their electrochemical transduction. Such devices include enzyme biosensors; DNA, aptamer, and protein sensors; immunosensors; and biochips. The last decade has seen truly phenomenal growth in the field of electrochemical (mainly amperometric) biosensors with applications in different areas, including food and beverage analysis (e.g., presence of genetically modified organisms [GMOs]), medical applications (e.g., diagnosis of genetic disorders), detection of pathogens, forensic applications, drug response measurement, cancer diagnosis, and environmental issues [1–4]. Concerning the

Electrochemical Biosensors
Edited by Serge Cosnier

ISBN 978-981-4411-46-2 (Hardcover), 978-981-4411-47-9 (eBook)
www.panstanford.com

last of these, as a result of human and technological development, hundreds of new chemicals of unknown toxicity and effects on human health are released into the environment every year. The increasing number of pollutants to monitor and the alarming health and environmental consequences involved result in increasing scientific and social concerns in this area. For example, a growing number of initiatives and legislative actions for stricter environmental pollution control, with particular emphasis on water quality control for sustainable use of water, has been adopted in recent years. The sustainable use of water, in fact, is desirable in terms of addressing the growing human impact and increasing demand for drinking water. In this context, the contamination of water by river water and groundwater discharge and widespread use of herbicides, insecticides, fungicides, and chemicals like cadmium, lead, mercury, polychlorinated biphenyls (PCBs), pentachlorophenol (PCP), DDT, and dioxins requires techniques to quantify and evaluate the impact of these pollutants on human health. Indeed, the scientific community has established correlations between these contaminants and a whole range of reproductive and other health problems, both in wildlife and in humans. In particular, synthetic hormones and certain chemical products used in industry and agriculture are capable of disrupting the endocrinal system. Although there are many tests, some of them are insufficiently sensitive. The stricter environmental control urgently calls for simple, fast, sensitive, selective, cost-effective, real-time, on-site, and field-portable monitoring technologies with negligible waste generation that correspond to the definition of amperometric biosensors.

Generally, the biological element is immobilized onto a transducer that can transform the molecular recognition process into a thermal, gravimetric, optical, or electrochemical signal proportional to the target concentration. Among the electrochemical transducing techniques (potentiometry, voltammetry, conductimetry, impedance spectrometry), amperometric measurement is a relatively mature field. The latter consists of holding the bioelectrode at a constant potential, while the current resulting from the oxidation or reduction of electroactive species is monitored. The latter can be generated or consumed by an enzymatic reaction used as a recognition process of a target or as a label for immunosensors or protein and DNA sensors.

4.2 Functioning Principles of Amperometric Enzyme Electrodes

More than four decades ago, Clark and Lyons reported the ingenious concept of intimately combining the recognition properties of an enzyme with the sensitivity of an amperometric transduction. This pioneering work led to the most successful electrochemical biosensor for the monitoring of blood glucose on the basis of glucose oxidase (GOx) [5].

This example is the source of a multitude of biosensors dedicated to glucose detection. The transduction step is normally based on the enzymatic production of hydrogen peroxide and its oxidative amperometric detection. GOx contains one tightly bound flavin adenine dinucleotide (FAD) unit per monomer. In a glucose biosensor, FAD oxidizes glucose and in the process is reduced to $FADH_2$. In the presence of oxygen, the active oxidized form is regenerated and hydrogen peroxide is produced. This naturally occurring "ping pong" mechanism is illustrated in Fig. 4.1.

Typically, the progress of the biochemical reaction is followed by oxidizing the peroxide produced, so the most common experiment involves poising the electrode at an anodic potential (0.6–0.7 V vs. a saturated calomel electrode [SCE]) and measuring the current response to glucose additions under stirred conditions. The latter are employed to eliminate the possibility of the rate being limited by substrate diffusion from bulk solution. It can be shown that the recorded current varies with the substrate concentration according to

$$I_S \frac{I_{max}[S]}{K_M + [S]}$$

where I_S is the steady-state current flowing, I_{max} is the maximum current under saturated substrate conditions [6], and $[S]$ is the

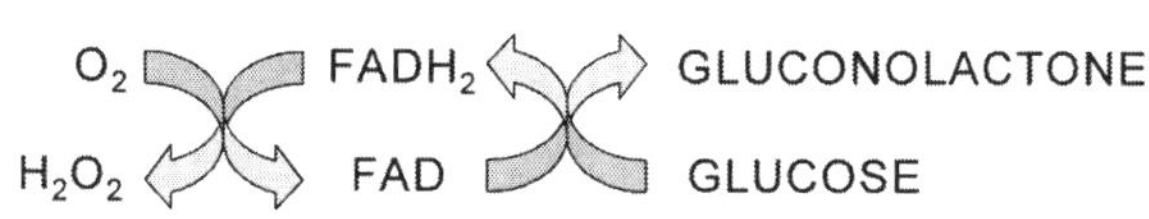

Figure 4.1 Ping-pong mechanism for the oxidation of glucose, catalyzed by FAD.

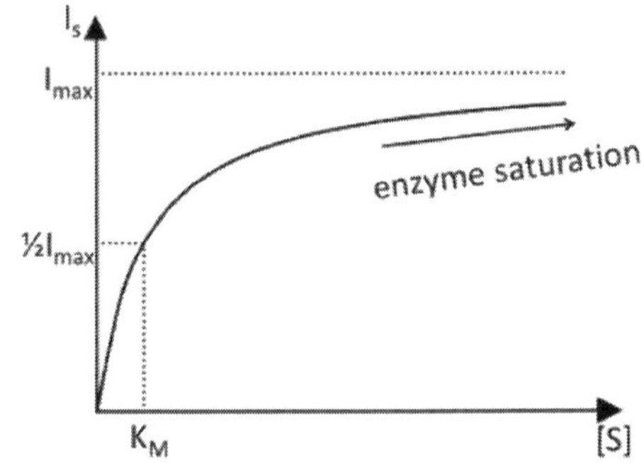

Figure 4.2 Schematic description of the amperometric biosensor response as a function of the substrate concentration.

substrate concentration. The term K_M is the apparent Michaelis–Menten constant, which has units of concentration. It should be noted that the caveat "apparent" is used because this constant characterizes the enzyme *electrode* and does not represent an intrinsic property of the enzyme itself. Its value may be affected by the diffusional constraints to which the substrate or product is subjected within the immobilized film. This equation predicts that the current is initially proportional to the substrate concentration, tending toward (but never reaching) an upper limit at larger $[S]$ values. The result is a rectangular hyperbola, as illustrated in Fig. 4.2.

Note from the equation that $(K_M, 1/2\ I_{max})$ is a point on the curve, so K_M can be estimated simply by taking concentration value corresponding to half the maximum current. The first part of the calibration curve corresponds to the concentration range where the enzymatic reaction rate increases linearly with substrate concentration. This linear region, where the reaction is of the first order with respect to the substrate concentration, defines the useful concentration range of the biosensor. The sensitivity of the biosensor is given by the slope of the initial linear part of the calibration curve, and its detection limit is conventionally calculated by dividing the sensitivity into thrice the standard deviation of the current residuals [7]. However, practical detection limits, based on a signal-to-noise ratio of 3, are thought to be more important. The calibration curve then bends gradually at higher concentrations, reaching a pseudo-plateau. This phenomenon reflects the maximum enzyme rate under saturated substrate conditions and corresponds to a maximum current response that is generally proportional to the amount of enzyme immobilized on the electrode surface. The comparison of biosensor performance is conventionally based on sensitivity, linear range, and detection limit.

This concept of enzyme electrodes was quickly extended to other oxidases that catalyze the oxidation of substrates such as oligosaccharides, alcohols, or neurotransmitters (dopamine, adrenaline, choline, and glutamate), leading to a wide range of amperometric biosensors.

In a similar way, many amperometric biosensors have been developed with dehydrogenases as the biomolecular recognition protein. These enzymes assume the target oxidation or reduction with the concomitant reduction or oxidation of their coenzyme factor: $NAD(P)^{+}/NAD(P)H$. Usually, dehydrogenase electrodes are employed for detecting a target through its oxidation by transferring two electrons and one proton to the oxidized form of the cofactor ($NAD(P)^{+}$). The amperometric signal is, thus, based on the oxidation of NAD(P)H. The reverse approach, namely, the use of a dehydrogenase biosensor for the detection of a substrate via its reduction, has been scarcely reported [8]. This is mainly due to the fact that the electrochemical oxidation of NAD(P)H always leads to the enzymatically active oxidized form, whereas the electrochemical reduction of $NAD(P)^{+}$ may provide dimers or other enantiomeric NAD(P)H forms that cannot be used again by the enzyme.

In parallel with these two major kinds of enzyme biosensors, a wide range of enzymes that catalyze the production or consumption of electroactive species has also been exploited for the development of enzyme electrodes. For instance, laccase, tyrosinase, and peroxidase catalyze the oxidation of phenolic compounds, while phosphatases and hydrolases generate phenol derivatives, the amperometric signal being based on the reduction of quinoid products or the oxidation of phenolic compounds, respectively. For example, the detection of parathion, an organophosphorus pesticide, has been achieved by a hydrolase electrode. This biosensor catalyzes the hydrolysis of parathion into nitrophenol, which is subsequently detected through its amperometric oxidation at the electrode

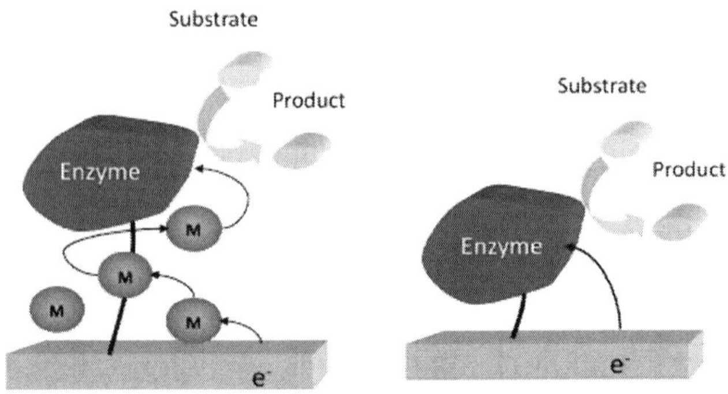

Figure 4.3 Schematic representation of mediated and direct electrical wiring of an immobilized enzyme.

surface [9]. Such a conventional approach provides detection limits lower than 1 μg/L.

Different strategies for amperometric detection have been reported involving either the oxidation or reduction of one component of the enzymatic reaction or an electron exchange with the prosthetic site of the enzyme (Fig. 4.3). The electrical connection of an oxidase may enhance its lifetime by suppressing the enzymatic generation of H_2O_2 and its deleterious effect. The enzyme wiring may also improve the selectivity of the amperometric transduction step and hence the efficiency of the biosensor. Generally, amperometric biosensors are restricted to enzymes able to catalyze the formation of electroactive species, such as hydrogen peroxide and NAD(P)H. However, the biosensor specificity is markedly reduced by the poor selectivity of the amperometric transduction step based on the oxidative detection of these electroactive species. The presence of electrical communication between the immobilized enzyme and the electrode surface should decrease the oxidation potential and hence avoid potential interferences with easily oxidizable metabolites such as uric acid or ascorbate present in physiological fluids. This electrical wiring also permits the establishment of amperometric communication with enzymes that are not able to catalyze the production or consumption of electroactive species.

The enzyme wiring is classified as direct electron transfer (DET) or mediated electron transfer if this process is performed via the use of redox mediators. The latter are commonly small redox-active molecules with good electron transfer kinetics that present a compatible geometrical configuration for the enzymatic active site. They are used as the electron donor/acceptor between the enzyme and the electrode surface, and their redox potential has to be as close as possible to the redox potential of the active center of the enzyme. Electron hopping between immobilized redox centers, thus, ensures the electron transport between the enzymes and the electrode surface.

4.2.1 *Enzyme Electrodes Based on Mediated Electron Transfer*

Usually, most oxidoreductases have their prosthetic group deeply buried within the protein shell. The spatial separation between this redox site and the electrode surface constitutes, in fact, a barrier to electron exchange. Generally, the electrical wiring of an enzyme is performed by small and mobile redox mediators. The latter can be freely diffusing in solution or grafted onto the electrode matrix or the biomolecule itself by a long spacer arm. The immobilization of redox mediators has also been accomplished by electrochemical polymerization or through their inclusion into electrode materials by electrostatic interactions or steric constraints.

The easier and more widely used approach to connect enzymes immobilized in organic or inorganic materials is their coentrapment with redox mediators [10]. For instance, the simultaneous immobilization of a pyrroloquinoline quinone and oxidases during the electrogeneration of polypyrrole films provides an amperometric biosensor operating at a low redox potential (−110 mV vs. Ag/AgCl) and hence without interferences due to easily oxidizable substances [11].

Biosensor fabrication based on inorganic gels consists of the adsorption of an enzyme clay aqueous mixture onto an electrode surface followed by chemical crosslinking by glutaraldehyde of the entrapped proteins. Moreover, the ion exchange properties of synthetic clays have been exploited to incorporate negatively or positively charged redox mediators within this matrix containing enzymes. In particular, synthetic anionic clays like layered double hydroxides constitute an attractive material for the design of amperometric biosensors based on wired enzymes. Thanks to their anion exchange properties, negatively charged redox mediators like 2,2′-azinobis 3-ethylbenzothiazoline-6-sulphonate (ABTS), anthraquinone mono- and disulphonate, or ferrocene sulphonate can be intercalated into the interlayer spacing of anionic clays. The incorporated redox mediators act as electron shuttles between the electrode and the prosthetic center of enzymes. The anion exchange properties of synthetic clays have been thus exploited to immobilize simultaneously peroxidase and ABTS, providing an efficient amperometric biosensor for H_2O_2 at 0.0 V [12]. The latter, based on the electrical wiring of peroxidase, provides an excellent sensitivity and detection limit, namely, 443 mA M^{-1} cm^{-2} and 10 nM, respectively. In addition, the coimmobilization of peroxidase and GOx leads to the selective detection of glucose. This biosensor was also employed for the sensitive determination of cyanide via its noncompetitive inhibiting action on peroxidase, providing a sensitive detection limit of 10^{-9} M [13]. In the same vein, a clay-ABTS matrix was employed for the immobilization and wiring of laccase. Taking into account that laccase catalyzes the four-electron reduction of oxygen into water by oxidizing ABTS anions, the resulting bioelectrode allows the amperometric monitoring of oxygen in the concentration range of 6×10^{-8} to 4×10^{-6} M [14]. Nevertheless, it should be noted that one of the limitations of enzyme–clay electrodes lies in the nonconductive nature of the clay nanoparticles, the electron transfer being assumed by an electronhopping mechanism between intercalated redox species.

However, the major drawback of the incorporation of redox mediators is their low mechanical stability. The latter leads to a slow but steady release of these mediators in solution, leading to a decrease in the efficiency of the enzyme connection and thus to a decrease in the biosensor response. An alternative consists of covalently linking the redox mediator to the matrix. Thus, osmium-based hydrogels were developed and now constitute one of the most powerful host redox matrixes for the electrical connection of enzymes. One major advantage of polypyridyl osmium complexes lies in the easy tuning of the redox potential of the Os center by modulating the *N*-heterocyclic ligand structure with electron-donating or electron-withdrawing groups. The Os complex is generally linked to a polyvinylpyridine (PVP) backbone, which is subsequently deposited on the electrode surface in presence of enzymes. The latter are entrapped inside the osmium gel by using poly(ethylene glycol) diglycidyl ether. For instance, the electrical wiring of enzymes such as GOx, bilirubin oxidase, or peroxidase was carried out with an osmium–polyvinyl pyridine gel polymer. An amperometric biosensor, based on osmium polymer gel and gluthatione sulhydryl oxidase combined with peroxidase, was elaborated for the detection of gluthatione, an indicator of oxidative stress [15]. The biosensor presents a bilayer configuration with an external layer containing gluthatione oxidase, while the internal layer consists of peroxidase wired by the osmium complex. Its functioning principle was based on the enzymatic oxidation of gluthatione linked to the production of hydrogen peroxide followed by the catalytic reduction of hydrogen peroxide by the wired peroxidase. The resulting amperometric biosensor exhibits a sensitive detection limit of 500 nM at a low operating potential.

Another way to use osmium gel consists of the electrodeposition of the redox polymer by controlled potential electrolysis at −1.4 V versus Ag/AgCl in phosphate buffer. A nanostructured multilayered biosensor for glucose was thus fabricated by successive layer-by-layer deposition of carbon nanotubes, osmium polymer, and GOx [16]. As expected, the redox signal of the Os complex increases with the number of deposited layers and leads to an anodic catalytic current at 0.3 V in the presence of glucose.

It should be noted that the sensitivity exhibited by this amperometric biosensor based on wired GOx (16.4 mA M^{-1} cm^{-2}) appears really attractive compared to the highest sensitivities reported for conventional glucose biosensors based on the amperometric detection of hydrogen peroxide, namely, 82–106 mA M^{-1} cm^{-2} [16, 17].

Another approach concerns the development of electropolymerized films functionalized by redox moieties. The covalent binding of redox mediators to the polymer skeleton has been initiated by the functionalization of ferrocene and osmium groups redox by an electropolymerizable pyrrolic group [18].

4.2.2 *Enzyme Electrodes Based on Direct Electron Transfer*

Generally, DET is observed for enzymes that have a redox site involved in the catalytic process or the prosthetic site itself, localized near the surface of the surrounding protein. Moreover, even if the active center of the enzyme is ideally located near the protein shell or on its surface, the enzyme immobilization should lead to a correct orientation of the enzyme with respect to the surface, while not impeding the access of substrate to the enzyme. Enzymes also can contain several redox systems that constitute an electroactive chain from the prosthetic site located inside the protein to a redox group located at the periphery. For instance, cellobiose dehydrogenase, which catalyzes the oxidation of carbohydrates like cellobiose, glucose or lactose, contains a dehydrogenase domain with a FAD and a cytochrome domain with a heme b. The enzymatic oxidation of carbohydrates involves the reduction of FAD. Then the electrons are donated to the heme group by internal electron transfer and injected to the electrode surface by DET [19].

Conducting polymers, polyelectrolytes, self-assembled layers, ionic liquids, and nanoparticles are used to establish direct transfer of electrons between the enzyme and an electrode. For instance, the use of metal nanoparticles constitutes a flexible way to entrap enzymes at electrode surfaces. Thanks to the easy functionalization of gold nanoparticles via thiol derivatives, these nanoparticles were widely used for the elaboration of enzymatic biosensors [20, 21]. In particular, polymer films like polyaniline or chitosan were combined with gold nanoparticles to produce hybrid matrixes, leading to the establishement of DET with redox enzymes such as bilirubin oxidase or glucose dehydrogenase [22].

However, most oxidoreductases do not communicate easily with the electrode surface because, as mentioned earlier, their prosthetic group is deeply buried within the protein shell. In this context, DET has been achieved using nanomaterials like carbon nanotubes or graphite nanoparticles. Because of their high conductivity and high specific surface area (up to 1000 m^2g^{-1}) combined with mechanical strength, carbon nanotubes are promising materials for the development of nanostructured deposits exhibiting direct electrical connections with enzymes adsorbed, entrapped, or covalently linked to the nanotube deposit. Given their diameter on the nanometer scale, the similarity in size with the protein allows very close proximity between the conductive material and the protein. This may involve a distortion of the protein or a penetration that will lead to the establishment of the electron transfer.

Given the multitude of publications describing the phenomenon of DET, it is necessary to clearly redefine the basic criteria for asserting the presence of DET.

For cyclic voltammetry experiments, the establishment of DET must lead to the appearance of a reversible signal at the same potential as that given for the enzyme prosthetic site. A potential difference between these two values may reflect denaturation of the protein. This then leads to the observation of the redox couple of the prosthetic system of the enzyme but not its catalytic activity. For example, the immobilization of GOx produces sometimes a denaturation of the protein, leading to the observation of the redox couple of the FAD. This was observed for GOx mixed and entrapped with multiwalled carbon nanotubes previously dispersed with a surfactant in water [23]. The cyclic voltammogram of the resulting electrode displays a couple of redox peaks at −0.466 V versus an SCE, indicating a two-electron transfer coupled with a two-proton exchange. However, no catalytic current was reported in the presence of glucose, although the biosensor exhibits an activity for glucose oxidation with a freely diffusing redox mediator, namely, ferrocene monocarboxylic acid. This demonstrates that the immobilized enzyme retains its catalytic activity, but the observed electroactive FAD centers are unable to catalyze the oxidation of glucose. As a consequence, the real DET with the enzyme should be characterized by the appearance of a catalytic current at the potential of the redox couple in the presence of the enzyme substrate.

4.3 Bioelectrodes Based on Multienzyme Configurations

A combination of several enzymes has been developed to extend the scope of detectable substrates or to optimize the response of the biosensor by improving the selectivity of the electrochemical transduction or by amplifying it. Indeed, the selectivity of the biosensor response obviously depends on the selectivity of the molecular recognition by the biomolecule, but it can be degraded by an electrochemical detection step that is not selective.

4.3.1 *Extension of the Substrate Range via Enzyme Association*

To increase the scope of detectable substrates, a combination of esterases and oxidases has been widely used. This strategy consists of combining an enzyme that recognizes the target substrate with an enzyme or a cascade of enzymes, producing an electrochemically oxidizable or reducible compound from the product of the first enzymatic reaction. In the biomedical field, for example, many configurations of biosensors for the determination of cholesterol or neurotransmitters, such as acetylcholine, have been made by association of cholesterol esterase and cholesterol oxidase or acetylcholine esterase and choline oxidase. In the same vein, more complex configurations involving three or more enzymes have been described.

Biosensors used for the determination of phosphate are often based on the use of phosphate as a cosubstrate of oxidases, such as nucleoside phosphorylase, maltose phosphorylase, or pyruvate oxidase. A combination of the three enzymes (maltose phosphorylase, mutarotase, and GOx) generates one glucose molecule per phosphate, glucose being then oxidized to gluconic acid and H_2O_2 [24].

$$\text{Maltose} + \text{Phosphate} \rightarrow \alpha\text{-D-glucose} + \beta\text{-D-glucose} - 1$$

$$\text{phosphate (maltose phosphorylase)}$$

$$\alpha\text{-D-glucose} \rightarrow \beta\text{-D-glucose (mutarotase)}$$

$$\beta\text{-D-glucose} + O_2 \rightarrow H_2O_2 + \beta\text{-D-gluconic acid (GOx)}$$

Dale et al. reported the spatially controlled immobilization of three enzymes, xanthine oxidase, purine nucleoside phosphorylase, and adenosine deaminase, on a microelectrode to detect adenosine, an important agent of the peripheral and central nervous systems. The cascade enzyme reactions provided, thus, one molecule of H_2O_2 for one of adenosine [25].

4.3.2 *Improvement of the Selectivity of the Amperometric Transduction Step*

Another motivation concerns selectivity improvement of the electrochemical transduction step. The functioning principle of biosensors based on oxidases is generally based on the oxidation of H_2O_2 to the surface of a platinum electrode at a potential between 0.5 V and 0.7 V versus an SCE. Thus, one of the major challenges of amperometric biosensors based on H_2O_2 oxidation is the elimination of interferences due to more easily oxidizable species present, for instance, in blood and serum in the case of the detection of glucose or neurotransmitters like glutamate. As a consequence, hydrogen peroxide has been detected through the use of peroxidases immobilized on the electrode surface. Peroxidase was electrically connected to the electrode, allowing a sensing operation at substantially milder potentials [26–28].

For instance, the immobilization of a peroxidase within a laponite gel doped by graphite nanoparticles and its electrical wiring by an electropolymerised film of 3,4-dihydroxybenzaldehyde was accomplished for the amperometric determination of H_2O_2 via its reduction at 0.0 V versus an SCE. GOx was immobilized on the preceding biosensor, allowing the interference-free detection of glucose.

The improvement of the selectivity of the transduction step was also undertaken via the design of bienzymatic systems able to replace elegantly detection by oxidation of a substrate by detection by reduction. This concept has, for example, allowed the determination of salicylate in the absence of any interference.

Conventionally, an amperometric biosensor for salicylate was fabricated by the immobilization of salicylate hydroxylase on an electrode surface. In the presence of dihydronicotinamide adenine dinucleotide (NADH) coenzyme and oxygen, this enzyme catalyzes the following reaction:

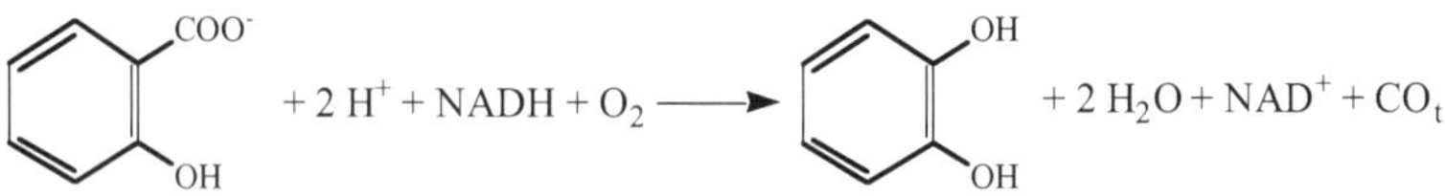

The amperometric determination of salicylate was consequently carried out at 0.4 V versus Ag/AgCl via the oxidation of the enzymatically generated catechol. Since polyphenol oxidase catalyzes the oxidation of phenol and diphenol derivatives to *o*-quinones, while oxygen was reduced to water, this enzyme was combined with salicylate hydroxylase. The resulting bienzyme electrode allows the detection of salicylate through the reduction of *o*-quinone enzymatically generated from catechol. Consequently, the amperometric detection of salicylate was performed at −0.2 V. This functioning potential allows the elimination of the classical interferences due to easily oxidizable compounds like urate, ascorbate, or acetaminophen present in physiological fluids [29].

In the same vein, polyphenol oxidase and alkaline phosphatase, which exhibit complementary activities, were associated to detect phenyl phosphate. Owing to its phosphohydrolytic activity, alkaline phosphatase catalyzes the hydrolysis of phenyl phosphate into phenol. Therefore, the amperometric detection of phenyl phosphate was generally assayed by oxidizing the enzymatically generated phenol at 0.6–0.7 V versus Ag/AgCl. The functioning principle of the bienzyme electrode is based on the hydrolysis of phenyl phosphate by alkaline phosphatase, followed by the oxidation of phenol by polyphenol oxidase to *o*-quinone, which is then electrochemically reduced at −0.2 V. In contrast to the monoenzyme electrode, the transduction step occurs via a reduction process, thus circumventing interferences but also the problem of electrode fouling due to the electro-oxidation of phenol and its related polymerization [30]. The coimmobilization of polyphenol oxidase, ascorbate oxidase, and GOx on a platinum electrode constituted also an approach for the selective detection of glucose via the amperometric oxidation of H_2O_2. Indeed, the common interfering species like ascorbate and acetaminophen were enzymatically oxidized before reaching the electrode surface.

4.4 Amplification of the Biosensor Response

Substrate-recycling reactions have gained considerable interest in terms of sensitivity enhancement. In this context, the Scheller group has thus described up to 5 orders of magnitude of signal amplification [31]. For instance, a tyrosinase and a glucose dehydrogenase were combined onto a Clark electrode for the sensitive monitoring of phenol derivatives. Tyrosinase assumes the molecular recognition of phenols and then catalyzes their oxidation into *o*-quinoid products. The transduction step consists of the amperometric detection of oxygen, which is the cosubstrate of the tyrosinase reaction, and hence its concentration is directly related to those of targets. Glucose dehydrogenase used these generated quinones as redox mediators for the oxidation of glucose to glunonic acid. This enzymatic reaction regenerates the *o*-dihydroxybenzene derivatives, which are tyrosinase substrates, inducing thus an amplification factor of the biosensor response between 150 and 450 for chlorophenol, *p*-cresol, phenol, and catechol [32]. Similarly, the association of glucose dehydrogenase with laccase instead of tyrosinase provided an extremely sensitive sensor for neurotransmitters reaching the nanomolar concentration range like 0.5 nM for adrenaline [33]. This strategy was extended to the amplification of the response of immunoassays. For instance, immunodetection of cocaine was performed by recognition of this target by a polyclonal anticocaine antibody labeled with an alkaline phosphatase. The transduction was based on the amperometric detection of phenol produced by the alkaline phosphatase from phenyl phosphate. This transduction step involved the enzymatic oxidation of phenol to *o*-quinone by an immobilized tyrosinase coupled to a glucose dehydrogenase that used the generated quinone to catalyze the glucose oxidation. The association of these two immobilized enzymes induced an amplified consumption of oxygen. Its amperometric detection via a Clark-type oxygen electrode leads to a low detection limit for cocaine, namely, 380 pM [34].

Another approach of enzyme amplification concerns substrate recycling via the coimmmobilization of glutamate oxidase and glutamate-pyruvatetransaminase (GPT), the product of the first enzyme reaction being the substrate for the second enzyme. The glutamate oxidase catalyzes the oxidation of glutamate to 2-oxo-glutarate, while GPT transforms 2-oxo-glutarate into glutamate in the presence of L-analine. As a consequence, the resulting glutamate biosensor based on the amperometric detection of H_2O_2 led to a 1000-fold increase in sensitivity for the bienzyme configuration compared to the monoenzyme one, the detection limit being subnanomolar (0.2 nM) [35].

The combination of several enzymes working in cascade can also be considered to amplify the signal by recovering more electrons for a single substrate molecule. This strategy was recently used in the field of enzymatic biofuel cells to boost the power density of biofuel cells. For instance, the oxidation of ethanol by alcohol dehydrogenase was employed for biofuel cells and amperometric biosensors involving commonly two electrons per molecule. In contrast, the oxidation of ethanol by an alcohol dehydrogenase then an aldehyde dehydrogenase and finally by a formate dehydrogenase led to CO_2 with six electrons per molecule of ethanol. In particular, Minteer's group has immobilized six enzymes enabling the complete oxidation of ethanol to carbon dioxide and water with the aim to increase biofuel cell energy. It appears that the power density was increased 8.71-fold compared to a single enzyme (alcohol dehydrogenase)-based ethanol/air biofuel cell [36].

An amplification process based on substrate recycling was nicely illustrated by the enzymatic oxidation of catechol followed by the electroreduction of the generated quinone into the initial catechol molecule. In the same vein, an original amplification for the detection of flavins based also on a substrate recycling was reported [37]. In the presence of reduced pyridine nucleotides, flavin reductase catalyses the reduction of flavins into dihydroflavins that can be electrochemically reoxidised. As a consequence, the immobilisation of flavin reductase on an electrode surface leads to an amperometric biosensor for the determination of flavins such as riboflavin, FMN, and FAD. Since the amperometric detection of the enzymatically generated dihydroflavin regenerates the enzyme substrate, an amplification process occurs which yields detection limits in the nanomolar range (Fig. 4.4).

$$\textit{Flavin reductase}: \mathrm{NAD(P)H{+}H^{+}{+}Flavin \rightarrow NAD(P)^{+}{+}Fihydroflavin}$$

$$\textit{Electrode}: \mathrm{Dihydroflavin \rightarrow 2H^{+} + 2e^{-} + Flavin}$$

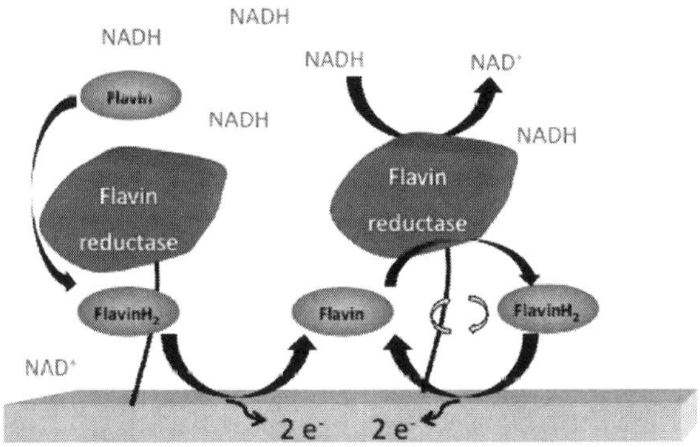

Figure 4.4 Amplification process of the biosensor response based on an electrochemical recycling of the enzyme substrate illustrated with flavin reductase.

It should be noted that flavin reductase was successfully combined with dehydrogenase to develop amperometric biosensors focused on dehydrogenase substrates. The functioning principle of the resulting bienzyme electrode is based on the enzymatic oxidation of the dehydrogenase substrate with the concomitant production of NADH. In the presence of excess flavin, NADH is then detected through its oxidation by flavin reductase and the formation of dihydroflavin. The latter is reoxidized at −0.1 V versus an SCE, a negative potential that allows the system to circumvent interferences from easily oxidizable metabolites [38].

4.5 New Concept of Bienzyme Sensors Displaying a Positive Amperometric Response to the Inhibitive Effect

New bienzymatic configurations have also been successfully applied to the development of biosensor designs dedicated to the detection of an inhibitor by the appearance of an electrochemical signal in contrast to the disappearance of a signal typically observed in the case of a biosensor based on a phenomenon of inhibition. Generally, biosensor design was based on the antagonist activities of two enzymes separately immobilized. The outer enzyme layer consumes the substrate of the enzyme located in the inner layer, decreasing or suppressing its electrochemical signal. Biosensors based on the principle of an enzyme inhibition process were applied to a wide range of significant pollutants such as organophosphorous pesticides, organochlorine pesticides, derivatives of insecticides, heavy metals, and glycoalkaloids. The injection of an inhibitor like a pollutant that inhibits selectively the outer enzyme layer leads thus to the appearance of a biosensor signal.

This concept can be applied to the detection of all oxidase inhibitors and was illustrated with the detection of Hg^{2+} using a combination of laccase and GOx [39].

The bienzyme sensor was composed of an inner layer entrapping laccase wired by ABTS and an outer layer containing GOx. In the presence of glucose, the latter catalyzes the oxidation of this sugar with the reduction of O_2 to H_2O, inducing a depletion of O_2 in the inner layer. Taking into account that the electrically wired laccase catalyzes the reduction of O_2 to H_2O_2, the activity of the GOx should suppress the biosensor response. The injection of Hg^{2+} induces the specific inhibition of GOx activity and hence blocks the O_2 consumption. Oxygen can thus reach the inner layer, leading to the appearance of an amperometric signal due to the four-electron reduction of oxygen to water by the electrically wired laccase (Fig. 4.5). In addition, this amperometric signal increases with the increase in inhibitor concentration. The same strategy based on two

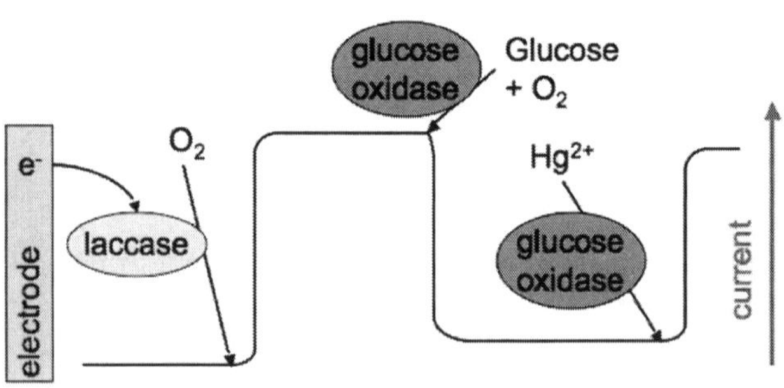

Figure 4.5 Schematic description of the bienzyme electrode response to successively oxygen, glucose, and Hg^{2+} at 0.2 V versus an SCE.

enzymes involving no initial biosensor response and an increase instead of a decrease in the biosensor signal in the presence of an enzyme inhibitor was used for the development of a nitrite biosensor. The latter was composed of two separated enzyme layers. Catalase located in the outer layer catalyzes the decomposition of hydrogen peroxide to H_2O and O_2, while the inner layer consisted of peroxidase electrically wired by clay–ABTS nanoparticles to the electrode. In the presence of H_2O_2, the expected electroenzymatic reduction by the wired peroxidase is theoretically suppressed by the H_2O_2 decomposition by catalase. As a specific inhibitor of catalase, nitrite blocks the H_2O_2 consumption by catalase, inducing thus an increase in the amperometric signal of the H_2O_2 reduction at 0 V by the wired peroxidase [40].

4.6 DNA Sensors and Immunosensors Based on Enzyme Labeling for Amperometric Transduction of the Biological Sensor Response

All the principles of amperometric transduction of enzymatic reactions developed in the preceding paragraphs are obviously applicable to the establishment of an amperometric transduction for immunosensors and DNA sensors. Constraints that relate to the amperometric detection method concern the anchoring layer of biomolecules (DNA, antibody, or antigen). By porous or hydrophilic character, the latter must facilitate the permeation of the electroactive product of the enzymatic reaction through the biological layer to the electrode surface. In addition, this anchoring layer must ensure immobilization of biomolecules as close as possible to the surface so that marking by an enzyme induces a position of this enzyme marker as close as possible to the sensing surface.

4.6.1 *Immunosensors Based on the Electroenzymatic Response*

Immunosensors are of great interest in clinical analysis because they are specific, simple, and rapid, with small sample requirements compared to conventional immunoassay techniques. They are designed to detect the binding event between antibody and antigen after the immobilization of one of the partners of the immunoreaction by covalent binding, affinity interactions, or physical adsorption on the electrode surfaces. The immobilized sensing element can be either an antibody to monitor the presence of a viral antigen or an antigen (whole virus, chemically neutralized virus, phage-displayed epitopic antigen, etc.) to detect the presence of antibodies.

With the aim to develop amperometric transduction based on an enzyme reaction, the immunoreaction was detected by the use of secondary antibodies labeled with enzymes. This enzyme labeling of the immunosensor surface allows the production of electroactive species that are amperometrically monitored at the electrode surface. The resulting signal is commonly proportional to the amount of antibody target anchored on the transducer surface. Obviously, all the schemes previously developed for enzyme electrodes may be applied to the detection of the immunoreaction (Fig. 4.6).

Different enzyme markers may be thus used for an immunoreaction. The nature of the electrode surface can influence the choice of the enzyme marker to exploit for the secondary antibody. Thus, oxidases such as GOx can be used with gold or platinum electrodes for the detection of H_2O_2. On the contrary, electrodes based on graphite or glassy carbon require the use of peroxidases or phosphatases or even polyphenol oxidases.

The sensitivity of such an approach obviously depends on the specific activity of the enzyme as well as on the electrochemical reaction involved by the enzymatic product and its nature (molecular size and electrostatic charge). Indeed, the immunosensor performance is also dependent on the accessibility to the electrode surface for the generated electroactive species. The latter point is controlled by the steric constraints generated by the immobilization procedure of the immunoreagent used as a probe and hence by the bulkiness of the diffusing electroactive species [41].

For instance, amperometric immunosensors for detection of the antibody directed against the cholera toxin were developed by immobilization of the cholera toxin onto an electrode surface. After immunoreaction, the corresponding target (anticholera toxin)

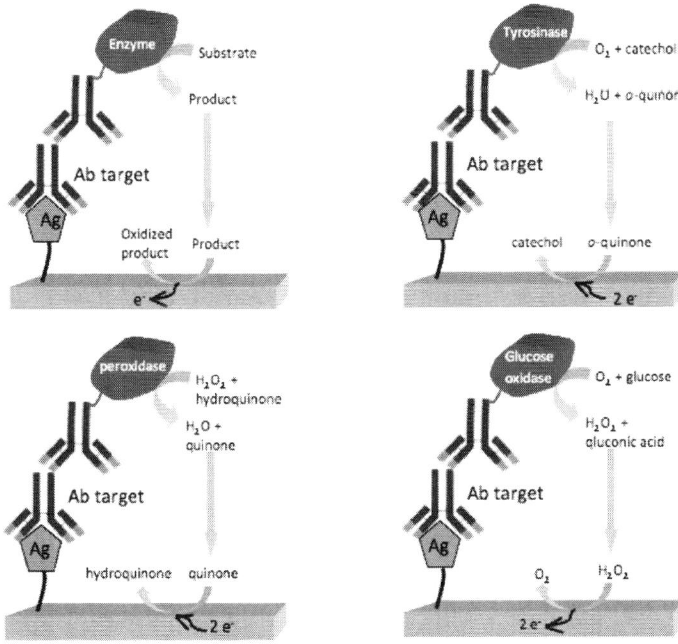

Figure 4.6 Schematic representation of the immunosensor design after its labeling by a secondary antibody conjugated with an enzyme. Examples of enzyme labeling and related amperometric signals based on oxidation or reduction of enzymatic products.

antibody was detected through its labeling by an enzyme marker. This step was performed by recognition of the immobilized anticholera toxin antibody either by a secondary antibody labeled with a peroxidase or by a secondary antibody linked to an avidin, followed by incubation with biotinylated GOx or biotinylated polyphenol oxidase [42].

Owing to the large range of peroxidase substrates, various electrochemical reactions may be envisioned with a peroxidase marker. Thus, three electroactive systems (ferrocyanide/H_2O_2, ferrocene dicarboxilic acid/H_2O_2, and hydroquinone/H_2O_2) were examined, the most sensitive being hydroquinone/H_2O_2. Concerning GOx and polyphenol oxidase, their advantage is the easy diffusion of H_2O_2 and the signal amplification by recycling, respectively. Comparison of the electroenzymatic performances of the three enzyme configurations shows that the more sensitive amperometric immunosensor was based on peroxidase using hydroquinone/H_2O_2 substrates, the detection limit being 50 ng/mL of anticholera toxin antibody. By addition of H_2O_2, peroxidase catalyzed the oxidaton of hydroquinone into quinone, which is then reduced at the electrode surface at −0.1 V versus an SCE.

A similar electroenzymatic system was reported as a transduction tool for an amperometric immunosensor devoted to the detection of an antibody against West Nile virus. West Nile virus belongs to the *Flaviridae* family and can cause meningitis or encephalitis. Since there are no effective therapies or vaccines against West Nile virus infection, the prevention of West Nile virus invasion is an important public health concern. The primary tool for diagnosing West Nile virus infection is the detection of West Nile virus–specific antibodies in serum, plasma, or cerebrospinal fluid. The amperometric immunosensor configuration was based on entrapment into a functionalized polypyrrole film of phages previously modified by an additional peptide sequence of West Nile virus. After detection of the antibody target and then incubation with a secondary peroxidase-labeled antibody, the immunosensors were applied to the amperometric determination of the antibody target via the reduction of the enzymatically generated quinone in the presence of hydroquinone and H_2O_2. This immunosensor configuration allows the detection of different antibody dilutions until a titer of $1{:}10^7$ [43].

4.6.2 *DNA Sensors Based on the Electroenzymatic Response*

In the last decade, hybridization of DNA became increasingly important in the diagnosis and management of infectious diseases. Therefore, considerable research has been done to develop DNA sensors for rapid detection of the hybridization event [44, 45]. Concerning transduction of the hybridization into an electrical signal, indirect methods based on postfunctionalization of the immobilized duplex resulting from the hybridization of the immobilized probe with the DNA target are generally employed. Typically, amperometric DNA sensors use well-defined sequences of single strands as a biological receptor. These DNA probes are immobilized on an electrode surface, where hybridization with the unknown single-stranded target DNA occurs. Thanks to complementary base recognition, detection and identification are possible, for example, by measuring current changes due to the hybridization process with the DNA target. In view of detecting extremely low concentrations of DNA, signal amplification can be performed using the labeling of the duplex with active markers such as enzymes. The use of enzyme markers constitutes an attractive strategy for amplification of the hybridization event via the catalytic production of redox species (or light for optical DNA sensors or arrays). For instance, alkaline phosphatase, GOx, or horseradish peroxidase were commonly used for the fabrication of amperometric DNA sensors, providing extremely sensitive detection limits, namely, 80 aM [46–50]. Usually, the detection of duplex formation is carried out through the reduction or oxidation of enzymatically generated redox species that diffuse from the immobilized enzyme label to the electrode surface.

The labeling step can be carried out by biotinylation of the DNA sequences contained in the analyte before the hybridization. Then, the biotin label will be used for the successive specific anchoring of avidin and biotinylated enzymes. For instance, rapid diagnosis of hepatitis C virus was performed by an amperometric DNA sensor. The latter was prepared by entrapment of streptavidin in thin films of siloxane-poly(propylene oxide) hybrids deposited on a graphite electrode followed by the affinity anchoring of a biotinylated 18-mer probe. The RNA of hepatitis C virus from serum was submitted to a reverse transcriptase–linked polymerase chain reaction (PCR) to produce biotinylated cDNA as a target. Its hybridization onto the sensor surface was followed by its labeling by avidin and then biotinylated peroxidase. The enzymatic response was recorded by using H_2O_2 and KI, the amount of enzymatically generated I_2 being detected via its amperometric reduction at −0.45 V versus Ag/AgCl [51]. In the same vein, the amperometric detection of specific DNA sequences of human cytomegalovirus was investigated with an amperometric sensor involving the biotinylation of the DNA target by PCR amplification with biotinylated primers and its direct immobilization on neutravidin-coated electrodes. The resulting electrode was then exposed to complementary oligonucleotides labeled with digoxigenin probes and then incubated with an antidigoxigenin–peroxidase conjugate. The determination of the amplified viral DNA sequence was performed by recording the amperometric response of the DNA sensor in the presence of H_2O_2 and $[OsIII(bpy)_2pyCl]^{2+}$ as peroxidase substrates [49].

Another approach involves postfunctionalization of the immobilized duplex resulting from the hybridization of the immobilized probe with the DNA target. Taking into account that the hybridization probe is commonly a short oligonucleotide sequence (10–30 bases), the resulting duplex can undergo an additional hybridization process between the long DNA target and a complementary biotinylated oligonucleotide. This step is followed by the specific attachment of a biotinylated enzyme via an avidin bridge (Fig. 4.7). For example, the construction of an amperometric DNA sensor for the detection of West Nile virus was realized by

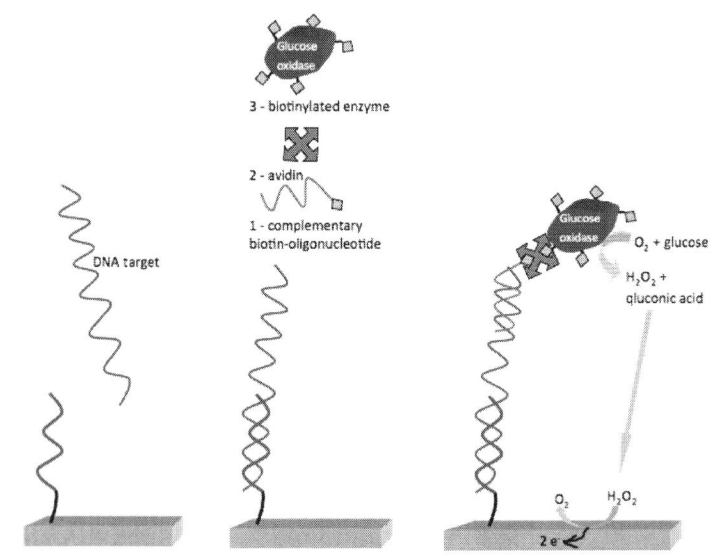

Figure 4.7 Schematic representation of the electroenzymatic transduction of a DNA sensor by the detection of the DNA target by hybridization; its labeling by incubation successively with a biotinylated complementary DNA, avidin, and a biotinylated enzyme; and finally electro-oxidation of the enzymatically produced H_2O_2.

the covalent binding of a 21-mer oligonucleotide probe onto a polypyrrole film electropolymerized on a platinum electrode [52]. After incubation with the target analyte, the duplex formation on the sensor surface was exposed to an additional hybridization process with a complementary biotinylated 15-mer DNA, followed by the successive attachment by affinity interactions of avidin and then biotinylated GOx. The quantity of the duplex was thus determined at 0.6 V by amperometric detection of H_2O_2 generated by the enzyme marker in the presence of glucose. This leads to an extremely sensitive detection limit, namely, 1 fg/mL, of West Nile virus DNA.

4.7 Future Directions and Challenges

In the area of amperometric biosensors, the exploitation of nanomaterials reflects a growing importance due to their ability to enhance the performance of amperometric transduction. Of particular interest are nanomaterials like graphene, silicon nanowires, thin-walled graphitic nanocages, or carbon nanotubes decorated with silicon dioxide or magnetic nanoparticles that confer numerous morphologies to the conductive sensing surface and can be mass-produced. Owing to the extended specific surface area, the latter enhance the immobilization capabilities of interfaces for biological materials. Moreover, the combination of different organic, inorganic, or magnetic nanomaterials is a promising area of research due to expected synergy effects resulting from these composites. Nanostructuration of interfaces by layer-by-layer and Langmuir–Blodgett methods emphasize the control of molecular architectures. These methods allow the combination of biomolecules and metallic nanoparticles or organic catalysts, while maintaining the biological activity of the immobilized biomolecules. Thanks to the control at the molecular level between enzymes and metallic or carbon nanoparticles, such approaches must play an important role in the development of new composite nanostructured films devoted to amperometric biosensing. It is expected that such 3D composite configuration will constitute an avenue for producing highly sensitive amperometric biosensors. In parallel, the conductive nature of some nanomaterials constitutes an attractive means for increasing the signal response of the resulting biological electrodes. Furthermore, the ability of nanomaterials such as carbon nanotubes for enabling DET with enzymes also represents an interesting potential, which is the subject of a growing research area.

Recently, the electrospinning method has been actively explored due to the simplicity of the process. An electric force on induced charges on the polymer liquid overcomes surface tension, inducing thus the ejection of a thin polymer jet. The charged jet is elongated and accelerated by the electric field and forms by deposition on an electrode surface a nanofibrous material that can be used for a broad range of applications. The latter provides a 3D surface network structure based on electrospun nanofiber coatings. Besides their high specific surface area due to a large surface area-to-volume ratio, electrospun fibers present huge active sites for further interaction or attachment. Thus, an electrospun poly(acrylonitrile-co-acrylic acid) nanofiber deposit was successfully applied to the immobilization of a tris(2,2′-bipyridyl)ruthenium(II) complex via electrostatic interaction between a cationic luminescence probe glassy carbon electrode (GCE) and negatively charged nanofibers, providing an electrochemiluminescent sensor [53]. The polymer nanofibers also offer the possibility of various biological modifications. For instance, TiO_2 nanofibers that exhibited catalytic electroactivity toward the oxidation of H_2O_2 were thus deposited on a platinum electrode, leading to a 30% increase in current response to this metabolite. The immobilization of oxidases by entrapment within chitosan spread onto these nanofiber coatings leads to a high-performance amperometric glucose biosensor [54]. It is thus expected that electrospun nanofibers may constitute a promising platform for the development of sensitive amperometric biosensors.

References

1. Viswanathan, S., Radecka, H., Radecki, J. (2009) Electrochemical biosensors for food analysis, *MonatshChem*, **140**, 891–899.
2. Erdem, A., Ozsoz, M. (2002) Electrochemical DNA biosensors based on DNA-drug interactions, *Electroanalysis*, **14**, 965–974.

3. Wang, J. (2006) Electrochemical biosensors: towards point-of-care cancer diagnostics, *Biosens. Bioelectron.*, **21**, 1887–1892.
4. Kimmel, D. W., Leblanc, G., Meschievitz, M. E., Cliffel, D. E. (2012) Electrochemical sensors and biosensors, *Anal. Chem.*, **84**, 685–707.
5. Clark, L. C. (Jr.), Lyons, C. (1962) Electrode systems for continuous monitoring in cardiovascular surgery, *Ann. NY Acad. Sci.*, **102**, 29–45.
6. Kamin, R. A., Wilson, G. S. (1980) Rotating ring-disk enzyme electrode for biocatalysis kinetic studies and characterisation of the immobilised enzyme layer, *Anal. Chem.*, **52**, 1198–1205.
7. Brett, C. M. A., Brett, A. M. O. (1998) *Electroanalysis* (Oxford University Press), 80–81.
8. Suaud-Chagny, M. F., Gonon, F. G. (1986) Immobilization of lactate dehydrogenase on a pyrolytic carbon fiber microelectrode, *Anal. Chem.*, **58**, 412–415.
9. Sacks, V., Eshkenazi, I., Neufeld, T., Dosoretz, C., Rishpon, J. (2000) Immobilised parathion hydrolase: an amperometric sensor for parathion, *Anal. Chem.*, **72**, 2055–2058.
10. Schuhmann, W. (1995) Conducting polymer based amperometric enzyme electrodes, *Mikrochim. Acta*, **121**, 1–29.
11. Loughram, M. G., Hall, J. M., Turner, A. P. F. (1996) Development of a pyrroloquinoline quinone (PQQ) mediated glucose oxidase enzyme electrode for detection of glucose in fruit juice, *Electroanalysis*, **8**, 870–875.
12. Shan, D., Cosnier, S., Mousty, C. (2003) HRP wiring by redox active layered double hydroxides: application to the mediated H_2O_2 and glucose detection, *Anal. Lett.*, **36**, 907–920
13. Shan, D., Cosnier, S., Mousty, C. (2004) HRP/[Zn–Cr–ABTS] redox clay-based biosensor: design and optimization for cyanide detection, *Biosens. Bioelectron*, **20**, 390–396.
14. Mousty, C., Cosnier, S., Vieille, L. (2007) Laccase immobilization in redox active layered double hydroxides: a reagentless amperometric biosensor, *Biosens. Bioelectron.*, **22**, 1733–1738.
15. Mao, L., Yamamoto, K. (2000) Amperometric biosensor for gluthatione based on an osmium-polyvinylpyridine gel polymer and gluthatione sulfhydryl oxidase, *Electroanalysis*, **12**, 577–582.
16. Gao, Q., Guo, Y., Liu, J., Yuan, X., QI, H., Zhang, C. (2011) A biosensor prepared by co-entrapment of glucose oxidase and a carbon nanotube within an electrochemically deposited redox polymer multilayer, *Bioelectrochemistry*, **81**, 109–113.

17. Cosnier, S., Mousty, C., Gondran, C., Lepellec, A. (2006) Entrapment of enzyme within organic and inorganic materials for biosensor applications: comparative study, *Mater. Sci. Eng. C*, **26**, 442–447.

18. Foulds, N. C., Lowe, C. R. (1988) Immobilization of glucose oxidase in ferrocene-modified pyrrole polymers, *Anal. Chem.*, **60**, 2473–2478.

19. Schulz, C., Ludwig, R., Milcheelsen, P., Toscano, M., Gorton, L. (2012) Enhancement of enzymatic activity and catalytic current of cellobiose dehydrogenase by calcium ions, *Electrochem. Commun.*, **17**, 71–74.

20. Yi, X., Huang-Xian, J., Hong-Yuan, C. (2000) Direct electrochemistry of horseradish peroxidase immobilized on a colloid/cysteamine-modified gold electrode, *Anal. Biochem.*, **278**, 22–28.

21. Liu, S., Ju, H. (2003) Reagentless glucose biosensor based on DET of glucose oxidase immobilized on colloidal gold modified carbon paste electrode, *Biosens. Bioelectron.*, **19**, 177–183.

22. Yehezkeli, O., Tel-Vered, R., Raichlin, S., Willner, I. (2011) Nano-engineered flavin-dependent glucose dehydrogenase/gold nanoparticle-modified electrodes for glucose sensing and biofuel cell applications, *ACS Nano*, **5**, 2385–2391.

23. Cai, C., Chen, J. (2004) DET of glucose oxidase promoted by carbon nanotubes, *Anal. Biochem.*, **332**, 75–83.

24. Mousty, C., Cosnier, S., Shan, D., Mu, S. (2001) Trienzymatic biosensor for the determination of inorganic phosphate, *Anal. Chim. Acta*, **443**, 1–8.

25. Llaudet, E., Botting, N. P., Crayston, J. A., Dale, N. (2003) A three-enzyme microelectrode sensor for detecting purine release from central nervous system, *Biosens. Bioelectron.*, **18**, 43–52.

26. Yang, Y., Mu, S. (1997) Bioelectrochemical responses of the polyaniline horseradish peroxidase electrodes, *J. Electroanal. Chem.*, **432**, 71–78.

27. Bartlett, P. N., Birkin, P. R., Wang, J. H., Palmisano, F., De Benedetto, G. (1998) An enzme switch employing direct electrochemical communication between horseradish peroxidase and a poly(aniline) film, *Anal. Chem.*, **70**, 3685–3694.

28. Ruzgas, T., Csöregi, E., Emnéus, J., Gorton, L., Marko-Varga, G. (1996) Peroxidase-modified electrodes: fundamentals and application, *Anal. Chim. Acta*, **330**, 123–138.

29. Cosnier, S., Gondran, C., Watelet, J.-C. (2001) A polypyrrole-bienzyme electrode (salicylatehydroxylase-polyphenoloxidase) for the interference-free determination of salicylate, *Electroanalysis*, **13**, 906–910.

30. Cosnier, S., Gondran, C., Watelet, J.-C., De Giovani, W. F., Furriel, R. P. M., Leone, F. A. (1998) A bienzyme electrode (alkaline phosphatase-polyphenoloxidase) for the amperometric determination of phosphate, *Anal. Chem.,* **70**, 3952–3956.

31. Wollenberger, U., Schubert, F., Pfeiffer, D., Scheller, F. W. (1993) Enhancing biosensor performance using multi-enzyme systems, *TIBTECH,* **11**, 265–262.

32. Makower, A., Eremenko, A. V., Streffer, Wollenberger, U., Scheller, F. W. (1996) Tyrosinase-glucose dehydrogenase substrate-recycling biosensor: a highly-sensitive measurement of phenolic compounds, *J. Chem. Tech. Biotechnol.*, **65**, 39–44.

33. Lisdat, F., Wollenberger, U., Makower, A., Hortnagl, H., Pfeiffer, D., Scheller, F. W. (1997) Catecholamine detection using enzymatic amplification, *Biosens. Bioelectron.*, **12**, 1199–1211.

34. Bauer, C. G., Eremenko, A., Kuhn, A., Kurzinger, K., Makower, A., Scheller, F. W. (1998) Automated amplified flow immunoassay for cocaine, *Anal. Chem.*, **70**, 4624–4630.

35. Yao, T., Yamamoto, H., Wasa, T. (1990) L-glutamate enzyme electrode involving amplification by substrate recycling, *Anal. Chim. Acta*, 236, 437–440.

36. Sokic-Lazic, D., Minteer, S. D. (2008) Citric acid cycle biomimic on a carbon electrode, *Biosens. Bioelectron.*, **24**, 939–944.

37. Cosnier, S., Fontecave, M., Limosin, D., Nivière, V. (1997) A poly(amphiphilic pyrrole)-flavin reductase electrode for amperometric determination of flavins, *Anal. Chem.*, **69**, 3095–3099.

38. Cosnier, S., Fontecave, Innocent, C., Nivière, V. (1997) An original electroenzymatic system: flavin reductase-riboflavin for the improvement of dehydrogenase-based biosensors. application to the amperometric detection of lactate, *Electroanalysis*, **69**, 685–688.

39. Cosnier, S., Mousty, C., Guelorget, A., Sanchez-Paniagua Lopez, M., Shan, D. (2011) A fast and direct amperometric determination of Hg^{2+} by a bienzyme electrode based on the competitive activities of glucose oxidase and laccase, *Electroanalysis*, **23**, 1776–1779.

40. Chen, H., Mousty, C., Chen, L., Cosnier, S. (2008) A new approach for nitrite determination based on a HRP/catalase biosensor, *Mater Sci. Eng. C,* **28**, 726–730.

41. Ionescu, R., Gondran, C., Gheber, L., Cosnier, S., Marks, R. (2004) Construction of amperometric immunosensors based on the electrogen-

eration of a permeable biotinylated polypyrrole film, *Anal. Chem.*, **76**, 6808–6813.

42. Ionescu, R., Gondran, C., Gheber, L., Cosnier, S., Marks, R. (2005) Comparison between the performances of amperometric immunosensors for cholera antitoxin based on three enzyme markers, *Talanta*, **66**, 15–20.
43. Ionescu, R., Hermann, S., Cosnier, S., Marks, R. (2007) Amperometric immunosensor for the detection of anti-West Nile virus IgG, *Anal. Chem.*, **79**, 8662–8668.
44. Odenthal, C., Gooding, J. J. (2007) An introduction to electrochemical DNA biosensors, *Analyst*, **132**, 603–610.
45. Cosnier, S., Mailley, P. (2008) Recent advances in DNA sensors, *Analyst*, **133**, 984–991.
46. Elsholz, B., Worl, R., Blohm, L., et al. (2006) Automated detection and quantitation of bacterial RNA by using electrical microarrays, *Anal. Chem.*, **78**, 4794–4802.
47. Kavanagh, P., Leech, D. (2006) Redox polymer and probe DNA tethered to gold electrodes for enzyme-amplified amperometric detection of DNA hybridization, *Anal. Chem.*, **78**, 2710–2716.
48. Miranda-Castro, R., de-los-Santos-Alvarez, P., Lobo-Castanon, M. J. A., Miranda-Ordieres, J., Tunon-Blanco, P. (2007) Hairpin-DNA probe for enzyme-amplified electrochemical detection of legionella pneumophila, *Anal. Chem.*, **79**, 4050–4055.
49. Djellouli, N., Rochelet-Dequaire, M., Limoges, B., Druet, M., Brossier, P. (2007) Evaluation of the analytical performances of avidin-modified carbon sensors based on a mediated horseradish peroxidase enzyme label and their application to the amperometric detection of nucleic acid, *Biosens. Bioelectron.*, **22**, 2906–2913.
50. Zhang, L. Y., Wan, Y., Zhang, J., et al. (2012) The enzyme-amplified amperometric DNA sensor using an electrodeposited polymer redox mediator, *Sci. China B Chem.*, **52**, 746–750.
51. Riccardi, C. D., Dahmouche, K., Santilli, C. V., da Costa, P. I., Yamanaka, H. (2006) Immobilization of streptavidin in sol-gel films: application on the diagnosis of hepatitis C virus, *Talanta*, **70**, 637–643.
52. Ionescu, R., Hermann, S., Cosnier, S., Marks, R. (2006) A polypyrrole cDNA electrode for the amperometric detection of the West Nile Virus, *Electrochem. Commun.*, **8**, 1741–1748.

53. Shan, D., Cosnier, S., Xue, H.-G. (2010) Enhanced solid-state electrochemiluminescence of tris(2,2-bipyridyl)ruthenium(II) incorporated into electrospun nanofibrous mat, *Anal. Chem.*, **82**, 5892–5896.

54. Tang, H., Yan, F., Tai, Q., Chan, H. L. W. (2010) The improvement of glucose bioelectrocatalytic properties of platinum electrodes modified with electrospun TiO_2 nanofibers, *Biosens. Bioelectron.*, **25**, 1646–1651.

Chapter 5

Electrochemical Biosensors Based on Carbon Nanotubes and Nanohybrids: From Fundamental to Biological Architectures

Youxing Fang and Erkang Wang

State Key Laboratory of Electroanalytical Chemistry,
Changchun Institute of Applied Chemistry, Chinese Academy of Sciences, Changchun, Jilin 130022, China
ekwang@ciac.ac.cn

5.1 Introduction

Carbon materials possess valuable properties for the design of electrodes used in electroanalytical chemistry, because of their relatively wide potential windows in aqueous media, low cost, and relative chemical inertness in most electrolyte solutions. There are several available microstructures of carbon materials, such as diamond, amorphous powders, glassy carbon (GC), carbon fiber, graphite, nanotubes, and graphene [1]. Among all these carbon materials, carbon nanotubes (CNTs) have been attracting extensive interest and becoming a popular component, owing to their large

Electrochemical Biosensors
Edited by Serge Cosnier

ISBN 978-981-4411-46-2 (Hardcover), 978-981-4411-47-9 (eBook)
www.panstanford.com

surface area and low-dimensional nature for diverse applications in the fields of sensors, electronics, biomedicine, and engineering [2, 3]. CNTs are divided into single-walled carbon nanotubes (SWCNTs) and multiwalled carbon nanotubes (MWCNTs): SWCNTs can be considered one rolled-up sheet of graphene, while MWCNTs are concentrical tubes of rolled-up graphene [3]. It should be noted that the electronic properties and detailed electrochemical activity of pristine SWCNT and MWCNT are different [3]. In most case, the functionalized/hybrid CNTs are often employed for the fabrication of the electrochemical and electronics sensors. Compared with the electronics sensors, the request of functionalized/hybrid CNTs in electrochemical sensors are not rigorous as the electronics except for the CNT-field effect transistor (FET). In a typical case, one-dimensional CNTs are lain on the substrate and build a conducting network, leading to increased electrode surface area and enhanced electron transfer, while the huge impenetrate hollow network is in favor of exchanges of electrochemical active materials. However, it should be pointed out that electrochemical sensors based on SWCNT-FETs request semiconducting SWCNTs; hence elimination of metallic SWCNTs is a necessary procedure.

To endow CNTs with specific functions and take advantage of their superior electrochemical and electronic properties, a wide range of functional groups/materials have been used to integrate with CNTs for electrochemical analysis, especially for bioanalysis. On the one hand, by solely using pristine CNTs it may be difficult to achieve selectivity for a biosensor, especially for an immunosensor, because the bare CNT surface hardly specifically adsorbs or bonds the target. On the other hand, the functional components, like oxide nanoparticles (NPs) and enzymes, probably are poorly electroconductive for an electrochemical sensor with low sensitivity, unless they are incorporated with the conducting components, which can decrease the overpotential and increase the current density. As a result, CNT hybrids have been widely developed and employed in the field of electrochemical assays for a variety of biological analytes discussed below.

Inspired by the features of CNTs and their hybrids, electrochemical biosensors have been flourishing recently, and many research papers have reported the use of CNTs for the modification of

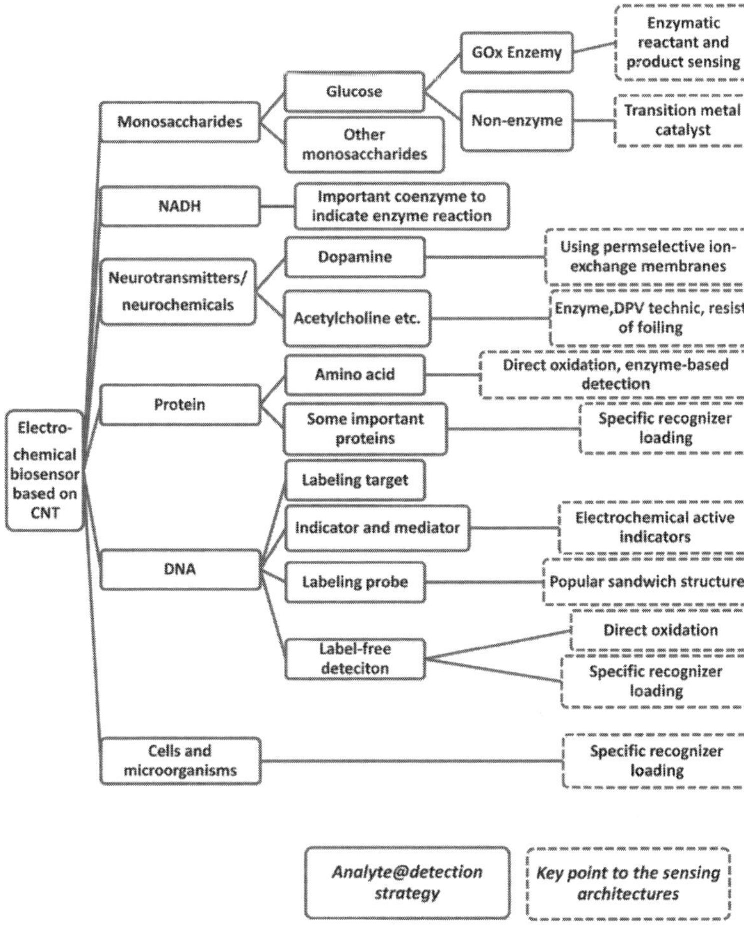

Figure 5.1 Electrochemical biosensor based on CNTs with different sensing strategies.

interfaces for enhanced electrochemical signaling. In this chapter, we will address the research accomplishments that have led to powerful CNT-based electrochemical biosensors with different surface architectures sorted by analytes (Fig. 5.1) and examine challenges and future prospects. However, due to the explosion of publications in this field, it is difficult to browse all the published works about CNTs and their hybrids as an electrochemical platform for biosensing. We apologize to the authors of much excellent research work that, due to the large activity in this important field, we have unintentionally left them out.

5.2 Synthesis and Functionalization of CNTs

Arc discharge, laser ablation, and chemical vapor deposition (CVD) are the three most important techniques employed for CNT synthesis [4]. High-temperature preparation of the former two techniques was first used to produce CNTs, but nowadays these methods have been replaced by low-temperature CVD techniques, since the orientation, alignment, nanotube length, diameter, purity, and density of CNTs can be precisely controlled in the latter [5]. The CVD method is the most commercial-scale synthesis with the goal of mass production, and CNTs by CVD are widely used in many fields, especially electrochemistry. Detailed advanced CNT synthesis can be found in the review about CNT synthesis [4, 5] and properties [6]. Since the synthesis of CNTs using metal nanoparticles (NPs) as catalyst, CNTs remain trace metal impurities, which may play an important role in the electrocatalytic property of CNTs [7, 8]. Moreover, the edge-plane like sites at the CNT open end also contribute it electrochemical catalytic activity [9].

Functionalized CNTs are apt to conjugate with different recognition molecules as well as incorporate with other functional materials (such as noble metal NPs, conducting polymers, proteins, etc.) for electrochemical bioanalysis. To meet specific requirements of an electrochemical biosensor, CNTs can be functionalized by both noncovalent procedures and covalent methods. Noncovalent linkages between CNTs and other species are mainly achieved through hydrophobic and π–π interactions [10]. Chen et al. reported that noncovalent functionalization involves a bifunctional molecule, 1-pyrenebutanoic acid, succinimidyl ester, irreversibly adsorbed onto the inherently hydrophobic surfaces of SWCNTs due to π–π stacking [11]. The resulting functionalization of SWCNTs with terminated succinimidyl ester groups is highly reactive to nucleophilic substitution by amines that exist in abundance on the surface of most proteins, which enables the immobilization of a wide range of biomolecules on the sidewalls of SWCNTs. CNTs capped by biological molecules (such as carbohydrates [12], DNA [13], peptides [14]) are also demonstrated to disperse in aqueous solution. For example, bundled SWCNTs are effectively dispersed in water under sonication in the presence of single-stranded DNA (ssDNA), because ssDNA can helically wrap CNTs through π–π stacking suggested by molecular modeling results [13].

The most common treatment of CNTs under strong oxidation conditions (such as strong acid [15] and ozone [16] treatment) introduces a variety of oxygenated groups, such as carbonyl, carboxyl, and hydroxyl groups, serving as the most common pre-procedure for covalent functionalization of CNTs. The carboxylates can be used to link to the amino groups of biomolecules or proteins using a well-known carbodi-imide procedure. The oxidized CNTs react with 1-ethyl-3-(3-dimethylaminopropyl) carbodiimide (EDC) or (*N*,*N*'-dicyclohexylcarbodiimide) (DCC) under ambient conditions, producing a stable active ester in the presence of *N*-hydroxysuccinimide (NHS). Then, the active ester is reacted with the amine groups of a biomolecule (DNA, enzyme, etc.) to form an amide bond. For instance, the treated CNTs are covalently bonded with both the amino groups on DNA bases [17] and amino-terminated DNA strands [18]. In addition, a direct addition reaction is also available for the covalent functionalization of CNTs due to unsaturated π bonds in CNTs [10].

5.3 Fabrication of CNT-Based Biosensors

Now the nonaligned CNT biosensors are well and widely developed because of the commercial-scale availability of CNTs. The most convenient method for fabricating CNT-based electrochemical sensors is preparing the solvent dispersion of CNTs with or without the aid of additives, followed by casting resulting dispersion at the surface of the electrode. Apart from the simplicity and low cost of this method, the functional additives for the nonaligned CNT biosensors may bring additional merits, such as reinforced membrane strength, ion selectivities, anchoring sites of the probe, enhanced electrochemical activities, etc. Another important technique is self-assembly immobilization, which generally involves self-organizing the functionalized CNTs with other components by different driving forces. For example, the interaction between positively charged poly(diallyl dimethyl ammonium) chloride (PDDA) and negatively charged oxidized CNTs was used to fabricate multilayer films for the determination of dimethyl ammonium [19], while the π–π interactions between CNTs and PDDA may be also responsible for the assembly, as results demonstrated by Yang et al. [20]. Additionally, other protocols for the nonaligned CNT biosensors can be achieved by CNTs incorporated paste electrodes and electropolymerization immobilization, which was well summarized by Hu.

The CNTs by CVD can be directly constructed to the aligned CNT electrodes. On the basis of their different densities, aligned CNT electrodes possess electrochemical properties resembling conventional-scale electrodes at high density or microelectrode arrays at low density [21]. For the high-density aligned CNTs, both unmodified [22, 23] and modified [24, 25] CNTs are available for electrochemical bioassays. Similar to nonaligned CNTs, functional additives are adoptable for modified aligned CNTs. For instance, Gao et al. reported the electrochemical polymerization of pyrrole in the presences of glucose oxidase (GOx) at the surface of aligned CNTs, and the resulting electrode was used to detect glucose by monitoring the intermediate of H_2O_2 [25]. On the other hand, the aligned CNTs are considered as microelectrodes if the spacing is sufficiently larger than the diameter of the nanotubes to avoid diffusion layer overlap of neighboring CNTs. Controlling the density of the metal catalyst for CNT growth is the key to the CNT microelectrodes. For example, random electrodeposited Ni NPs on a metal substrate [26] and patterned Ni catalysts by ion beam sputtering [27] are used to grow CNTs, which are further employed as the work electrode for biosensors.

5.4 Biomolecular Sensing Monosaccharides

5.4.1 *Glucose*

Glucose monitoring deserves a great deal of attention since diabetes management is of utmost concern for health care personnel, individuals, and society as it is taking a heavy economic toll. Fast quantitative determination of glucose is important in the field of clinical chemistry and food analysis [28]. Therefore, nearly all the sensing concepts are initially tested for the development of glucose sensors as it has a huge commercial impact. Two types of CNT electrochemical-sensing strategies, based on the presence and absence of an enzyme as the biorecognition element, have been used for the detection of glucose.

5.4.1.1 Glucose sensors with enzymes

Glucose biosensors have long been developed by use of GOx, which is a flavin-dependent oxidase with specific activity for D-glucose. It catalyzes the oxidation of glucose according to the following reaction:

$$\text{Glucose} + O_2 \xrightarrow{\text{glucose oxidase}} \text{Gluconic acid} + H_2O_2$$

It is well known that GOx catalyzes the oxidation of glucose to gluconolactone in the presence of dissolved oxygen. So, monitoring either the amount of used oxygen or the production of H_2O_2 classical can determine the glucose concentration [29]. Unfortunately, the amperometric determination of H_2O_2 at high anodic potential [30] causes the interference from ascorbic acid (AA), uric acid (UA), and acetaminophen. Thus, it is preferable to introduce novel nanomaterials to the direct electrochemistry of H_2O_2 at a low potential [31–33]. With a high surface area, a high electrocatalytic effect, and a fast electron transfer rate [34], CNTs are the one of the most popular nanomaterials for the supported and electron transducers in the field of glucose biosensing. For example, Wang et al. reported the CNT-/nafion-/GOx-coated electrode for the electrocatalytic detection of hydrogen peroxide with the permselectivity of nafion, offering selective low-potential (−0.05 V vs. Ag/AgCl) biosensing of glucose [35]. Ferrocene (FCA)-modified MWCNT nanocomposite electrodes with GOx are prepared and employed for glucose sensing, taking FCA as a mediator of electron transfer and MWCNTs as the conductor to enhance the enzymatic response of glucose oxidation [36].

Particular attention has been given to metal NPs incorporated with CNTs for enhanced amperometric biosensing of glucose. For the sensitive detection of glucose with fast and steady responses, noble metal NPs (such as Ag, Au, Pt, and Pd) have been widely used because of their superior chemical stability and high chemical catalytic activity. The resulting nanohybrids offer enhanced electrocatalytic activity toward the hydrogen peroxide product or offer direct electron transfer of GOx. For example, a Ag NPs/CNT/chitosan film composite [37] was coated on a GC electrode by Lin et al. This hybrid film achieves a detection limit of 0.1 μM with a linear range from 0.5 to 50 μM. Pt NPs were also introduced by sol–gel [38] and covalent linkage [39] to CNTs for improved glucose enzyme electrodes. Pd NPs and GOx are codeposited onto a nafion-solubilized CNT film and used for glucose sensing [40]. This hybrid electrode has eliminated interference from UA and AA, with increased storage stability and performance. Another composite consisting of SWCNTs, nafion, and Pt NPs with a size of 2–3 nm is very sensitive toward H_2O_2 with a low detection limit of 25 nM. After modifying GOx on the resulting electrode, they constructed a GC-based biosensor that responds even more sensitively to glucose than the GC/GOx electrode modified by Pt NPs or CNTs alone, with a detection limit of 0.5 μM. Claussen et al. developed a glucose biosensor by immobilizing GOx on networks of SWCNTs decorated with Au-coated Pd (Au/Pd) nanocubes [41]. They grew vertical SWCNTs in porous anodic alumina (PAA), where the individual tubes were lain on the PAA surface, and when they grow overrun the pore. Then, the Au/Pd nanocubes were in situ fabricated on the SWCNTs. The hybrid electrode exhibits a low diction limit of H_2O_2 and glucose (with GOx) of 2.3 nM and 1.3 μM, respectively.

A polymer/CNT composite is another class of useful material to immobilize GOx for a glucose biosensor. Apart from direct mixing, chemical and electrochemical polymerizations have been mainly used to fabricate CNT/polymer composites [42–44]. An oxidant is needed in the chemical approach, which affects the properties of the resulting product. For electrochemical polymerization, electrochemical variables can control the polymer morphology and other characteristics of the polymer coating [45]. Zou et al. fabricated polyaniline (PANI)–Prussian blue composite via a spontaneous redox reaction and then mixed it with MWCNTs to obtain the dispersion of the three components [46]. A biosensing electrode for detecting glucose was simply made by casting the above mixture and a GOx solution to a GC electrode, which has a detection limit of 0.01 mM. The CNT/epoxy composites loaded with GOx were also used for the glucose biosensor [47]. Jia et al. have combined the functionalized polyethylenimine, Au NPs, and CNTs as a hybrid matrix for the immobilization of GOx for glucose detection [48]. A polypyrrole (PPy)–MWCNT/GOx nanobiocomposite film was prepared by electro-oxidation of pyrrole in an aqueous solution containing appropriate amounts of functionalized MWCNTs and GOx through Tsai et al.'s work [49]. The amperometric responses vary proportionately to the concentration of hydrogen peroxide at the PPy-MWCNTs nanocomposites, indicating that the electroanalytical PPy–MWCNTs–GOx nanocomposite film was highly sensitive and suitable for the glucose biosensor based on GOx function. Another interesting sensing platform uses two enzymes. For example, Sheng et al. developed a special method based on covalent attachment of two enzymes, GOx and horseradish peroxidase (HRP), onto carboxylic-derived MWCNTs for the deposition of electroactive PANI under ambient conditions (Fig. 5.2a). They can detect the glucose

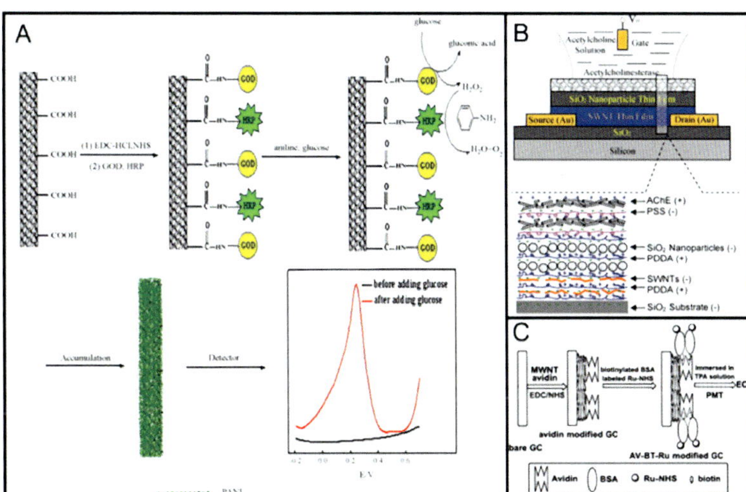

Figure 5.2 CNTs incorporated with enzymes for the electrochemical biosensors of glucose (A, [50]), acetylcholine (B, [89]) and albumin (C, [125]). Reproduced with permission from Elsevier.

concentration by monitoring the amount of PANI deposited on the enzymatic electrode [50].

5.4.1.2 Glucose sensors without enzymes

Although enzyme-based biosensors offer high selectivity and sensitivity and GOx is quite stable compared to other enzymes, their practical application is limited by the serious problem of lacking long-time stability originated from the intrinsic nature of enzymes. To solve this problem, nonenzymatic glucose sensors have also been explored to directly oxidize glucose in the sample. Apart from the few special nonenzyme glucose biosensors, all the catalysts are dependent on the presence of transition metals [51]. The nonenzymatic electrocatalytic system mostly involves metals (including metal hybrids, alloys, and complex) and metal oxides. Especially, the introduction of CNTs to these catalysts can be a powerful way to increase the surface area of the electrode and enhance electrocatalysis. For example, CNTs decorated with Pt [52, 53], Pd [54], Cu [55], and bimetallic Pt [56, 57] NPs reported enhanced electrochemical responses due to CNTs and metal NPs. In addition to metal catalysts, some metal oxides, including Cu_2O [58], NiO [59], and MnO_2 [60], have been used in conjunction with CNTs to produce electrocatalytic glucose sensors. Notably, Chen et al. report the electrodeposition of MnO_2 on vertically aligned MWCNTs. The MnO_2/MWNTs electrode displayed high electrocatalytic activity toward the oxidation of glucose in an alkaline solution, showing about a 0.30 V negative shift in peak potential with oxidation starting at ca. −0.20 V (vs. Ag/AgCl) compared to a bare MWCNT electrode. This MnO_2/MWCNT electrode is highly resistant against poisoning by chloride ions and avoids interference from the oxidation of common interfering species such as AA, dopamine (DA), and UA.

5.4.2 *Other Monosaccharides*

Wang's group reported that CNT-modified GC electrodes realized a variety of monosaccharide detection, including fructose, mannose, xylose, maltose, and arabinose, in 2004 [61]. Antiochia et al. described that a mediated carbon nanotube paste (CNTP) was electropolymerized by 3,4-dihydroxybenzaldehyde, followed by immobilizing d-fructose dehydrogenase for the fructose amperometric biosensor with a detection limit of 1 μM [62]. Another d-fructose sensor was made by Masato et al. [63]. They directly grew CNTs on a Pt electrode, after which they immersed the electrode in a d-fructose dehydrogenase solution for a sensor electrode. The calibration range for the fructose concentration was up to ca. 40 mM, and the detection limit was ca. 5 mM. Galactose sensing can be achieved by covalent immobilized galactose oxidase on a chemically crosslinked SWCNT–chitosan matrix [64], while the casting galactose oxidase on the MWCNT-modified electrode is realized the galactose sensing [65].

5.4.3 *Dihydronicotinamide Adenine Dinucleotide*

Dihydronicotinamide adenine dinucleotide (NADH) coenzyme oxidation at the electrode surface has received considerable interest due to its significance both as a cofactor in many naturally occurring enzymatic reactions and for its role in the electron transfer chain in biological systems. NADH acts as an electron carrier [66]. NADH is a vital cofactor in over 300 dehydrogenase-based enzymatic reactions and is the reduced form of NAD^+. NAD^+ can accept two electrons and a proton to produce NADH. For NADH sensing, three main protocols can be adopted: surface-modified electrodes for NADH sensing, surface redox-mediated NADH probes, and bulk-modified electrodes for the electrocatalytic oxidation of NADH [67]. CNTs, as widespread nanomaterials used in electroanalysis, particularly involve all the sections for NADH sensing.

Zhang et al. reported a facilitated CNT-/chitosan film–modified GC electrode for the electro-oxidation of NADH [68]. This GC/CNT/chitosan sensor for NADH required 0.3 V less overpotential than the bare GC electrode. The wide linear range is obtained from 5 μM to 0.3 mM with a detection limit of 3 μM at an applied potential of 0.4 V. Subsequently, the glucose dehydrogenase can be introduced to the CNT/chitosan film by covalent linkages for a facile glucose sensor. Recently, Bai et al. reported ionic liquid (IL)-modified CNTs for NADH and O_2 sensing, followed by constructing a glucose biosensor based on glucose dehydrogenase [69]. Other efforts have been concentrated on the simultaneous use of CNTs together with organic mediators in order to obtain highly electroactive surfaces at low overpotential (mediators) and being less prone to fouling (CNTs) [67]. For example, Meldola's blue [70], toluidine blue O [71], Variamine blue [72], and the oxide of serotonin [73] have been used as mediators for NADH sensing on the CNT electrode. Except for the mediator strategy, CNT-based nanocomposites for NADH sensors, such as NPs [74, 75], resin [76], and polymers [77], have been also reported.

5.4.4 *Neurotransmitters/Neurochemicals*

5.4.4.1 Dopamine

DA is one of the most important catecholamines. It is a significant neurotransmitter and plays a vital role in the central nervous system. A low level of DA may cause neurological disorders, such as Parkinson's disease and schizophrenia [78]. Therefore, it is important to determine the concentration of this neurochemical in the clinic. DA can be determined by electrochemical methods because it is an electrochemical active compound. CNTs can be adopted for DA electrochemical sensors to increase sensitivity and eliminate interference by other species. Unfortunately, the DA concentration is relatively low (0.01 μM) in biologic samples and always coexists with some interferences, such as AA, UA, etc., for electrochemical determination [78]. A common strategy to eliminate interferences is to cover the electrode with permselective ion exchange membranes such as nafion [79] and poly(3,5-dihydroxy benzoic acid) [80], while the use of ion exchange membranes cannot solve the sensitivity problem. Thus, CNTs have been adopted for DA electrochemical sensors to increase sensitivity. To date, quite a few works have been published for sensitively determining DA in the presence of AA and UA using both nafion and CNTs [81, 82]. The selectivity issue can be also addressed by the immobilization of tyrosinase on a SWCNT–PPy composite [83] and tiron-doped PPy/CNTs [84], because of selective oxidation of DA by an enzyme for the former case and the repulsive interaction with anionic sites of the film for the latter case. DA can be also simultaneously detected in the presence of other analytes like neurochemicals [85], amino acid [86], and drugs [87].

5.4.4.2 Other important neurochemicals

Cui's group reports several acetylcholine sensors based on self-assembly of CNTs and acetylcholinesterase (AChE) molecules on the substrate (Fig. 5.2b). The thin-film FET of CNTs can reach high sensitivity at nanomolar levels [88–90], while a similar detection method without CNTs has relatively low sensitivity [91]. A CNT-FET uses alternating layers of AChE and poly(sodium 4-styrenesulfonate) (PSS) as the sensing component. AChE catalyzes the hydrolysis reaction of acetylcholine to release H^+ ions, which change the gate voltage and the conductance of the device. Jordi et al. reported a potentiometric ion-selective electrode for an acetylcholine sensor. They used an octaamide cavitand as a molecular recognition element and a CNT as a solid transducer material to detect choline and derivatives. The detection limit was 0.4 μM [92].

Epinephrine (EP) and norepinephrine (NEP) are two neurotransmitters with similar structures and are difficult to distinguish by common electrochemical methods, and only a single voltammetric method has been reported to date for the simultaneous determination of EP and NEP [93]. Rajendra et al. fabricated an MWCNT-modified edge-plane pyrolytic graphite electrode for simultaneous determination of EP and NEP, and the oxidation peaks of EP and NEP have been well separated at ~150 and ~215 mV, respectively [93]. Linear calibration curves were obtained for EP and NEP in the range of 0.5–100 nM with limits of detection 0.15 nM and 0.09 nM, respectively; thereafter they detect the blood sample successfully. A CNTP electrode of 2-(4-oxo-3-phenyl-3,4-dihydro-quinazolinyl)-*N*'-phenyl-hydra-zinecarbothioamide was also used to detect EP and NEP simultaneously with a potential difference of 0.240 V [94]. As another example for a bioassay of real samples, EP in human plasma was successfully detected by using the an edge-plane pyrolytic graphite electrode modified with MWCNTs [95].

Serotonin is electrochemically active, but detection is more difficult than DA because reactive species formed after oxidation can adsorb on the electrode, decreasing sensitivity over time by a fouling effect [96]. Venton et al. observed that CNT-modified electrodes were fouled significantly less by serotonin [97]. They distinguished serotonin and DA using the reduction peaks, which were separated by 200 mV. Sun et al. also reported an electrochemical sensor based on a CNT–IL composite for the simultaneous determination of serotonin and DA, with a detection limit of 8 nM for serotonin and of 60 nM for DA [85]. Anther interesting work has been published by Aleix et al., who compared three electrodes (GC, boron-doped diamond, and CNT network) for serotonin sensors [98]. They observed that the CNT network was most sensitive to serotonin, with the lowest detection limit of 10 nm, while the fast-scan cyclic voltammetry (CV) technique will benefit significantly from the reduced background currents of the CNT electrode by the surface-fouling effect.

5.4.5 *Proteins*

Proteins are essential components of organisms and thus are involved in many biological functions. There are increasing demands for ultrasensitive protein detection because many important protein biomarkers are present at ultralow levels, especially during the early stages of disease [99]. Electrochemical-based protein sensors offer selectivity, sensitivity, and reliability with low cost, making them very attractive tools for protein detection. CNTs are incorporated because of faster electron transfer kinetics to provide a wire to the redox site of a protein [100].

5.4.5.1 Amino acids

L-cysteine can be detected by using a CNT electrode loaded with Pt NPs, which can be either electrodeposited [101] or deposited using sputtering [102]. Shen et al. fabricated a chromium hexacyanoferrate/MWCNT composite for the amperometric sensor of L-cysteine without interference from tryptophane (Trp) and tyrosine (Tyr). However, although a low detection limit (10 nM) was obtained, the applied potential was relatively high (0.84 V) [103]. Moreover, by virtue of oxygen-containing moieties, the nitric acid–treated CNT-based electrode can also successfully detect L-cysteine [104]. A boron-doped carbon nanotube (BCNT)-modified GC electrode was constructed for the detection of L-cysteine. The work revealed that the BCNT/GC electrode possesses much higher sensitivity (25.3 $\pm$ 1.20 nA·mM^{-1}) than the CNT/GC electrode (1.16 $\pm$ 0.07 nA·mM^{-1}), probably being attributed to more edge-plane sites and functional groups on the surface of BCNTs [105].

A chemically modified electrode composed of MWCNTs/4-aminobenzeresulfonic acid has been used for electrocatalytic oxidation of tyrosine by CV and differential pulse voltammetry (DPV), making it suitable for submicromolar detection of tyrosine [106]. Okuno et al. developed a label-free electrochemical immunosensor of prostate-specific antigen (PSA) using T-PSA-mAb immobilized on an SWCNT electrode with good selectivity [107]. The current signals were derived from the oxidation of Tyr and Trp residues, and the detection limit for T-PSA was 0.25 ng/mL. The chiral discrimination of between D- and L-tryptophan was also attempted through the oxidation current on the L-alanine ethyl ester–modified MWCNT by Kang et al., despite the poorly separated oxidation peaks (by ~8 mV) of D- and L-tryptophan [108].

Electrochemical determination of glutamate can be achieved by using glutamate dehydrogenase, which produces NADH for electrochemical responses at the electrode. With this sensing protocol, introducing CNTs can efficiently increase the sensitivity of the electrochemical biosensor of NADH and glutamate. The mediator molecules such as thionine [109] and Meldola's blue [110], as well as Pt NPs [111], have been incorporated with CNTs to detect NADH at the micromolar level.

5.4.5.2 Some important proteins

Cid et al. have developed two human immunoglobulin G (IgG) electrochemical biosensors by using a FET based on a network of SWCNTs [112, 113]. IgG antibodies are the basic elements of the recognition layer due to antigen–antibody reactions, and IgG antibodies can be loaded on SWCNTs by nonspecifical adsorption [112] and covalent linkages [113]. An IgG sensor based on a CNT-FET was demonstrated by Kim et al. [114]. They reduced the receptor size on the CNT-FET using the fragment of an IgG antibody to make the target approaching to CNTs (Fig. 5.3a). The results show that CNT-FET biosensors using a whole antibody had very low sensitivity (detection limit was ~1000 ng/mL), whereas those based on small fragments of IgG antibodies could detect 1 pg/mL (~7 fM level). An immunosensor of mouse IgG was also accessible by monitoring the electrical impedance change of a nanotube array electrode with immobilized mouse IgG antibodies [115]. Aptamers can also replace antibodies to react with antigens and make the specific interaction reaction within the Debye length of CNTs. Maehashi et al. presented IgE biosensors based on aptamer-modified CNT-FETs with detection limits of 25 pM [116].

Streptavidin was successfully detected by coating CNTs with a lipid bilayer to immobilize membrane proteins to a CNT-FET [117]. Alternatively, a probe molecule of biotin can be covalently linked to a CNT-FET, and the biological recognition reaction between the streptavidin and biotin causes conductance changes [118, 119].

C-reactive protein (CRP) is a blood test that indicates inflammation and is a marker for some diseases. Anti-CRP was orientedly bound to MWCNTs on a screen-printed carbon electrode, while a layer of HRP was finally constructed to serve as the tag for the H_2O_2-involved detection strategy [120]. An enzyme-free immunosensor of CRP was fabricated by use of *N,N'*-bis-(2-hydroxy-methylene)-o phenylenediamine cobalt (CoRb), which can replace peroxidase for the above detection strategy [121]. Another difference is that the CoRb was constructed at the inner layer of the electrode in contrast to the above enzyme-based strategy. Both of them achieved the detection limits at the nanogram per milliliter level, meeting the requested threshold of the care test.

Albumin is one of the most abundant proteins in the body, and consequently it is an important test analyte. Human serum albumin (HSA) can be detected by amperometric enzyme-linked immunoassays based on vertically aligned arrays of SWCNTs. HSA antibodies (anti-HSA) were attached to the carboxylated ends of nanotube forests and then were selected to react with HAS by an antigen–antibody reaction. Another layer of HRP-conjugated anti-HSA was coated on top of the electrode, which can produce HAS-dependent current response of the current of H_2O_2 reduction [122].

CNT-FETs are also used for HAS sensors. Abe et al. reported two similar HSA immunosensors with detection limits at the nanomolar lever, and selectivity was achieved by the antigen–antibody reaction as well [123, 124]. CNTs modified electrode is also employed for the fabrication of electrochemiluminesce (ECL) sensors. Fang et al. [125] achieved the detection of the bovine serum albumin (BSA) with double protein labeling protocol with a biotin label for biorecognition and ruthenium label for ECL detection (Fig. 5.2c). Avidin was bound to the carboxylated MWCNTs for biorecognition for selective determination of BSA. The MWCNTs can accelerate the electron transfer of the probe ion from the bulk solution to the surface of the electrode.

5.4.6 *DNA Sensors*

DNA sensors have grown rapidly because of their importance in scientific and medical fields, such as detection of dangerous toxins, disease diagnosis, drug discovery, and forensic investigations. Electrochemical methods of DNA detection present the extremely sensitive yet inexpensive and robust way without the need of the sample pretreatment and derivatization when used in the above areas [126]. The combination of unique electric properties, high surface area, hollow geometry, and useful mechanical properties of CNTs with DNA hybridization provides the possibility of constructing DNA biosensors with high sensitivity, simplicity, and multiplexing.

A strategy for DNA detection relies on chemical labeling of target DNA sequences with specific functionality species as signal reporter such as redox-active molecules and enzymes. The peroxidase-labeled target DNA can be detected by the simple amperometric method [127, 128]. A biotinylated target DNA at the nanomolar level was successfully detected on the aligned CNT electrode with a linear response from 50 to 200 nM, whereas the label-free method showed a linear response between 0.5 and 10 μM [129]. He and Dai reported that the ssDNA chains grafted onto aligned CNTs interacted with FCA-labeled complementary DNA; thus the redox response of FCA can be recorded at the CNT electrode, whereas the noncomplementary DNA chains did not show any redox response of FCA. In addition, with other target DNA chains containing complementary sequences of both grafted DNA and related noncomplementary DNA chains, a strong redox response was also obtained [130]. The enzyme-based electrochemical bioassay of alkaline phosphatase–labeled target DNA was also achieved on CNT-modified GC transducers, and the alkaline phosphatase was able to catalyze the α-naphthyl phosphate hydrolysis reaction to produce the electroactive product of α-naphthol [131]. Herein the CNTs can not only promote electrocatalytic action but also decrease the surface-fouling effect by the oxidation product of phenol.

Electrochemical indicators are important for DNA biosensors, and the indicators can generate electrochemical signals (changes) correlated with the concentration of DNA. CNTs wrapped by chitosan film were immobilized on the surface of graphite to build a DNA sensor of salmon sperm DNA using methylene blue (MB) as the electroactive indicator, which can be inserted into the double-stranded DNA [132]. CNTs can increase the electroactive surface of the electrode as well as act as a bridge to accelerate the rate of heterogeneous electron transfer between the electrode and the redox-active MB. A low detection limit of 0.252 nM for fish sperm DNA was achieved, while HSA didn't interfere with the detection. A paste electrode assembled by MWCNTs and immobilizing the DNA probe within electropolymerized PPy was also realized for electrochemical detection of DNA hybridization using ethidium bromide (EB) as the mediator [133]. Similar detection with an indicator of daunomycin was presented by Zhu et al., who combined CNTs with Pt NPs to detect the hybridization of DNA by DPV measurement [134].

As these indicators were just intercalated within the double helix of DNA during hybridization, the noncovalent bond with DNA is relatively weak with nonspecific adsorption, which may limit the sensitivity of the detection. Enzyme/electroactive molecule–labeled probe DNA can avoid nonspecific adsorption. Wang et al. adopted a sandwich hybridization gene-sensing structure for the selective and sensitive DNA sensor by using a DNA probe (DNA probe 1) immobilized on magnetic beads, the target DNA, and another DNA probe (DNA probe 2) conjugated on a SWCNT/alkaline phosphatase adduct (Fig. 5.4) [135]. CNTs play a dual amplification role in both recognition and transduction events, as carriers for enzyme tags and for accumulating the product of the enzymatic reaction. An electrochemical DNA sensor was realized by oxidizing the product of the enzymatic reaction, which is associated with DNA target concentration. A similar sandwich structure for DNA sensing was reported by Yang et al. They used FCA-modified SWCNTs conjugated with DNA probe 2 to record the electrochemical signal of H_2O_2, and the middle layer of target DNA concentration can be derived as well [136].

The other interest of DNA electrochemical sensors was focused on the label-free protocol without using indicators or functional enzymes [137]. DNA hybridization can be detected by examining direct oxidation of guanine or adenine [138]—the simplest way to directly detect the DNA strand. The generated current can be detected using voltammetric techniques with CNTs loaded with probe DNA [126, 139, 140]. However, the electrochemical signal is strongly dependent on the proportion of guanine or adenine bases in the nucleic acids, while the sensor is not reusable because of the destruction of the nucleic acid sequences. Then, it is possible to measure the impedance [141] to build the DNA sensor. This technique depends on changes in the electrical properties of the surface (capacitance and resistance), resulting solely from the presence of the target molecule. So the impedance sensing of DNA does not require labels and indicators. For example, the detection of specific sequences on hepatitis B virus using SWCNT-modified graphite loaded with complementary DNA probe (ssDNA) was achieved by monitoring the impedance changes before and after the hybridization [142]. Label-free detection of DNA hybridization using CNT-FETs has also been discussed and summarized [137, 143]. Martinez et al. synthesized a block copolymer that can bond noncovalently to CNTs and covalently link to the aminated ssDNA probe. The modified CNT-based FETs can detect the specific complementary DNA (Fig. 5.3b) [144]. In addition, functionalized with peptide nucleic acids, CNT-FETs can also recognize the complementary RNA oligomer (the hepatitis C virus RNA), enabling another label-free RNA sensor [145]. Specific recognition can be achieved by a functional peptide, MutS DNA mismatch repair protein, which interacts with the mismatched DNA and endows the selective response at the CNT-FETs [146]. Recently,

Sorgenfrei et al. took advantage of defect-dominated conductance in nanotubes to produce measurements of DNA hybridization kinetics. They used a point-defect CNT-FET to bind the single DNA probe, which induces conduction changes when it interacts with the complementary DNA target. Apart from the single-molecule-level detection, the kinetics of the system can be also studied as a function of temperature, allowing the measurement of rate constants, melting curves, and activation energies for different sequences and target concentrations [147].

5.5 Cells and Microorganism Sensing

When CNTs are modified with recognition functional species, cells can be trapped by the affinity between the cell surface and the CNTs, which is the principle of a CNT-based cell sensor. For example, tetrapeptide- or fibronectin-containing arginine–glycine–aspartic acid–serine (RGDS) was covalently bonded to SWCNTs. The conjugated RGDS showed a predominant ability to capture cells on the electrode surface by the specific combination of RGD domains with integrin receptors of the cell. Thus, the cells can be immobilized on the RGDS–SWCNTs for further construction of an electrochemical cell sensor [148–150]. Cancer cells were

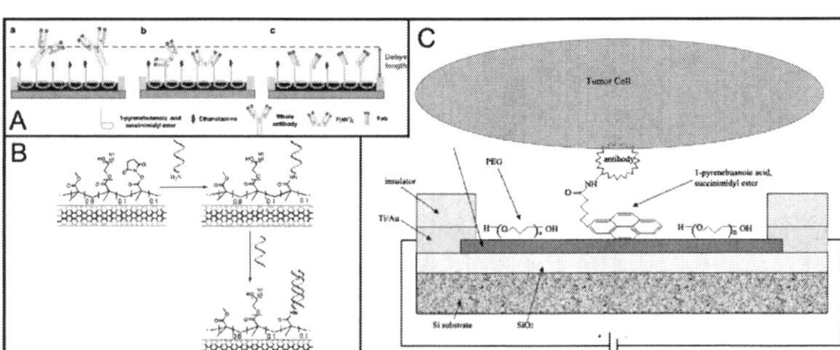

Figure 5.3 Electrochemical sensors by use of CNT-FET. (A) protein. Reproduced from Ref. [114] with permission from Elsevier. (B) DNA. Reproduced from Ref. [144] with permission from the American Chemical Society. (C) Cancer cell. Reproduced from Ref. [152] with permission from the Institute of Physics Publishing.

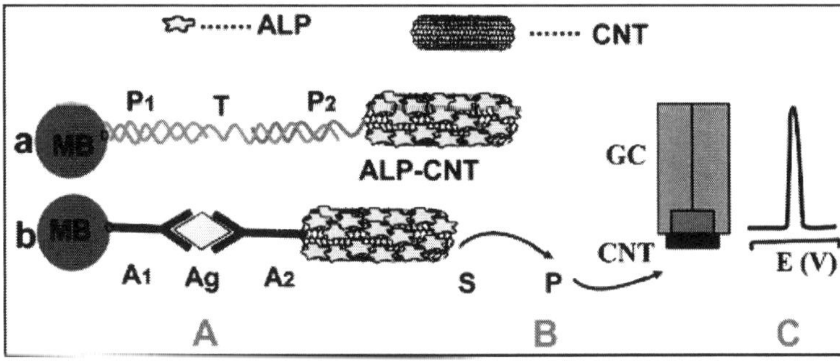

Figure 5.4 An analytical protocol of sandwich structures for electrochemical DNA sensors. Reproduced from Ref. [135] with permission from the American Chemical Society.

also detected on a folate nanoprobe–modified CNT electrode by electrochemical impedance spectroscopy [151]. By the antibody-modified CNT-FET approach, the breast cancer cell sensor successfully worked with high sensitivity and selectivity (Fig. 5.3c) [152]. An interesting work recognizes three different leukemia cancer cells by their different electrochemical behaviors [153]. Microorganisms can be also detected by a similar way. In two respective works, *Candida albicans* [154], *Escherichia coli* O157:H7, and bacteriophage T7 [155] sensors were used on specific antibody-modified CNTs. Gustavo et al. reported that selectively binding living bacteria to a aptamer–CNT electrode causes potential changes for a bacteria electrochemical sensor [156].

5.6 Conclusion and Prospects

CNTs possess many superior electrochemical properties for modified electrodes, such as enhanced detection sensitivity, electrocatalytic effects, and reduced fouling, which provide an attractive feature in the advancement of electrochemical biosensors. In addition, the high surface area of CNTs is also beneficial for the attachment of functional groups/molecules for biosensing goals. Thus, CNTs and their hybrids and the general use of CNTs with a modified surface combined with NPs, polymers, enzymes, DNA etc., offer an enormous choice for constructing an electrochemical biosensor via diverse sensing strategies. However, there are still some critical challenges to implementing CNTs into commercial available biosensors that can be widely used [100]. For example, the cytotoxicity and biodegradation of CNTs concerned with human heath may hinder the bioassay in vivo. The manufacture of commercial viable biosensors with high consistency and cost-effectiveness cannot compete with other commercially available products. Additionally, graphene may serve as a strong competitor with similar or higher electrochemical properties in contrast to CNTs.

Although the design and fabrication of electrochemical biosensors using CNTs are extensive and intensive evolving in the laboratory in virtue of the rapid development of nanoscience, the use of CNTs for in situ/in vivo biodetection still needs a lot of work for biomedical applications that potentially reward opportunities to develop novel CNTs hybrids and lead to more exciting discoveries.

Acknowledgments

This work was supported by the National Natural Science Foundation of China with grants Nos. 21427811 and 21190040, and the Instrument Developing Project of the Chinese Academy of Sciences (no. YZ201203).

References

1. Huang, J. S., Liu, Y., You, T. Y. (2010) Carbon nanofiber based electrochemical biosensors: a review, *Anal. Methods*, **2**, 202–211.
2. Niyogi, S., Hamon, M. A., Hu, H., Zhao, B., Bhowmik, P., Sen, R., Itkis, M. E., Haddon, R. C. (2002). Chemistry of single-walled carbon nanotubes, *Acc. Chem. Res.*, **35**, 1105–1113.
3. Tasis, D., Tagmatarchis, N., Bianco, A., Prato, M. (2006) Chemistry of carbon nanotubes, *Chem. Rev.*, **106**, 1105–1136.
4. Kingston, C. T., Simard, B. (2003) Fabrication of carbon nanotubes, *Anal. Lett.*, **36**, 3119–3145.
5. Prasek, J., Drbohlavova, J., Chomoucka, J., et al. (2011) Methods for carbon nanotubes synthesis-review, *J. Mater. Chem.*, **21**, 15872–15884.

6. Terrones, M. (2004) Carbon nanotubes: synthesis and properties, electronic devices and other emerging applications, *Int. Mater. Rev.*, **49**, 325–377.
7. Banks, C. E., Crossley, A., Salter, C., Wilkins, S. J., Compton, R. G. (2006) Carbon nanotubes contain metal impurities which are responsible for the "electrocatalysis" seen at some nanotube-modified electrodes, *Angew. Chem., Int. Ed.*, **45**, 2533–2537.
8. Sljukic, B., Banks, C. E., Compton, R. G. (2006) Iron oxide particles are the active sites for hydrogen peroxide sensing at multiwalled carbon nanotube modified electrodes, *Nano Lett.*, **6**, 1556–1558.
9. Banks, C. E., Davies, T. J., Wildgoose, G. G., Compton, R. G. (2005) Electrocatalysis at graphite and carbon nanotube modified electrodes: edge-plane sites and tube ends are the reactive sites, *Chem. Commun.*, 829–841.
10. He, P., Dai, L. (2006) Carbon nanotube biosensors, in *BioMEMS and Biomedical Nanotechnology*, Vol. 1 (Springer, New York), 171–201.
11. Chen, R. J., Zhang, Y. G., Wang, D. W, Dai, H. J. (2001) Noncovalent sidewall functionalization of single-walled carbon nanotubes for protein immobilization, *J. Am. Chem. Soc.*, **123**, 3838–3839.
12. Star, A., Steuerman, D. W., Heath, J. R., Stoddart, J. F. (2002) Starched carbon nanotubes, *Angew. Chem., Int. Ed.*, **41**, 2508–2512.
13. Zheng, M., Jagota, A., Semke, E. D., et al. (2003) DNA-assisted dispersion and separation of carbon nanotubes, *Nat. Mater.*, **2**, 338–342.
14. Dieckmann, G. R., Dalton, A. B., Johnson, P. A., et al. (2003) Controlled assembly of carbon nanotubes by designed amphiphilic peptide helices, *J. Am. Chem. Soc.*, **125**, 1770–1777.
15. Hirsch, A., Vostrowsky, O. (2005) Functionalization of carbon nanotubes, in *Functional Molecular Nanostructures*, Vol. 245, Ed. A. D. Schlüter (Springer, Berlin/Heidelberg), 193–237.
16. Mawhinney, D. B., Naumenko, V., Kuznetsova, A., Yates, J. T., Liu, J., Smalley, R. E. (2000) Infrared spectral evidence for the etching of carbon nanotubes: ozone oxidation at 298 K, *J. Am. Chem. Soc.*, **122**, 2383–2384.
17. Guo, M. L., Chen, J. H., Liu, D. Y., Nie, L. H, Yao, S. Z. (2004) Electrochemical characteristics of the immobilization of calf thymus DNA molecules on multi-walled carbon nanotubes, *Bioelectrochemistry*, **62**, 29–35.
18. Dwyer, C., Guthold, M., Falvo, M., Washburn, S., Superfine, R., Erie, D. (2002) DNA-functionalized single-walled carbon nanotubes, *Nanotechnology*, **13**, 601–604.

19. Zhang, M. N., Gong, K. P., Zhang, H. W., Mao, L. Q. (2005) Layer-by-layer assembled carbon nanotubes for selective determination of dopamine in the presence of ascorbic acid, *Biosens. Bioelectron.*, **20**, 1270–1276.
20. Yang, D. Q., Rochette, J. F., Sacher, E. (2005) Spectroscopic evidence for pi-pi interaction between poly(diallyl dimethylammonium) chloride and multiwalled carbon nanotubes, *J. Phys. Chem. B*, **109**, 4481–4484.
21. Hu, C., Hu, S. (2009) Carbon nanotube-based electrochemical sensors: principles and applications in biomedical systems, *J. Sens.*, **2009**, Article ID: 187615.
22. Ye, J. S., Wen, Y., De Zhang, W., Gan, L. M., Xu, G. Q, Sheu, F. S. (2003) Selective voltammetric detection of uric acid in the presence of ascorbic acid at well-aligned carbon nanotube electrode, *Electroanalysis*, **15**, 1693–1698.
23. Ye, J. S., Wen, Y., Zhang, W. D., Gan, L. M., Xu, G. Q, Sheu, F. S. (2004) Nonenzymatic glucose detection using multi-walled carbon nanotube electrodes, *Electrochem. Commun.*, **6**, 66–70.
24. Soundarrajan, P., Patil, A., Dai, L. M. (2003) Surface modification of aligned carbon nanotube arrays for electrochemical sensing applications, *J. Vac. Sci. Technol. A*, **21**, 1198–1201.
25. Gao, M., Dai, L. M., Wallace, G. G. (2003) Biosensors based on aligned carbon nanotubes coated with inherently conducting polymers, *Electroanalysis*, **15**, 1089–1094.
26. Lin, Y. H., Lu, F., Tu, Y, Ren, Z. F. (2004) Glucose biosensors based on carbon nanotube nanoelectrode ensembles, *Nano Lett.*, **4**, 191–195.
27. Koehne, J., Li, J., Cassell, A. M., et al. (2004) The fabrication and electrochemical characterization of carbon nanotube nanoelectrode arrays, *J. Mater. Chem.*, **14**, 676–684.
28. Malhotra, B. D., Chaubey, A. (2003) Biosensors for clinical diagnostics industry, *Sens. Actuators, B*, **91**, 117–127.
29. Wang, J. (2001) Glucose biosensors: 40 years of advances and challenges, *Electroanalysis*, **13**, 983–988.
30. Ianniello, R. M., Yacynych, A. M. (1981) Immobilized enzyme chemically modified electrode as an amperometric sensor, *Anal. Chem.*, **53**, 2090–2095.
31. Wang, J. (2008) Electrochemical glucose biosensors, *Chem. Rev.*, **108**, 814–825.
32. Harper, A., Anderson, M. R. (2010) Electrochemical glucose sensors-developments using electrostatic assembly and carbon nanotubes for biosensor construction, *Sensors*, **10**, 8248–8274.

33. Rahman, M. M., Ahammad, A. J. S., Jin, J.-H., Ahn, S. J., Lee, J.-J. (2010) A comprehensive review of glucose biosensors based on nanostructured metal oxides, *Sensors*, **10**, 4855–4886.
34. Wang, J. (2005) Carbon-nanotube based electrochemical biosensors: a review, *Electroanalysis*, **17**, 7–14.
35. Wang, J., Musameh, M., Lin, Y. H. (2003) Solubilization of carbon nanotubes by nafion toward the preparation of amperometric biosensors, *J. Am. Chem. Soc.*, **125**, 2408–2409.
36. Qiu, J. D., Zhou, W. M., Guo, J., Wang, R., Liang, R. P. (2009) Amperometric sensor based on ferrocene-modified multiwalled carbon nanotube nanocomposites as electron mediator for the determination of glucose, *Anal. Biochem.*, **385**, 264–269.
37. Lin, J., He, C., Zhao, Y., Zhang, S. (2009) One-step synthesis of silver nanoparticles/carbon nanotubes/chitosan film and its application in glucose biosensor, *Sens. Actuators, B*, **137**, 768–773.
38. Yang, M. H., Yang, Y. H., Liu, Y. L., Shen, G. L., Yu, R. Q. (2006) Platinum nanoparticles-doped sol-gel/carbon nanotubes composite electrochemical sensors and biosensors, *Biosens. Bioelectron.*, **21**, 1125–1131.
39. Chu, X., Duan, D., Shen, G., Yu, R. (2007) Amperometric glucose biosensor based on electrodeposition of platinum nanoparticles onto covalently immobilized carbon nanotube electrode, *Talanta*, **71**, 2040–2047.
40. Lim, S. H., Wei, J., Lin, J. Y., Li, Q. T., KuaYou, J. (2005) A glucose biosensor based on electrodeposition of palladium nanoparticles and glucose oxidase onto nafion-solubilized carbon nanotube electrode, *Biosens. Bioelectron.*, **20**, 2341–2346.
41. Claussen, J. C., Franklin, A. D., ul Haque, A., Porterfield, D. M., Fisher, T. S. (2009) Electrochemical biosensor of nanocube-augmented carbon nanotube networks, *ACS Nano*, **3**, 37–44.
42. Baibarac, M., Gomez-Romero, P. (2006) Nanocomposites based on conducting polymers and carbon nanotubes: from fancy materials to functional applications, *J. Nanosci. Nanotechnol.*, **6**, 289–302.
43. Yogeswaran, U., Chen, S.-M. (2008) Recent trends in the application of carbon nanotubes-polymer composite modified electrodes for biosensors: a review, *Anal. Lett.*, **41**, 210–243.
44. Moniruzzaman, M., Winey, K. I. (2006) Polymer nanocomposites containing carbon nanotubes, *Macromolecules*, **39**, 5194–5205.

45. Tiwari, I., Singh, K. P., Singh, M. (2009) An insight review on the application of polymer-carbon nanotubes based composite material in sensor technology, *Russ. J. Gen. Chem.*, **79**, 2685–2694.
46. Zou, Y., Sun, L.-X., Xu, F. (2007) Biosensor based on polyaniline-prussian blue/multi-walled carbon nanotubes hybrid composites, *Biosens. Bioelectron.*, **22**, 2669–2674.
47. Perez, B., Pumera, M., del Valle, M., Merkoci, A., Alegret, S. (2005) Glucose biosensor based on carbon nanotube epoxy composites, *J. Nanosci. Nanotechnol.*, **5**, 1694–1698.
48. Jia, F., Shan, C., Li, F., Niu, L. (2008) Carbon nanotube/gold nanoparticles/polyethylenimine-functionalized ionic liquid thin film composites for glucose biosensing, *Biosens. Bioelectron.*, **24**, 945–950.
49. Tsai, Y. C., Li, S. C., Liao, S. W. (2006) Electrodeposition of polypyrrole-multiwalled carbon nanotube-glucose oxidase nanobiocomposite film for the detection of glucose, *Biosens. Bioelectron.*, **22**, 495–500.
50. Sheng, Q., Zheng, J. (2009) Bienzyme system for the biocatalyzed deposition of polyaniline templated by multiwalled carbon nanotubes: a biosensor design, *Biosens. Bioelectron.*, **24**, 1621–1628.
51. Toghill, K. E., Compton, R. G. (2010) Electrochemical non-enzymatic glucose sensors: a perspective and an evaluation, *Int. J. Electrochem. Sci.*, **5**, 1246–1301.
52. Li, L.-H., Zhang, W.-D. (2008) Preparation of carbon nanotubes supported platinum nanoparticles by an organic colloidal process for nonenzymatic glucose sensing, *Microchim. Acta*, **163**, 305–311.
53. Yoon, S. W., Kim, S. Y., Park, J., Park, C. J., Lee, C. J. (2005) Electronic structure and field emission of muldwalled carbon nanotubes depending on growth temperature, *J. Phys. Chem. B*, **109**, 20403–20406.
54. Meng, L., Jin, J., Yang, G., Lu, T., Zhang, H, Cai, C. (2009) Nonenzymatic electrochemical detection of glucose based on palladium-single–walled carbon nanotube hybrid nanostructures, *Anal. Chem.*, **81**, 7271–7280.
55. Male, K. B., Hrapovic, S., Liu, Y. L., Wang, D. S, Luong, J. H. T. (2004) Electrochemical detection of carbohydrates using copper nanoparticles and carbon nanotubes, *Anal. Chim. Acta*, **516**, 35–41.
56. Li, L.-H., Zhang, W.-D., Ye, J.-S. (2008) Electrocatalytic oxidation of glucose at carbon nanotubes supported PtRu nanoparticles and its detection, *Electroanalysis*, **20**, 2212–2216.
57. Cui, H.-F., Ye, J.-S., Zhang, W.-D., Li, C.-M., Luong, J. H. T., Sheu, F.-S. (2007) Selective and sensitive electrochemical detection of glucose

in neutral solution using platinum-lead alloy nanoparticle/carbon nanotube nanocomposites, *Anal. Chim. Acta*, **594**, 175–183.

58. Zhang, X., Wang, G., Zhang, W., Wei, Y., Fang, B. (2009) Fixure-reduce method for the synthesis of Cu_2O/MWCNTs nanocomposites and its application as enzyme-free glucose sensor, *Biosens. Bioelectron.*, **24**, 3395–3398.
59. Zhang, W. D., Chen, J., Jiang, L. C., Yu, Y. X., Zhang, J. Q. (2010) A highly sensitive nonenzymatic glucose sensor based on NiO-modified multi-walled carbon nanotubes, *Microchim. Acta*, **168**, 259–265.
60. Chen, J., Zhang, W.-D., Ye, J.-S. (2008) Nonenzymatic electrochemical glucose sensor based on MnO_2/MWNTs nanocomposite, *Electrochem. Commun.*, **10**, 1268–1271.
61. Deo, R. P., Wang, J. (2004) Electrochemical detection of carbohydrates at carbon-nanotube modified glassy-carbon electrodes, *Electrochem. Commun.*, **6**, 284–287.
62. Antiochia, R., Lavagnini, I., Magno, F. (2004) Amperometric mediated carbon nanotube paste biosensor for fructose determination, *Anal. Lett.*, **37**, 1657–1669.
63. Tominaga, M., Nomura, S., Taniguchi, I. (2009) D-Fructose detection based on the direct heterogeneous electron transfer reaction of fructose dehydrogenase adsorbed onto multi-walled carbon nanotubes synthesized on platinum electrode, *Biosens. Bioelectron.*, **24**, 1184–1188.
64. Tkac, J., Whittaker, J. W., Ruzgas, T. (2007) The use of single walled carbon nanotubes dispersed in a chitosan matrix for preparation of a galactose biosensor, *Biosens. Bioelectron.*, **22**, 1820–1824.
65. Fu, H., Huang, H., Cai, Q. (2007) Study on multi-walled carbon nanotubes modified amperemetric galactose biosensor, *Transducer Microsyst. Technol.*, **26**, 67–69 (in Chinese).
66. Gorton, L. (2002) Electrochemistry of $NAD(P)^+$/NAD(P)H, in *Encyclopedia of Electrochemistry*, Vol. 9, Eds. A. J. Bard, M. Stratmann, G. S. Wilson (Wiley-VCH, New York), 67–143.
67. Radoi, A., Compagnone, D. (2009) Recent advances in NADH electrochemical sensing design, *Bioelectrochemistry*, **76**, 126–134.
68. Zhang, M. G., Smith, A., Gorski, W. (2004) Carbon nanotube-chitosan system for electrochemical sensing based on dehydrogenase enzymes, *Anal. Chem.*, **76**, 5045–5050.
69. Bai, L., Wen, D., Yin, J., Deng, L., Zhu, C., Dong, S., Carbon nanotubes-ionic liquid nanocomposites sensing platform for NADH oxidation and oxygen, glucose detection in blood, *Talanta*, **91**, 110–115.

70. Zhu, L. D., Zhai, J. G., Yang, R. L., Tian, C. Y., Guo, L. P. (2007) Electrocatalytic oxidation of NADH with Meldola's blue functionalized carbon nanotubes electrodes, *Biosens. Bioelectron.*, **22**, 2768–2773.

71. Zeng, J. X., Wei, W. Z., Wu, L., Liu, X. Y., Liu, K., Li, Y. (2006) Fabrication of poly(toluidine blue O)/carbon nanotube composite nanowires and its stable low-potential detection of NADH, *J. Electroanal. Chem.*, **595**, 152–160.

72. Radoi, A., Compagnone, D., Valcarcel, M. A., et al. (2008) Detection of NADH via electrocatalytic oxidation at single-walled carbon nanotubes modified with Variamine blue, *Electrochim. Acta*, **53**, 2161–2169.

73. Raj, C. R., Chakraborty, S. (2006) Carbon nanotubes-polymer-redox mediator hybrid thin film for electrocatalytic sensing, *Biosens. Bioelectron.*, **22**, 700–706.

74. Deng, L., Wang, Y. Z., Shang, L., Wen, D., Wang, F. A., Dong, S. J. (2008) A sensitive NADH and glucose biosensor tuned by visible light based on thionine bridged carbon nanotubes and gold nanoparticles multilayer, *Biosens. Bioelectron.*, **24**, 951–957.

75. Manso, J., Mena, M. L., Yanez-Sedeno, P., Pingarron, J. M. (2008) Alcohol dehydrogenase amperometric biosensor based on a colloidal gold-carbon nanotubes composite electrode, *Electrochim. Acta*, **53**, 4007–4012.

76. Pumera, M., Merkoci, A., Alegret, S. (2006) Carbon nanotube-epoxy composites for electrochemical sensing, *Sens. Actuators, B*, **113**, 617–622.

77. Agui, L., Pena-Farfal, C., Yanez-Sedeno, P., Pingarron, J. M. (2007) Poly-(3-methylthiophene)/carbon nanotubes hybrid composite-modified electrodes, *Electrochim. Acta*, **52**, 7946–7952.

78. Wightman, R. M., May, L. J., Michael, A. C. (1988) Detection of dopamine dynamics in the brain, *Anal. Chem.*, **60**, 769A–779A.

79. Jeong, H., Jeon, S. (2008) Determination of dopamine in the presence of ascorbic acid by nafion and single-walled carbon nanotube film modified on carbon fiber microelectrode, *Sensors*, **8**, 6924–6935.

80. Zhou, X., Zheng, N., Hou, S., Li, X., Yuan, Z. (2010) Selective determination of dopamine in the presence of ascorbic acid at a multi-wall carbon nanotube-poly(3,5-dihydroxy benzoic acid) film modified electrode, *J. Electroanal. Chem.*, **642**, 30–34.

81. Hocevar, S. B., Wang, J., Deo, R. P., Musameh, M., Ogorevc, B. (2005) Carbon nanotube modified microelectrode for enhanced voltammetric detection of dopamine in the presence of ascorbate, *Electroanalysis*, **17**, 417–422.

82. Umasankar, Y., Thiagarajan, S., Chen, S. M. (2009) Nanocomposite of functionalized multiwall carbon nanotubes with nafion, nano platinum, and nano gold biosensing film for simultaneous determination of ascorbic acid, epinephrine, and uric acid, *Anal. Biochem.*, **390**, 224–224.

83. Min, K., Yoo, Y. J. (2009) Amperometric detection of dopamine based on tyrosinase-SWNTs-Ppy composite electrode, *Talanta*, **80**, 1007–1011.

84. Li, J., Wei, W., Luo, S. (2010) A novel one-step electrochemical codeposition of carbon nanotubes-DNA hybrids and tiron doped polypyrrole for selective and sensitive determination of dopamine, *Microchim. Acta*, **171**, 109–116.

85. Sun, Y., Fei, J., Hou, J., Zhang, Q., Liu, Y, Hu, B. (2009) Simultaneous determination of dopamine and serotonin using a carbon nanotubes-ionic liquid gel modified glassy carbon electrode, *Microchim. Acta*, **165**, 373–379.

86. Noroozifar, M., Khorasani-Motlagh, M., Akbari, R., Parizi, M. B. (2011) Simultaneous and sensitive determination of a quaternary mixture of AA, DA, UA and Trp using a modified GCE by iron ion-doped natrolite zeolite-multiwall carbon nanotube, *Biosens. Bioelectron.*, **28**, 56–63.

87. Zare, H. R., Nasirizadeh, N. (2009) Application of hematoxylin multi-wall carbon nanotube modified carbon paste electrode as a chemical sensor for simultaneous determination of dopamine and acetaminophen, *Int. J. Electrochem. Sci.*, **4**, 1691–1705.

88. Xue, W., Cui, T. (2008) A high-resolution amperometric acetylcholine sensor based on nano-assembled carbon nanotube and acetylcholinesterase thin films, *J. Nano Res.*, **1**, 1–9.

89. Xue, W., Cui, T. (2008) A thin-film transistor based acetylcholine sensor using self-assembled carbon nanotubes and SiO_2 nanoparticles, *Sens. Actuators, B*, **134**, 981–987.

90. Xue, W., Cui, T. (2007) A high-resolution amperometric acetylcholine biosensor based on nano self-assembly of carbon nanotubes, in *Proceedings of the IEEE Twentieth Annual International Conference on Micro Electro Mechanical Systems, Vols 1 and 2*, 766–769.

91. Kharitonov, A. B., Zayats, M., Lichtenstein, A., Katz, E., Willner, I. (2000) Enzyme monolayer-functionalized field-effect transistors for biosensor applications, *Sens. Actuators, B*, **70**, 222–231.

92. Ampurdanes, J., Crespo, G. A., Maroto, A., Angeles Sarmentero, M., Ballester, P., Xavier Rius, F. (2009) Determination of choline and derivatives with a solid-contact ion-selective electrode based on

octaamide cavitand and carbon nanotubes, *Biosens. Bioelectron.*, **25**, 344–349.

93. Goyal, R. N., Bishnoi, S. (2011) Simultaneous determination of epinephrine and norepinephrine in human blood plasma and urine samples using nanotubes modified edge plane pyrolytic graphite electrode, *Talanta*, **84**, 78–83.
94. Beitollahi, H., Karimi-Maleh, H., Khabazzadeh, H. (2008) Nanomolar and selective determination of epinephrine in the presence of norepinephrine using carbon paste electrode modified with carbon nanotubes and novel 2-(4-oxo-3-phenyl-3,4-dihydro-quinazolinyl)-N'-phenyl-hydrazinecarbothioamide, *Anal. Chem.*, **80**, 9848–9851.
95. Goyal, R. N., Bishnoi, S. (2011) A novel multi-walled carbon nanotube modified sensor for the selective determination of epinephrine in smokers, *Electrochim. Acta*, **56**, 2717–2724.
96. Jackson, B. P., Dietz, S. M., Wightman, R. M. (1995) Fast-scan cyclic voltammetry of 5-hydroxytryptamine, *Anal. Chem.*, **67**, 1115–1120.
97. Swamy, B. E. K., Venton, B. J. (2007) Carbon nanotube-modified microelectrodes for simultaneous detection of dopamine and serotonin in vivo, *Analyst*, **132**, 876–884.
98. Guell, A. G., Meadows, K. E., Unwin, P. R., Macpherson, J. V. (2010) Trace voltammetric detection of serotonin at carbon electrodes: comparison of glassy carbon, boron doped diamond and carbon nanotube network electrodes, *Phys. Chem. Chem. Phys.*, **12**, 10108–10114.
99. Zhang, H. Q., Zhao, Q., Li, X. F., Le, X. C. (2007) Ultrasensitive assays for proteins, *Analyst*, **132**, 724–737.
100. Jacobs, C. B., Peairs, M. J., Venton, B. J. (2010) Review: carbon nanotube based electrochemical sensors for biomolecules, *Anal. Chim. Acta*, **662**, 105–127.
101. Fei, S. D., Chen, J. H., Yao, S. Z., Deng, G. H., He, D. L., Kuang, Y. F. (2005) Electrochemical behavior of L-cysteine and its detection at carbon nanotube electrode modified with platinum, *Anal. Biochem.*, **339**, 29–35.
102. Ye, M. L., Xu, B., Zhang, W. D. (2011) Sputtering deposition of Pt nanoparticles on vertically aligned multiwalled carbon nanotubes for sensing L-cysteine, *Microchim. Acta*, **172**, 439–446.
103. Shen, R. X., Li, X. Z., Lium, G. L., Ji, Y. L., Wang, G. F., Fang, B. (2010) Synthesis and characterization of chromium hexacyanoferrate/multiwalled carbon nanotube composite and its biosensing for L-cysteine, *Electroanalysis*, **22**, 2383–2388.

104. Gong, K. P., Dong, Y., Xiong, S. X., Chen, Y., Mao, L. Q. (2004) Novel electrochemical method for sensitive determination of homocysteine with carbon nanotube-based electrodes, *Biosens. Bioelectron.*, **20**, 253–259.

105. Deng, C. Y., Chen, J. H., Chen, X. L., Wang, M. D., Nie, Z., Yao, S. Z. (2009) Electrochemical detection of L-cysteine using a boron-doped carbon nanotube-modified electrode, *Electrochim. Acta*, **54**, 3298–3302.

106. Huang, K. J., Luo, D. F., Xie, W. Z., Yu, Y. S. (2008) Sensitive voltammetric determination of tyrosine using multi-walled carbon nanotubes/4-aminobenzeresulfonic acid film-coated glassy carbon electrode, *Colloids Surf. B*, **61**, 176–181.

107. Okuno, J., Maehashi, K., Kerman, K., Takamura, Y., Matsumoto, K., Tamiya, E. (2007) Label-free immunosensor for prostate-specific antigen based on single-walled carbon nanotube array-modified microelectrodes, *Biosens. Bioelectron.*, **22**, 2377–2381.

108. Kang, S. Z., Chen, H., Li, X. Q., Mu, J. (2010) Preparation of L-alanine ethyl ester modified multiwalled carbon nanotubes and their chiral discrimination between D- and L-tryptophan, *Diamond Relat. Mater.*, **19**, 1221–1224.

109. Meng, L., Wu, P., Chen, G. X., Cai, C. X., Sun, Y. M., Yuan, Z. H. (2009) Low-potential detection of glutamate based on the electrocatalytic oxidation of NADH at thionine/single-walled carbon nanotubes composite modified electrode, *Biosens. Bioelectron.*, **24**, 1751–1756.

110. Chakraborty, S., Raj, C. R. (2007) Amperometric biosensing of glutamate using carbon nanotube based electrode, *Electrochem. Commun.*, **9**, 1323–1330.

111. Tang, L. H., Zhu, Y. H., Xu, L. H., Yang, X. L., Li, C. Z. (2007) Amperometric glutamate biosensor based on self-assembling glutamate dehydrogenase and dendrimer-encapsulated platinum. nanoparticles onto carbon nanotubes, *Talanta*, **73**, 438–443.

112. Cid, C. C., Riu, J., Maroto, A., Rius, F. X. (2008) Carbon nanotube field effect transistors for the fast and selective detection of human immunoglobulin G, *Analyst*, **133**, 1005–1008.

113. Cid, C. C., Riu, J., Maroto, A., Rius, F. X. (2008) Detection of human immunoglobulin g at physiological conditions with chemically functionalizated carbon nanotube field effect transistors, *Curr. Nanosci.*, **4**, 314–317.

114. Kim, J. P., Lee, B. Y., Hong, S., Sim, S. J. (2008) Ultrasensitive carbon nanotube-based biosensors using antibody-binding fragments, *Anal. Biochem.*, **381**, 193–198.

115. Yun, Y., Bange, A., Heineman, W. R., et al. (2007) A nanotube array immunosensor for direct electrochemical detection of antigen-antibody binding, *Sens. Actuators, B*, **123**, 177–182.

116. Maehashi, K., Katsura, T., Kerman, K., Takamura, Y., Matsumoto, K., Tamiya, E. (2007) Label-free protein biosensor based on aptamer-modified carbon nanotube field-effect transistors, *Anal. Chem.*, **79**, 782–787.

117. Zhou, X. J., Moran-Mirabal, J. M., Craighead, H. G., McEuen, P. L. (2007) Supported lipid bilayer/carbon nanotube hybrids, *Nat. Nanotechnol.*, **2**, 185–190.

118. Hu, P., Fasoli, A., Park, J., et al. (2008) Self-assembled nanotube field-effect transistors for label-free protein biosensors, *J. Appl. Phys.*, **104**, 074310.

119. Hu, P. A., Zhang, J., Wen, Z. Z., Zhang, C. (2011) Network single-walled carbon nanotube biosensors for fast and highly sensitive detection of proteins, *Nanotechnology*, **22**, 335502.

120. Buch, M., Rishpon, J. (2008) An electrochemical immunosensor for c-reactive protein based on multi-walled carbon nanotube-modified electrodes, *Electroanalysis*, **20**, 2592–2594.

121. Zeng, S. L., Zhou, H. K., Gan, N., Cao, Y. T. (2011) A renewable c reactive protein amperometric immunosensor based on magnetic multiwalled carbon nanotubes probles modified electrode, *Appl. Mech. Mater.*, **80**, 452–456.

122. Yu, X., Kim, S. N., Papadimitrakopoulos, F., Rusling, J. F. (2005) Protein immunosensor using single-wall carbon nanotube forests with electrochemical detection of enzyme labels, *Mol. Biosyst.*, **1**, 70–78.

123. Abe, M., Murata, K., Ataka, T., Ifuku, Y., Matsumoto, K. (2009) Selective protein sensing using a carbon nanotube field-effect transistor, *J. Nanosci. Nanotechnol.*, **9**, 1947–1950.

124. Abe, M., Murata, K., Kojima, A., et al. (2007) Quantitative detection of protein using a top-gate carbon nanotube field effect transistor, *J. Phys. Chem. C*, **111**, 8667–8670.

125. Fang, L. Y., Lu, Z. Z., Wei, H., Wang, E. K. (2008) Quantitative electrochemiluminescence detection of proteins: avidin-based sensor and tris(2,2'-bipyridine) ruthenium(II) label, *Biosens. Bioelectron.*, **23**, 1645–1651.

126. Erdem, A., Papakonstantinou, P., Murphy, H. (2006) Direct DNA hybridization at disposable graphite electrodes modified with carbon nanotubes, *Anal. Chem.*, **78**, 6656–6659.

127. Caruana, D. J., Heller, A. (1999) Enzyme-amplified amperometric detection of hybridization and of a single base pair mutation in an 18-base oligonucleotide on a 7-μm diameter microelectrode, *J. Am. Chem. Soc.*, **121**, 769–774.

128. Zhang, Y. C., Kim, H. H., Heller, A. (2003) Enzyme-amplified amperometric detection of 3000 copies of DNA in a 10-μL droplet at 0.5 fm concentration, *Anal. Chem.*, **75**, 3267–3269.

129. Berti, F., Lozzi, L., Palchetti, I., Santucci, S., Marrazza, G. (2009) Aligned carbon nanotube thin films for DNA electrochemical sensing, *Electrochim. Acta*, **54**, 5035–5041.

130. He, P. G., Dai, L. M. (2004) Aligned carbon nanotube-DNA electrochemical sensors, *Chem. Commun.*, (3), 348–349.

131. Wang, J., Kawde, A. N., Jan, M. R. (2004) Carbon-nanotube-modified electrodes for amplified enzyme-based electrical detection of DNA hybridization, *Biosens. Bioelectron.*, **20**, 995–1000.

132. Li, J., Liu, Q., Liu, Y. J., Liu, S. C., Yao, S. Z. (2005) DNA biosensor based on chitosan film doped with carbon nanotubes, *Anal. Biochem.*, **346**, 107–114.

133. Qi, H., Li, X. X., Chen, P., Zhang, C. X. (2007) Electrochemical detection of DNA hybridization based on polypyrrole/ss-DNA/multi-wall carbon nanotubes paste electrode, *Talanta*, **72**, 1030–1035.

134. Zhu, N. N., Chang, Z., He, P. G., Fang, Y. Z. (2005) Electrochemical DNA biosensors based on platinum nanoparticles combined carbon nanotubes, *Anal. Chim. Acta*, **545**, 21–26.

135. Wang, J., Liu, G. D., Jan, M. R. (2004) Ultrasensitive electrical biosensing of proteins and DNA: carbon-nanotube derived amplification of the recognition and transduction events, *J. Am. Chem. Soc.*, **126**, 3010–3011.

136. Yang, X. Y., Lu, Y. H., Ma, Y. F., Liu, Z. F., Du, F., Chen, Y. S. (2007) DNA electrochemical sensor based on an adduct of single-walled carbon nanotubes and ferrocene, *Biotechnol. Lett.*, **29**, 1775–1779.

137. Balasubramanian, K. (2012) Label-free indicator-free nucleic acid biosensors using carbon nanotubes, *Eng. Life Sci.*, **12**, 121–130.

138. Shen, Q., Wang, X. M. (2009) Simultaneous determination of adenine, guanine and thymine based on beta-cyclodextrin/MWNTs modified electrode, *J. Electroanal. Chem.*, **632**, 149–153.

139. Muti, M., Kuralay, F., Erdem, A., Abaci, S., Yumak, T., Sinag, A. (2010) Tin oxide nanoparticles-polymer modified single-use sensors for electrochemical monitoring of label-free DNA hybridization, *Talanta*, **82**, 1680–1686.

140. Zhang, X. Z., Jiao, K., Liu, S. F., Hu, Y. W. (2009) Readily reusable electrochemical DNA hybridization biosensor based on the interaction of DNA with single-walled carbon nanotubes, *Anal. Chem.*, **81**, 6006–6012.
141. Bonanni, A., del Valle, M. (2010) Use of nanomaterials for impedimetric DNA sensors: a review, *Anal. Chim. Acta*, **678**, 7–17.
142. Caliskan, A., Erdem, A., Karadeniz, H. (2009) Direct DNA hybridization on the single-walled carbon nanotubes modified sensors detected by voltammetry and electrochemical impedance spectroscopy, *Electroanalysis*, **21**, 2116–2124.
143. Hu, P. A., Zhang, J., Li, L., Wang, Z. L., O'Neill, W., Estrela, P. (2010) Carbon nanostructure-based field-effect transistors for label-free chemical/biological sensors, *Sensors*, **10**, 5133–5159.
144. Martinez, M. T., Tseng, Y. C., Ormategui, N., Loinaz, I., Eritja, R., Bokor, J. (2009) Label-free DNA biosensors based on functionalized carbon nanotube field effect transistors, *Nano Lett.*, **9**, 530–536.
145. Dastagir, T., Forzani, E. S., Zhang, R., et al. (2007) Electrical detection of hepatitis C virus RNA on single wall carbon nanotube-field effect transistors, *Analyst*, **132**, 738–740.
146. Kim, S., Kim, T. G., Byon, H. R., Shin, H. J., Ban, C., Choi, H. C. (2009) Recognition of single mismatched DNA using MutS-immobilized carbon nanotube field effect transistor devices, *J. Phys. Chem. B*, **113**, 12164–12168.
147. Sorgenfrei, S., Chiu, C. Y., Gonzalez, R. L., et al. (2011) Label-free single-molecule detection of DNA-hybridization kinetics with a carbon nanotube field-effect transistor, *Nat. Nanotechnol.*, **6**, 125–131.
148. Cheng, W., Ding, L., Lei, J. P., Ding, S. J., Ju, H. X. (2008) Effective cell capture with tetrapeptide-functionalized carbon nanotubes and dual signal amplification for cytosensing and evaluation of cell surface carbohydrate, *Anal. Chem.*, **80**, 3867–3872.
149. Cheng, W., Ding, L., Ding, S. J., Yin, Y. B, Ju, H. X. (2009) A simple electrochemical cytosensor array for dynamic analysis of carcinoma cell surface glycans, *Angew. Chem., Int. Ed.*, **48**, 6465–6468.
150. Zhong, X., Qian, G. S., Xu, J. J., Chen, H. Y. (2010) Cytosensor constructed with a biomimetic fibronectin-functionalized carbon nanotubes on glassy carbon heated electrode, *J. Phys. Chem. C*, **114**, 19503–19508.
151. Zheng, T.-T., Zhang, R., Zou, L., Zhu, J.-J. (2012) A label-free cytosensor for the enhanced electrochemical detection of cancer cells using polydopamine-coated carbon nanotubes, *Analyst*, **137**, 1316–1318.

152. Shao, N., Wickstrom, E., Panchapakesan, B. (2008) Nanotube-antibody biosensor arrays for the detection of circulating breast cancer cells, *Nanotechnology*, **19**, 465101.

153. Shen, Q., You, S. K., Park, S. G., et al. (2008) Electrochemical biosensing for cancer cells based on TiO_2/CNT nanocomposites modified electrodes, *Electroanalysis*, **20**, 2526–2530.

154. Villamizar, R. A., Maroto, A., Rius, F. X. (2009) Improved detection of Candida albicans with carbon nanotube field-effect transistors, *Sens. Actuators, B*, **136**, 451–457.

155. Garcia-Aljaro, C., Cella, L. N., Shirale, D. J., et al. (2010) Carbon nanotubes-based chemiresistive biosensors for detection of microorganisms, *Biosens. Bioelectron.*, **26**, 1437–1441.

156. Zelada-Guillen, G. A., Riu, J., Duzgun, A., Rius, F. X. (2009) Immediate detection of living bacteria at ultralow concentrations using a carbon nanotube based potentiometric aptasensor, *Angew. Chem., Int. Ed.*, **48**, 7334–7337.

Chapter 6

Gold and Silver Nanoparticles for Electrochemical Immunosensors

Barry G. D. Haggett
Institute of Biomedical and Environmental Science and Technology, University of Bedfordshire, Luton, LU2 8DL, UK
barry.haggett@beds.ac.uk

6.1 Introduction

Gold and silver nanoparticles have attractive properties that are widely exploited in the development of electrochemical immunosensors. A wide variety of embodiments have been reported, but in the main, these can be categorized by where and how the nanoparticles are used.

6.1.1 *Nanoparticles*

Nanoparticles are commonly defined as particles with sizes in range of 1–100 nm. The properties of such particles can differ significantly from those of the corresponding bulk materials. In the most general terms, the surface-area-to-volume ratio of any nanoparticle is enormous compared to that of the bulk material,

Electrochemical Biosensors
Edited by Serge Cosnier

ISBN 978-981-4411-46-2 (Hardcover), 978-981-4411-47-9 (eBook)
www.panstanford.com

and for biosensor purposes, this provides a large surface area for immobilization of biomaterials. Gold and silver nanoparticles have additional attractive properties for electrochemical biosensing applications:

- Easy fabrication
- Good biocompatibility
- Excellent electrical conductivity
- Electrochemically active ions

Gold nanoparticles are prevalent in the development of immunosensing devices and have been widely reviewed [1–4], while silver nanoparticles have received comparatively little attention.

6.1.2 *Electrochemical Immunoassays*

Electrochemical immunoassays use voltammetry or potentiometry to measure the amount of analyte (antigen or antibody) bound in an antibody–antigen complex. The assays require one binding partner (the capture element) to be immobilized at a surface, for example:

- the surface of the electrode, where both the immunochemistry and the signal generation are to take place;
- another surface, such as the bottom of a multiwell plate, where the immunochemistry takes place prior to the measurement electrodes being introduced to the reaction products; or
- the surface of a magnetic particle used to separate antibody–antigen complexes from other sample components prior to being located at the electrode surface.

In assays utilizing a single antibody–antigen binding event, the capture element is exposed to the test sample to allow immunobinding to occur. Unbound materials are washed away, and the electrochemical signal is generated using an electroactive probe (Fig. 6.1a,b). In sandwich assays, captured antibody–antigen complexes are washed and exposed to a second, detection antibody to form either capture antibody–antigen detection antibody or capture antigen–antibody detection antibody "sandwiches." The

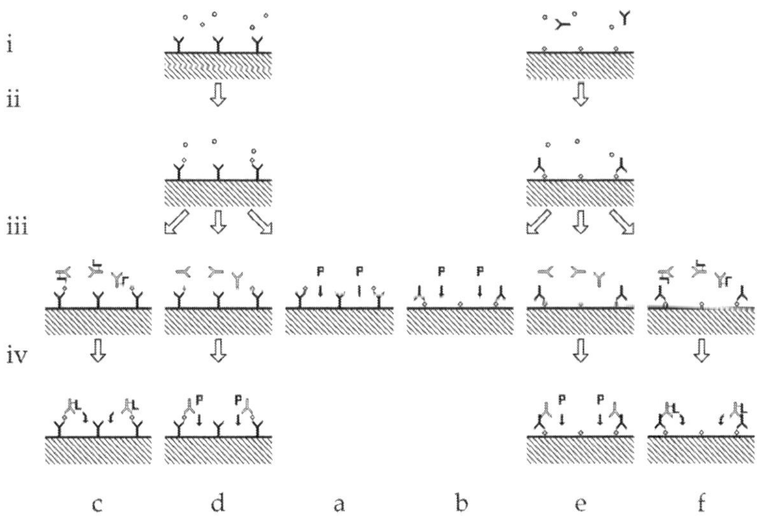

Figure 6.1 Schematic arrangements of the most usual electrochemical noncompetitive immunoassays. (a) Antigen assay using a capture antibody. (b) Antibody assay using a capture antigen. (c) Sandwich assay for an antigen using a labeled detection antibody. (d) Sandwich assay for an antigen using an electroactive probe. (e) Sandwich assay for an antibody using an electroactive probe. (f) Sandwich assay for an antibody using a labeled detection antibody. Measurement steps: (i) The capture element is exposed to the test sample; (ii) immunocomplexes are formed between the antibody and the antigen; (iii) the washed complexes are exposed to the electroactive probe or the detection antibody, as appropriate; and (iv) sandwich complexes are washed and the electrochemical signal generated using either an electroactive probe or a labeled detection antibody. The most usual arrangement is that the capture element is bound to the working electrode. Alternatively, the capture element may be bound to another surface, such as the bottom of a multiwell plate, or to magnetic particles, in which case the electrodes must be brought to the surface or a magnet used to locate the immunocomplexes at the working electrodes before measurement can occur. Key: (Y) antibodies; (◇) antigens; (O) other materials in the test sample; (L) enzyme or other label; and (P) electroactive probe that may be added to the test solution or incorporated into the working electrodes.

Table 6.1 Example electrochemical immunoassays utilizing gold and/or silver nanoparticles

Analyte	References	Analyte	References
α-1-Fetoprotein	[5–25]	Human lung cancer–associated antigen ENO1	[51]
Aflatoxin B1	[46–48]		
Albumin	[50]	Human serum albumin	[54]
Annexin II and MUC5AC	[52]	IgA and IgG anti-(tissue tranglutaminase) autoantibodies	[57]
Bacillus globigii	[53]		
Biotin	[55, 56]	IgG antibodies specific to *Echinococcus granulosus*	[61]
Brevetoxin B	[58]		
Carbohydrate antigen 199	[59, 60]	Inducible nitric oxide synthase	[64]
Carcinoma antigen 125	[60, 62, 63]	Interleukin	[89–93]
Carcinoma antigen 153	[60]	Kanamycin	[94]
Carcinoembryonic antigen	[65–88]	Microcystin-LR	[95]
Cardiac troponin I	[102]	Myeloperoxidase	[96, 97]
Casein	[104]	Myoglobin	[98–101]
Chloramphenicol	[106]	Neuron-specific enolase	[103]
Diuron	[107]	*N*-terminal pro-B-type natriuretic peptide	[105]
Escherichia coli O157:H7	[110, 111]		
Estradiol	[113, 114]	Ochratoxin	[108, 109]
Ferritin	[116]	*Plasmodium falciparum* histidine-rich protein-2	[112]
Hemoglobin HbA1c	[117]		
Hepatitis B surface antigen	[119–127]	Platelet-derived growth factor BB	[115]
HIV p24	[133]	Progastrin-releasing peptide	[118]
Human chorionic gonadotropin	[134, 135]	Progesterone	[128]
		Prolactin	[129]
Human epididymis-specific protein 4	[137]	Prostate-specific antigen	[89, 130–132]
		Pseudorabies virus antibody	[136]
Human IgE	[140]	α-Synuclein	[138]
Human IgG	[26–45]	Tumor necrosis factor α	[139]
Human IgM	[49]	Vascular endothelial growth factors	[141]

electrochemical signal is generated utilizing either a label on the detection antibody or a probe (Fig. 6.1c–f).

Gold and silver nanoparticles have been incorporated into electrochemical immunoassays for a wide variety of analytes, for example, Table 6.1. The majority of these assays were proteins such as α-fetoprotein (AFP) and carcinoembryonic antigen (CEA) using capture and/or detection antibodies (Fig. 6.1a,c,d). A minority of the reports described assays for antibodies (Fig. 6.1b,e,f) such as human immunoglobulins—particularly immunoglobulin G (IgG). In all of these assays, the metal nanoparticles can be deployed in one or more configurations.

6.1.3 *Use of Gold and Silver Nanoparticles*

For electrochemical immunoassays, gold and silver nanoparticles may be used to (1) modify the working electrode, (2) modify antibodies not bound to the working electrode, or (3) facilitate generation of a measurement signal at the working electrode. In each case, there is a multiplicity of different possible embodiments, and any given assay may use a combination of approaches. The variety of configurations adopted is exemplified by the range of assays developed for AFP (a tumor marker for hepatoblastoma, hepatocellular carcinoma, and germ cell tumors [142]), examples of which are summarized in Table 6.2.

6.2 Modification of the Working Electrode

6.2.1 *Use of Gold or Silver Nanoparticles*

Gold or silver nanoparticles may be localized at the working electrode by incorporation into the base electrode material (Fig. 6.2b), deposition on the base electrode surface (Fig. 6.2c), inclusion within a composite layer on the base electrode (Fig. 6.2d), or attachment to an intermediary layer (Fig. 6.2e). For research purposes, the working electrode is usually fabricated from monolithic gold or glassy carbon materials, but particularly for applied studies, the working electrode may be a carbon paste or screen-printed material.

Nanoparticles may be included within carbon paste electrodes or within pastes that are subsequently screen-printed and cured/dried to produce working electrodes (Fig. 6.2b). In these cases, the particles may be used to promote the conductivity of the base electrodes and/or link to antibodies (or other materials), where they are exposed at the electrode surfaces. More usually, metal nanoparticles are used to modify the surface of carbon paste electrodes [7, 49, 66, 72, 84, 147], as is the case with other types

Table 6.2 Example electrochemical immunoassays for α-fetoprotein

Base electrode	Preparation of electrode surface[a]	Preparation of unbound antibody[b]	Signal generation	Range /ng mL^{-1}	LOD[c] /ng mL^{-1}	Reference
CP (graphite, paraffin)	**Ai** Base electrode unmodified	**O** Au NP deposited on dendrimer-functionalized $CoFe_2O_4$ NP **Y** Anti-AFP deposited on magnetic core–shell–shell–gold NP	Functionalized NP located on base electrode with aid of magnet Change in open-circuit potential before and after addition of AFP	0.8–120	0.3	[143]
Screen-printed carbon	**Ai** Base electrode unmodified	CNT functionalized with hemin and poly(dimethyldiallylammonium chloride) **O** Au NP and Fe_3O_4 NP deposited on functionalized CNT **Y** Anti-AFP deposited on composite NP Surface blocked with BSA	AFP reacted with anti-AFP-functionalized composite NP Washed immunocomplex located on base electrode with aid of magnet Binding with AFP inhibiting hemin electrochemistry measured using DPV	0.1–200	0.04	[18]
GC	**Dii** Anti-AFP deposited on (Au NP in pores of mesoporous silica) and coated on base electrode	–	Au NP promoting electron transfer between base electrode and ferrocene monocarboxylic acid probe	1.0–90	0.2	[15]
Au	**Dii** HRP-labeled anti-AFP-functionalized Au NP–thionine–graphene–thionine–Au NP nanocomposite (blocked with BSA) deposited on base electrode	–	Binding with AFP inhibiting thionine-mediated measurement of HRP activity (H_2O_2 4.5 mM)	0.1–200	0.05	[6]

Au	**Di** Core–shell–shell Prussian blue–BSA–Au NP deposited on base electrode **Dii** Anti-AFP deposited on NP Surface blocked with BSA	–	Binding with AFP inhibiting Prussian blue electrochemistry measured by CV	0.02–200	0.006	[10]
CP (graphite-ionic liquid)	**Aiv** Amine-functionalized graphene deposited on base electrode **Ei** Au NP deposited on graphene **Eii** Anti-AFP deposited on NP Surface blocked with BSA	–	Binding with AFP inhibiting hexacyanoferrate (5 mM) electrochemistry measured using DPV	1–250	0.1	[11]
GC	**Di** Au NP in chitosan deposited on base electrode Thionine deposited on glutaraldehyde-activated film **Ei** Au NP deposited on film **Eii** Anti-AFP deposited on Au NP Surface blocked with BSA	–	Binding with AFP inhibiting thionine electrochemistry measured by CV	0.4–200	0.24	[17]
Screen-printed graphite	**Av** Anti-AFP crosslinked to chitosan film on base electrode	**O** Ag NP deposited on CNT **Y** Biotinylated anti-AFP attached to streptavidin-functionalized NP Surface blocked with BSA	Electrochemical stripping of Ag NP–promoted silver deposited from $AgNO_3$ (0.50 mM) and ascorbic acid	0.0001–5.0	0.00006	[13]

(*Contd.*)

Table 6.2 (*Contd.*)

Base electrode	Preparation of electrode surface[a]	Preparation of unbound antibody[b]	Signal generation	Range /ng mL^{-1}	LOD[c] /ng mL^{-1}	Reference
GC	**Ci** Au NP electrodeposited on base electrode **Ciii/ii** Anti-AFP attached to protein A–functionalized NP Surface crosslinked with dimethyl pimelimidate and then blocked with BSA	HRP-labeled anti-AFP	Measurement of HRP activity using thionine acetate (0.5 mM) as redox mediator		3.7	[9]
GC	**Ci** Au NP electrodeposited on base electrode **Ciii/ii** NP functionalized with dendrimer and conjugated to anti-AFP Surface blocked with BSA or StabilCoat®	HRP-labeled anti-AFP	Measurement of HRP activity using hydrogen peroxide (0.5 mM or 5 mM) and thionine acetate (0.5 mM) as redox mediator using either CV or double-step chronoamperometry	5–100[d] 50–500[d] 5–375[d]	3	[144]
Au	**Di** Au NP in chitosan deposited on base electrode	**O** Au NP deposited on composite Prussian blue–Fe_3O_4 NP	Prussian blue–mediated determination of GOD activity	0.014–142	0.007	[145]

	Av Anti-AFP covalently linked to chitosan	Additional gold deposited on Au–Prussian blue–Fe_3O_4 NP **Y/L** Anti-AFP, GOD, and HRP deposited on three-layer composite NP Surface blocked with BSA				
GC	**Aiv** Thionine electropolymerized on base electrode	**O** Au NP deposited on titania NP	Thionine-mediated measurement of HRP activity	0.001–200	0.0005	[22]
		Y HRP-labeled anti-AFP deposited on NP				
	Di Au NP–Ag NP–graphene hybrid nanosheets cast on thionine	Surface blocked with BSA				
	Dii Anti-AFP deposited on NP Surface blocked with BSA					
Screen-printed carbon	**Aiv** Prussian blue–poly (diallyldimethylammonium)–chitosan NP cast on base electrode	**O** Au NP deposited on poly(diallyldimethylammonium)-functionalized CNT	Prussian blue–mediated determination of GOD activity measured using DPV	0.0025–2.5	0.0022	[12]
	Ei Au NP deposited on composite layer	**Y/L** Anti-AFP and GOD deposited on NP				
	Eii Anti-AFP deposited on Au NP Surface blocked	Surface blocked with BSA				

(*Contd.*)

Table 6.2 (*Contd.*)

Base electrode	Preparation of electrode surface[a]	Preparation of unbound antibody[b]	Signal generation	Range /ng mL^{-1}	LOD[c] /ng mL^{-1}	Reference
GC	**Di** Core–shell–gold–Prussian blue (Au–PB) nanocomposite film electrochemically polymerized on base electrode	**O** Au NP deposited on chitosan-coated iridum oxide nanoparticles	*p*-Aminophenol (1.0 mM) oxidized to quinone imine by Prussian blue and reoxidized by base electrode *p*-Aminophenol regenerated by protons generated from water by iridium oxide NP	0.005–200	0.0005	[146]
		Y Anti-AFP deposited on NP				
	Ei Au NP deposited on chitosan film cast over Au–PB layer **Eii** Anti-AFP deposited on Au NP Surface blocked with BSA					
Au	**Aiv** CNT and BSA cast on base electrode **Ei/Di** Layer-by-layer assembly of Au NP and thionine **Dii** Anti-AFP deposited on film Surface blocked with BSA	**Y** HRP-labeled anti-AFP deposited on Au NP	Thionine-mediated measurement of HRP activity	0.25–45	0.05	[20]

AFP, α-fetoprotein; Ag NP, silver nanoparticle; Au NP, gold nanoparticle; BSA, bovine serum albumin; CNT, carbon nanotube; CP, carbon paste; CV, cyclic voltammetry; DPV, differential pulse voltammetry; GC, glassy carbon; GOD, glucose oxidase; HRP, horseradish peroxidase; NP, nanoparticle.

[a]Bold characters refer to the arrangement in Fig. 6.2.

[b]Bold characters indicate steps involving gold or silver nanoparticles: (Y) NP linked to antibody; (L) NP linked to label; and (O) NP linked to other species.

[c]LOD, limit of detection.

[d]Depends on the concentration of hydrogen peroxide.

[e]Range in serum with 5 mM H_2O_2.

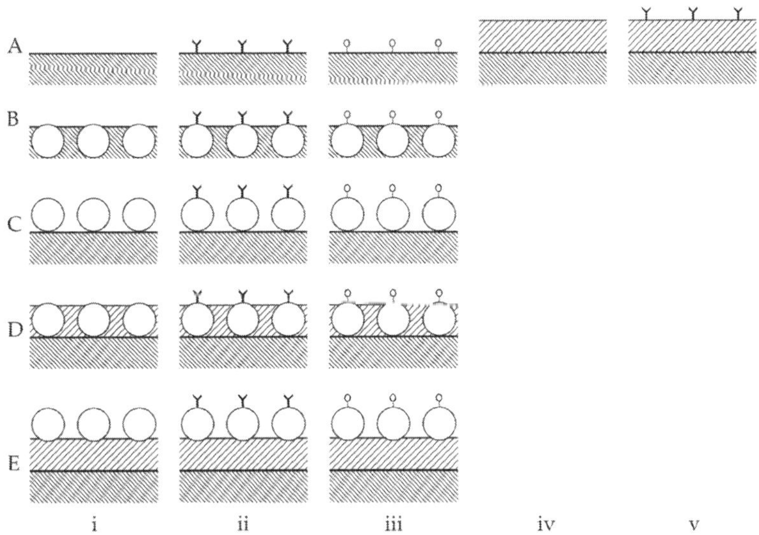

Figure 6.2 Schematics showing different configurations for modifying base electrodes with nanoparticles for immunoassays. (a) No nanoparticles. (b) Nanoparticles incorporated within the base electrode material. (c) Nanoparticles deposited on the base electrode. (d) Nanoparticles included within a composite layer on the base electrode. (e) Nanoparticles localized on an intermediary layer. In each case, the nanoparticles, or base electrode, may be (i) unmodified, (ii) functionalized with antibodies, or (iii) functionalized with material(s) other than antibodies. The base electrode may be modified with a film (iv) and functionalized with antibodies (v).

of working electrodes (Fig. 6.2c). Nanoparticles may be deposited on the unmodified electrode surface using electrodeposition. Gold nanoparticles deposited this way usually employ chloroauric acid. For example, Giannetto et al. coated glassy carbon electrodes using an applied potential of −0.4 V versus a Ag/AgCl/3 M KCl reference electrode in $HAuCl_4$(5 mM) solution with potassium nitrate (0.1 M) until the deposition charge was $\sim$1 mC mm^{-2} [9].

Gold nanoparticle suspensions are normally charged, so electrostatic attraction can be used to adsorb such particles to suitably modified electrode surfaces (Fig. 6.2e). Negatively charged gold nanoparticles (usually citrate stabilized) were adsorbed to base electrodes modified with, for example, amine-functionalized graphene [11], Prussian blue–poly(diallyldimethyl-ammonium)–chitosan nanoparticles [12], carbon nanotubes in bovine serum albumin (BSA) [20], or chitosan on core–shell–gold–Prussian blue nanoparticles [146] (Table 6.2).

Alternatively, nanoparticles may be incorporated within a composite layer on the electrode surface (Fig. 6.2d). Electrodeposition (−3.0 V vs. a saturated calomel electrode for 5 min in 0.1 M acetate buffer containing 0.1 M potassium chloride) was used to codeposit gold and chitosan onto a glassy carbon base electrode from a suspension of nanoparticles in a chitosan solution (0.5% w/w) [17]. A chitosan–gold nanoparticle hydrogel was also deposited onto a gold electrode [145]. Electrochemical polymerization was used to deposit a layer of core–shell–gold–Prussian blue nanoparticles on a glassy carbon electrode by cycling the base electrode between 0 and 100 mV in a potassium nitrate solution (0.1 M) containing chloroauric acid (1.5 mM) and potassium ferricyanide (1.0 mM) [146].

Electrostatic attraction was used to produce alternating layers of gold nanoparticles and thionine. First, BSA was used to entrap carbon nanotubes on a gold base electrode. Next, negatively charged nanoparticles were deposited by adsorption, and this was followed by deposition of positively charged thionine (from the acetate). Multiple layers ($n = 1$–4) of gold and thionine were assembled in this way [20].

A further strategy for incorporating metal nanoparticles into modified electrodes is to deposit preformed nanocomposite structures onto the electrode. Gold nanoparticles were deposited this way using gold–thionine–graphene–thionine–gold particles [22], core–shell–shell–Prussian blue–gold particles [10], and gold nanoparticles confined inside the pores of mesoporous silica particles [15]. A glassy carbon electrode was simultaneously modified with gold and silver nanoparticles by casting preformed gold–silver–graphene hybrid nanosheets onto an electropolymerized thionine layer [22].

6.3 Use of Antibodies and/or Other Materials

In addition to the possibilities for modifying the base electrode with nanoparticles, the electrode may incorporate antibodies

(Fig. 6.2ii,v), other materials (Fig. 6.2iii), or both. Antibodies may be conjugated to the base electrode via a short spacer (Fig. 6.2aii), attached to gold or silver nanoparticles immobilized at the working electrode (Fig. 6.2bii–eii), or attached to other materials (Fig. 6.2av).

6.3.1 *Antibodies Indirectly Attached to the Base Electrode*

Anti-α-fetoprotein (anti-AFP) was crosslinked to chitosan on a screen-printed graphite electrode [13] and to chitosan on a gold electrode [145] (Fig. 6.2av) and Table 6.2). Anti-AFP was conjugated to gold nanoparticles electrodeposited on glassy carbon and functionalized with protein A [9]. Giannetto et al. also used polyamidoaminic dendrimers to functionalize electrodeposited gold nanoparticles. The dendrimers were then conjugated to anti-AFP as capture antibodies used in a sandwich immunoassay [144] (Table 6.2b).

6.3.2 *Antibodies on Nanoparticles Attached to Intermediate Layers on the Base Electrode*

Anti-AFP was adsorbed to gold nanoparticles adsorbed to intermediate layers (Fig. 6.2ei): nanoparticles deposited on amine-functionalized graphene coating a carbon paste electrode [11], nanoparticles deposited on thionine conjugated to a chitosan film on a glassy carbon base electrode [17], and nanoparticles adsorbed to chitosan cast over a core–shell–gold–Prussian blue nanocomposite film [146] (Table 6.2).

6.3.3 *Antibodies on Nanoparticles Incorporated into Composite Layers on the Base Electrode*

Examples of antibodies adsorbed to nanoparticles incorporated into composite layers (Fig. 6.2di) include anti-AFP adsorbed to core–shell–shell–Prussian blue–BSA–gold nanoparticles deposited on a gold base electrode [10], gold nanoparticles confined inside pores of mesoporous silica and coated on a glassy carbon base electrode [15], and alternating layer-by-layer films of gold nanoparticles and thionine [20]. Finally, anti-AFP was adsorbed to both gold and silver nanoparticles incorporated into a graphene hybrid nanosheet [22] (Table 6.2).

6.4 Nanoparticle Modification of Antibodies/Antigens not Bound to the Working Electrode

Gold or silver nanoparticles may be linked to antibodies (or antigens) not bound at the working electrode surface. The biomaterials may be deposited on the nanoparticles using similar approaches to those used for binding immunomaterials to metal nanoparticles localized at the working electrode (Section 6.2.1). In short, there may be electrostatic attraction, direct conjugation with covalent bonding via a short spacer, or binding via a more complex molecular assembly.

In case the assay employs a single immunobinding reaction (Fig. 6.1a,b), the antibody (antigen) must be bound to some other surface—such as a multiwell plate, a microflow cell, or a magnetic bead—so that the resultant antibody-antigen complexes can be readily separated from unbound materials before the detection step. The detection step necessarily involves interaction of the working electrode with the immunocomplex, some agent attached to the complex, or a probe added to the test solution during the detection step.

In the case of sandwich assays, the unbound immunoreagent is the "detection" agent. In these cases, the unbound antibody/antigen usually carries a label (Fig. 6.1c,f), but in some cases, an electroactive probe may be added to the test solution during measurement or be incorporated in the working electrode assembly (Fig. 6.1d,e).

6.5 Antibodies Bound to Magnetic Composite Nanoparticles

Wang and Gan deposited anti-AFP on magnetic core–shell–shell–gold nanoparticles. The functionalized nanoparticles were located at a carbon paste working electrode with the aid of a magnet and the analytical signal generated by measuring the change in open-circuit potential before and after addition of AFP [143]. Meng et al. also used magnetic nanoparticles. In this case, immunocomplexes were formed on gold nanoparticles deposited on hemin-functionalized carbon nanotubes. Binding with AFP inhibited the hemin electrochemistry measured at the screen-printed carbon working electrode [18] (Table 6.2a).

6.6 Antibodies/Other Biomaterials Bound to Gold or Silver Nanoparticles

Su et al. used anti-AFP labeled with horseradish peroxidase to functionalize unbound gold nanoparticles. Sandwich immunocomplexes were formed with anti-AFP deposited on a modified working electrode incorporating thionine, and the analytical signal was provided by thionine-mediated measurement of the enzyme activity when hydrogen peroxide was added to the test solution [20]. A similar approach was used by Lai et al., who used gold nanoparticles functionalized with both anti-AFP and glucose oxidase. In this case, the analytical signal was provided by Prussian blue, incorporated in the working electrode, which mediated the oxidase enzyme activity, when glucose was added to sandwich immunocomplexes formed at the electrode surface [12]. An alternative mediated enzyme approach was used by Zhuo et al., who used a gold base electrode simply modified with anti-AFP conjugated to chitosan. The mediator (Prussian blue) was incorporated into Fe_3O_4–Prussian blue–gold composite nanoparticles functionalized with anti-AFP, horseradish peroxidase, and glucose oxidase [145].

Other arrangements were described by Tang et al., who used anti-AFP linked to gold nanoparticles bound to an iridium oxide core. The working electrode was modified with anti-AFP and Prussian blue, which oxidized *p*-aminophenol when it was added to the test solution [146]. CEA was assayed using gold–silver hollow nanospheres conjugated to glucose oxidase and anti-CEA. The analytical signal was generated by measuring Prussian blue–mediated glucose oxidase activity [86] (Fig. 6.3).

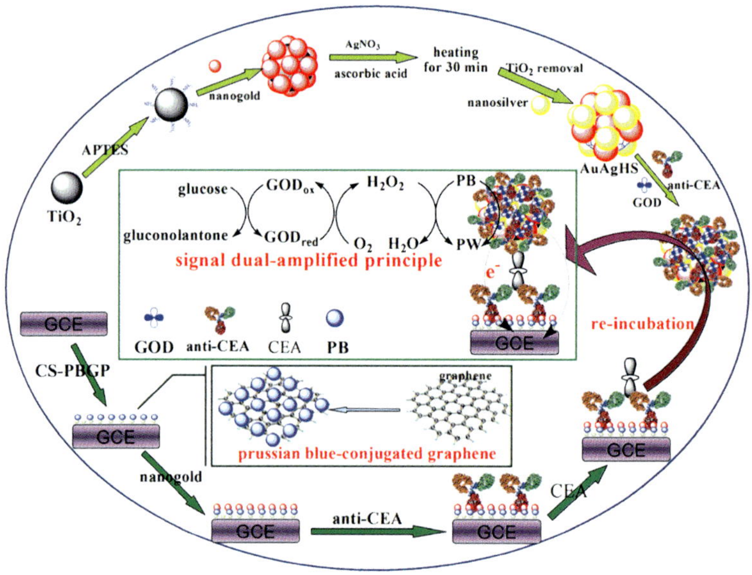

Figure 6.3 Determination of carcinoembryonic antigen using capture antibodies immobilized on nanogold particles deposited on Prussian blue-conjugated graphene and detection antibodies on gold–silver hollow nanospheres with glucose oxidase. Reprinted from Ref. [86], copyright 2011, with permission from Elsevier.

6.7 Generation of the Analytical Signal at the Working Electrode

Gold and/or silver nanoparticles used as labels can be used to enhance the sensitivity of bioassays by dissolving the metal and measuring the large number of metal ions released. Each immunobinding event can be amplified by the number of atoms (n) in the nanoparticle given by the volume of the nanoparticle and the density (ρ) and atomic weight (m_a) of the metal. Assuming that the particles are spherical, the factor n is given by Eq. 6.1:

$$n = \frac{4\pi r^3 \rho N_A}{3 m_a} \tag{6.1}$$

where r is the radius of the nanoparticle and N_A is Avogadro's constant. The molar volumes of metallic silver and gold are similar, and the amplification factor for both types of nanoparticles is about

2.5×10^2, 2.5×10^5, and 2.5×10^8 for 1 nm, 10 nm, and 100 nm radius particles, respectively. The amplification factor can be further elevated by enlarging the particles with deposition of gold or silver after formation of the immunocomplexes and before generation of the analytical signal.

6.8 Gold Nanoparticles

Captured gold nanoparticles can be dissolved either with or without preenlargement to provide gold(III) ions that may be measured electrochemically. Dequaire et al. dissolved captured gold nanoparticles in a mixture of hydrobromic acid (1 M) and bromine (0.1 mM) and used anodic stripping voltammetry on screen-printed carbon electrodes to measure the released gold(III) ions. The dynamic range for detection of goat IgG was 0.5–100 ng mL^{-1}, with a detection limit of 0.5 ng mL^{-1} (3 pM) [28].

Liao and Huang used formaldehyde/Au^{3+} to autocatalytically deposit gold on gold nanoparticles. Excess Au^{3+} was washed away and a mixture of bromine and hydrobromic acid used to dissolve the enlarged gold nanoparticles. The amount of gold deposited on the electrode was then measured using square-wave stripping voltammetry. This gave a linear dynamic range (semilog plot) of 1 pg mL^{-1}–500 pg mL^{-1} of rabbit IgG, with a detection limit of 0.25 pg mL^{-1} (1.6 fM) [148].

6.9 Gold Nanoparticles with Silver Deposition

Chu et al. (2005) described a sandwich immunoassay based on silver-enhanced gold nanoparticles. The immunocomplexes were incubated with hydroquinone and silver nitrate to precipitate silver on the gold labels. Subsequently, the deposited silver was dissolved with nitric acid and the metal ions determined using anodic stripping voltammetry. The detection limit for human IgG was estimated to be 1.0 ng mL^{-1} (6 pM) [149]. Lai et al. used alkaline phosphatase-labeled antibody-functionalized gold nanoparticles to catalyze electrodeposition of silver nanoparticles so as to generate electrochemical signals proportional to human and mouse IgG on dual working electrodes using anodic stripping voltammetry. The combination of enzyme and gold nanoparticle catalyzed silver deposition led to detection limits of 4.8 pg mL^{-1} and 6.1 pg mL^{-1} for human and mouse IgG, respectively [31]. Shim et al. also used deposition and stripping of silver to provide an analytical signal, but in this case, sandwich immunocomplexes were functionalized with a cysteamine–silver complex using the interaction between the free amine group of cysteamine and gold nanoparticles on the detection antibodies [34]. The amount of cysteamine attached was proportional to the concentration of IgG, with a limit of detection reported to be 0.4 fg mL^{-1}.

Lin et al. also used gold nanoparticles to catalyze silver deposition but with capture antibodies assembled on a graphene-coated glassy carbon electrode and detection antibodies on gold nanoparticles attached to poly(styrene-*co*-acrylic acid) microbeads. Silver stripping using linear sweep voltammetry gave a linear response to the logarithm of CEA concentration in the range of 0.5 pg mL^{-1} to 0.5 ng mL^{-1}, with a limit of detection of 0.12 pg mL^{-1} [59] (Fig. 6.4).

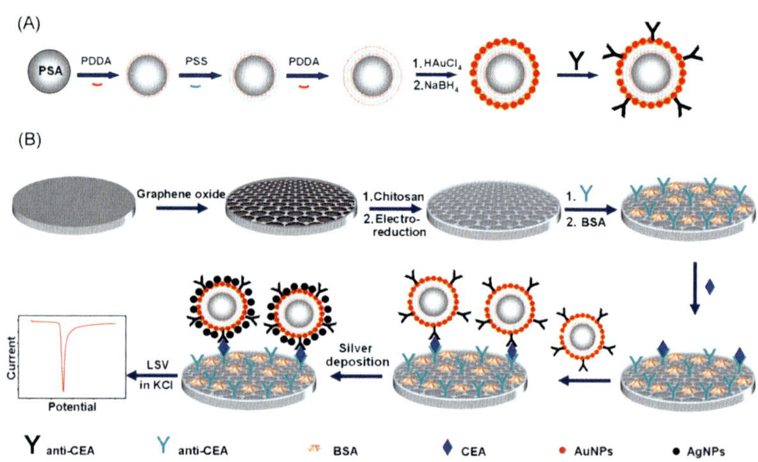

Figure 6.4 Schematic representation of (a) preparation of tracing tag and labeled detection antibody and (b) immunosensor fabrication and sandwich immunoassay. Reprinted from Ref. [59], copyright 2012, with permission from the American Chemical Society.

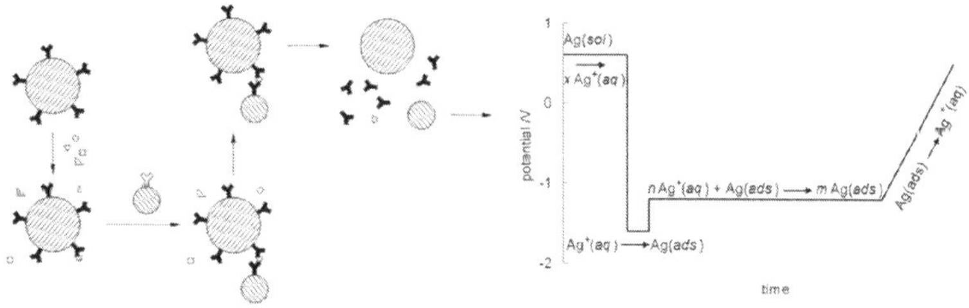

Figure 6.5 Schematic of an immunoassay using silver nanoparticles and magnetic beads. From top left: Magnetic beads conjugated to an antibody are exposed to an antigen to form antibody antigen complexes linked to a magnetic bead; in the second step, silver nanoparticles linked to a second antibody are added to the reaction mixture to form conjugates containing the antigen linked to both magnetic beads and silver nanoparticles; in the third step, a magnet is used to capture the magnetic beads, while unbound materials (including unbound silver nanoparticles) are washed from the system; and, in the fourth step, ammonium thiocyanate is added to the captured conjugates to break them up and stabilize silver. Finally, the silver nanoparticles are measured by (i) electrochemical dissolution to form silver ions, (ii) nucleation of silver on the electrode surface, (iii) accumulation of silver on the electrode surface, and (iv) rapid removal of silver from the surface to produce a measurable stripping charge. Adapted from Ref. [101]. Reproduced by permission of the PCCP Owner Societies.

6.10 Silver Nanoparticles

Captured silver nanoparticles can be used to generate an electrochemical signal, either with or without preenlargement, in a similar manner to gold nanoparticles but using milder reaction conditions. Szymanski et al. used a combination of magnetic beads and silver nanoparticles, each labeled with antibodies, to measure myoglobin using ammonium thiocyanate to dissolve the captured silver nanoparticles and stripping voltammetry to give a dynamic range spanning 0.2–20 ng mL^{-1} [100, 101] (Fig. 6.5).

Hao et al. reported an immunoassay that did not require dissolution of the silver nanoparticles before measurement [29]. Silver nanoparticles were activated with glutathione and conjugated to goat–anti-human IgG with glutaraldehyde. IgG was preadsorbed to screen-printed carbon working electrodes and incubated with the functionalized particles. Unbound materials were washed away and the remaining silver metal labeled immunocomplexes stripped using differential pulse voltammetry. The dynamic range was 1–1000 ng mL^{-1}, with a limit of detection of 0.4 ng mL^{-1}.

6.11 Conclusion

Gold and silver nanoparticles are widely used in electrochemical immunoassays, particularly due to their biocompatibility and large surface area, which enable high concentrations of antibodies, antigens, enzymes, and other materials to be immobilized at the nanoparticle surfaces. In addition, these particles provide good electrical conductivity and may be used to generate analytical signals using either indirect measurements (electrochemical deposition of metal ions formed by dissolution of the nanoparticles, with or without preenlargement) or direct measurements (electrochemical oxidation of metal nanoparticles).

Combinations of approaches for preparation and measurement can confer very low limits of detection on electrochemical immunoassays that enable useful measurements in clinical and other "real world" samples. Screen-printed and microfluidic devices have the potential for commercial exploitation, and time will tell which, if any, of the present technologies can be successfully commercialized.

References

1. Cao, X., Ye, Y., Liu, S. (2011) Gold nanoparticle-based signal amplification for biosensing, *Anal. Biochem.*, **417**, 1–16.
2. Li, Y., Schluesener, H., Xu, S. (2010) Gold nanoparticle-based biosensors, *Gold Bull. (Berlin, Ger.)*, **43**, 29–41.
3. Pingarrón, J. M., Yáñez-Sedeño, P., González-Cortés, A. (2008) Gold nanoparticle-based electrochemical biosensors, *Electrochim. Acta*, **53**, 5848–5866.
4. Vidotti, M., Carvalhal, R. F., Mendes, R. K., Ferreira, D. C. M., Kubota, L. T. (2011) Biosensors based on gold nanostructures, *J. Braz. Chem. Soc.*, **22**, 3–20.
5. Che, X., Yuan, R., Chai, Y., Li, J., Song, Z., Wang, J. (2010) Amperometric immunosensor for the determination of alpha-1-fetoprotein based on

multiwalled carbon nanotube-silver nanoparticle composite, *J. Colloid Interface Sci.*, **345**, 174–180.

6. Chen, H., Zhang, B., Cui, Y., Liu, B., Chen, G., Tang, D. (2011) One-step electrochemical immunoassay of biomarker based on nanogold-functionalized graphene sensing platform, *Anal. Methods*, **3**, 1615–1621.
7. Ding, C., Zhao, F., Ren, R., Lin, J. (2009) An electrochemical biosensor for α-fetoprotein based on carbon paste electrode constructed of room temperature ionic liquid and gold nanoparticles, *Talanta*, **78**, 1148–1154.
8. Gan, N., Wu, Y., Hu, F., Li, T., Zheng, L., Cao, Y. (2011) One novel nano magnetic Fe_3O_4/ZrO_2/nano Au composite membrane modified amperometric immunosensor for α-fetoprotein in human serum, *Int. J. Electrochem. Sci.*, **6**, 461–474.
9. Giannetto, M., Elviri, L., Careri, M., Mangia, A., Mori, G. (2011) A voltammetric immunosensor based on nanobiocomposite materials for the determination of alpha-fetoprotein in serum, *Biosens. Bioelectron.*, **26**, 2232–2236.
10. Hong, C., Yuan, R., Chai, Y., Zhuo, Y. (2008) Amperometric immunosensor for the determination of α-1-fetoprotein based on core-shell-shell Prussian blue-BSA-nanogold functionalized interface, *Electroanalysis*, **20**, 2185–2191.
11. Huang, K., Niu, D., Sun, J., Zhu, J. (2011) An electrochemical amperometric immunobiosensor for label-free detection of α-fetoprotein based on amine-functionalized graphene and gold nanoparticles modified carbon ionic liquid electrode, *J. Electroanal. Chem.*, **656**, 72–77.
12. Lai, G., Yan, F., Ju, H. (2009) Dual signal amplification of glucose oxidase-functionalized nanocomposites as a trace label for ultrasensitive simultaneous multiplexed electrochemical detection of tumor markers, *Anal. Chem.*, **81**, 9730–9736.
13. Lai, G., Wu, J., Ju, H., Yan, F. (2011) Streptavidin-functionalized silver-nanoparticle-enriched carbon nanotube tag for ultrasensitive multiplexed detection of tumor markers, *Adv. Funct. Mater.*, **21**, 2938–2943.
14. Liang, W., Yi, W., Li, S., et al. (2009) A novel, label-free immunosensor for the detection of α-fetoprotein using functionalised gold nanoparticles, *Clin. Biochem.*, **42**, 1524–1530.
15. Lin, J., Wei, Z., Chu, P. (2012) A label-free immunosensor by controlled fabrication of monoclonal antibodies and gold nanoparticles inside the mesopores, *Anal. Biochem.*, **421**, 97–102.

16. Liu, X., Wu, H., Zheng, Y., et al. (2010) A sensitive electrochemical immunosensor for α-fetoprotein detection with colloidal gold-based dentritical enzyme complex amplification, *Electroanalysis*, **22**, 244–250.
17. Liu, Y., Yuan, R., Chai, Y., Hong, C., Guan, S. (2010) Preparation of a composite film electrochemically deposited with chitosan and gold nanoparticles for the determination of α-1-fetoprotein, *Bioprocess Biosyst. Eng.*, **33**, 613–618.
18. Meng, L., Gan, N., Li, T., Cao, Y., Hu, F., Zheng, L. (2011) A three-dimensional, magnetic and electroactive nanoprobe for amperometric determination of tumor biomarkers, *Int. J. Mol. Sci.*, **12**, 362–375.
19. Song, Z., Yuan, R., Chai, Y., Wang, J., Che, X. (2010) Dual amplification strategy for the fabrication of highly sensitive amperometric immunosensor based on nanocomposite functionalized interface, *Sens. Actuators, B*, **145**, 817–825.
20. Su, B., Tang, J., Chen, H., Huang, J., Chen, G., Tang, D. (2010) Thionine/nanogold multilayer film for electrochemical immunoassay of alpha-fetoprotein in human serum using biofunctional double-codified gold nanoparticles, *Anal. Methods*, **2**, 1702–1709.
21. Su, B., Tang, J., Huang, J., et al. (2010) Graphene and nanogold-functionalized immunosensing interface with enhanced sensitivity for one-step electrochemical immunoassay of alpha-fetoprotein in human serum, *Electroanalysis*, **22**, 2720–2728.
22. Su, B., Tang, D., Li, Q., Tang, J., Chen, G. (2011) Gold–silver–graphene hybrid nanosheets-based sensors for sensitive amperometric immunoassay of alpha-fetoprotein using nanogold-enclosed titania nanoparticles as labels, *Anal. Chim. Acta*, **692**, 116–124.
23. Su, H., Yuan, R., Chai, Y., Zhuo, Y. (2012) Enzyme-nanoparticle conjugates at oil–water interface for amplification of electrochemical immunosensing, *Biosens. Bioelectron.*, **33**, 288–292.
24. Tan, L., Chen, Y., Yang, H., et al. (2009) Alpha-1-fetoprotein antibody functionalized Au nanoparticles: catalytic labels for the electrochemical detection of α-1-fetoprotein based on TiO_2 nanoparticles synthesized with ionic liquid, *Sens. Actuators, B*, **142**, 316–320.
25. Tang, J., Su, B., Tang, D., Chen, G. (2010) Conductive carbon nanoparticles-based electrochemical immunosensor with enhanced sensitivity for α-fetoprotein using irregular-shaped gold nanoparticles-labeled enzyme-linked antibodies as signal improvement, *Biosens. Bioelectron.*, **25**, 2657–2662.

26. Cui, R., Huang, H., Yin, Z., Gao, D., Zhu, J. (2008) Horseradish peroxidase-functionalized gold nanoparticle label for amplified immunoanalysis based on gold nanoparticles/carbon nanotubes hybrids modified biosensor, *Biosens. Bioelectron.*, **23**, 1666–1673.
27. Cui, R., Zhu, J. (2010) Fabrication of a novel electrochemical immunosensor based on the gold nanoparticles/colloidal carbon nanosphere hybrid material, *Electrochim. Acta*, **55**, 7814–7817.
28. Dequaire, M., Degrand, C., Limoges, B. (2000) An electrochemical metalloimmunoassay based on a colloidal gold label, *Anal. Chem.*, **72**, 5521–5528.
29. Hao, N., Li, H., long, Y., et al. (2011) An electrochemical immunosensing method based on silver nanoparticles, *J. Electroanal. Chem.*, **656**, 50–54.
30. Huang, K., Niu, D., Sun, J., Zhu, X., Zhu, J. (2010) Label-free amperometric immunobiosensor based on a gold colloid and Prussian blue nanocomposite film modified carbon ionic liquid electrode, *Anal. Bioanal. Chem.*, **397**, 3553–3561.
31. Lai, G., Yan, F., Wu, J., Leng, C., Ju, H. (2011) Ultrasensitive multiplexed immunoassay with electrochemical stripping analysis of silver nanoparticles catalytically deposited by gold nanoparticles and enzymatic reaction, *Anal. Chem.*, **83**, 2726–2732.
32. Leng, C., Lai, G., Yan, F., Ju, H. (2010) Gold nanoparticle as an electrochemical label for inherently crosstalk-free multiplexed immunoassay on a disposable chip, *Anal. Chim. Acta*, **666**, 97–101.
33. Liu, K., Zhang, J., Wang, C., Zhu, J. (2011) Graphene-assisted dual amplification strategy for the fabrication of sensitive amperometric immunosensor, *Biosens. Bioelectron.*, **26**, 3627–3632.
34. Noh, H., Rahman, M. A., Yang, J. E., Shim, Y. (2011) Ag(I)-cysteamine complex based electrochemical stripping immunoassay: ultrasensitive human IgG detection, *Biosens. Bioelectron.*, **26**, 4429–4435.
35. Qiu, L., Wang, C., Hu, P., Wu, Z., Shen, G., Yu, R. (2010) A label-free electrochemical immunoassay for IgG detection based on the electron transfer, *Talanta*, **83**, 42–47.
36. Tang, D., Niessner, R., Knopp, D. (2009) Flow-injection electrochemical immunosensor for the detection of human IgG based on glucose oxidase-derivated biomimetic interface, *Biosens. Bioelectron.*, **24**, 2125–2130.
37. Tang, D., Tang, J., Su, B., Chen, H., Huang, J., Chen, G. (2010) Highly sensitive electrochemical immunoassay for human IgG using double-

encoded magnetic redox-active nanoparticles, *Microchim. Acta*, **171**, 457–464.

38. Tang, J., Hu, R., Wu, Z., Shen, G., Yu, R. (2011) A highly sensitive electrochemical immunosensor based on coral-shaped AuNPs with CHITs inorganic–organic hybrid film, *Talanta*, **85**, 117–122.
39. Wang, G., Huang, H., Zhang, G., Zhang, X., Fang, B., Wang, L. (2010) Gold nanoparticles/l-cysteine/graphene composite based immobilization strategy for an electrochemical immunosensor, *Anal. Methods*, **2**, 1692–1697.
40. Wang, L., Jia, X., Zhou, Y., Xie, Q., Yao, S. (2010) Sandwich-type amperometric immunosensor for human immunoglobulin G using antibody-adsorbed Au/SiO_2 nanoparticles, *Microchim. Acta*, **168**, 245–251.
41. Wang, S., Wu, Z., Qu, F., Zhang, S., Shen, G., Yu, R. (2008) A novel electrochemical immunosensor based on ordered Au nano-prickle clusters, *Biosens. Bioelectron.*, **24**, 1020–1026.
42. Yang, Y., Dong, S., Shen, T., et al. (2011) Amplified immunosensing based on ionic liquid-doped chitosan film as a matrix and Au nanoparticle decorated graphene nanosheets as labels, *Electrochim. Acta*, **56**, 6021–6025.
43. Zhang, L., Liu, Y., Chen, T. (2009) Label-free amperometric immunosensor based on antibody immobilized on a positively charged gold nanoparticle/L-cysteine-modified gold electrode, *Microchim. Acta*, **164**, 161–166.
44. Zhang, S., Zheng, F., Wu, Z., Shen, G., Yu, R. (2008) Highly sensitive electrochemical detection of immunospecies based on combination of Fc label and PPD film/gold nanoparticle amplification, *Biosens. Bioelectron.*, **24**, 129–135.
45. Zhang, Y., Wang, H., Nie, J., Zhang, Y., Shen, G., Yu, R. (2009) Individually addressable microelectrode arrays fabricated with gold-coated pencil graphite particles for multiplexed and high sensitive impedance immunoassays, *Biosens. Bioelectron.*, **25**, 34–40.
46. Owino, J., Arotiba, O., Hendricks, N., et al. (2008) Electrochemical immunosensor based on polythionine/gold nanoparticles for the determination of aflatoxin B1, *Sensors*, **8**, 8262–8274.
47. Sun, A., Qi, Q., Dong, Z., Liang, K. (2008) An electrochemical enzyme immunoassay for aflatoxin B_1 based on bio-electrocatalytic reaction with room-temperature ionic liquid and nanoparticle-modified electrodes, *Sens. Instrum. Food Qual. Saf.*, **2**, 43–50.

48. Tang, D., Zhong, Z., Niessner, R., Knopp, D. (2009) Multifunctional magnetic bead-based electrochemical immunoassay for the detection of aflatoxin B1 in food, *Analyst*, **134**, 1554–1560.
49. Mao, X., Baloda, M., Gurung, A. S., Lin, Y., Liu, G. (2008) Multiplex electrochemical immunoassay using gold nanoparticle probes and immunochromatographic strips, *Electrochem. Commun.*, **10**, 1636–1640.
50. Omidfar, K., Dehdast, A., Zarei, H., Sourkohi, B. K., Larijani, B. (2011) Development of urinary albumin immunosensor based on colloidal AuNP and PVA, *Biosens. Bioelectron.*, **26**, 4177–4183.
51. Ho, J. A., Chang, H., Shih, N., et al. (2010) Diagnostic detection of human lung cancer-associated antigen using a gold nanoparticle-based electrochemical immunosensor, *Anal. Chem.*, **82**, 5944–5950.
52. Kim, D., Noh, H., Park, D. S., Ryu, S., Koo, J. S., Shim, Y. (2009) Immunosensors for detection of annexin II and MUC5AC for early diagnosis of lung cancer, *Biosens. Bioelectron.*, **25**, 456–462.
53. Mwilu, S. K., Aluoch, A. O., Miller, S., et al. (2009) Identification and quantitation of *Bacillus globigii* using metal enhanced electrochemical detection and capillary biosensor, *Anal. Chem.*, **81**, 7561–7570.
54. Ahirwal, G. K., Mitra, C. K. (2010) Gold nanoparticles based sandwich electrochemical immunosensor, *Biosens. Bioelectron.*, **25**, 2016–2020.
55. Ho, J. A., Chiu, J., Hong, J., Lin, C., Hwang, K., Hwu, J. R. (2009) Gold-nanostructured immunosensor for the electrochemical sensing of biotin based on liposomal competitive assay, *J. Nanosci. Nanotechnol.*, **9**, 2324–2329.
56. Ho, J. A., Hsu, W., Liao, W., et al. (2010) Ultrasensitive electrochemical detection of biotin using electrically addressable site-oriented antibody immobilization approach via aminophenyl boronic acid, *Biosens. Bioelectron.*, **26**, 1021–1027.
57. Neves, M. M. P. S., González-García, M. B., Nouws, H. P. A., Costa-García, A. (2012) Celiac disease detection using a transglutaminase electrochemical immunosensor fabricated on nanohybrid screen-printed carbon electrodes, *Biosens. Bioelectron.*, **31**, 95–100.
58. Tang, D., Tang, J., Su, B., Chen, G. (2011) Gold nanoparticles-decorated amine-terminated poly(amidoamine) dendrimer for sensitive electrochemical immunoassay of brevetoxins in food samples, *Biosens. Bioelectron.*, **26**, 2090–2096.

59. Lin, D., Wu, J., Wang, M., Yan, F., Ju, H. (2012) Triple signal amplification of graphene film, polybead carried gold nanoparticles as tracing tag and silver deposition for ultrasensitive electrochemical immunosensing, *Anal. Chem.*, **84**, 3662–3668.
60. Wu, J., Yan, F., Zhang, X., Yan, Y., Tang, J., Ju, H. (2008) Disposable reagentless electrochemical immunosensor array based on a biopolymer/sol-gel membrane for simultaneous measurement of several tumor markers, *Clin. Chem.*, **54**, 1481–1488.
61. Pereira, S. V., Bertolino, F. A., Messina, G. A., Raba, J. (2011) Microfluidic immunosensor with gold nanoparticle platform for the determination of immunoglobulin G anti-*Echinococcus granulosus* antibodies, *Anal. Biochem.*, **409**, 98–104.
62. Chen, S., Yuan, R., Chai, Y., Xu, Y., Min, L., Li, N. (2008) A new antibody immobilization technique based on organic polymers protected Prussian blue nanoparticles and gold colloidal nanoparticles for amperometric immunosensors, *Sens. Actuators, B*, **135**, 236–244.
63. Viswanathan, S., Rani, C., Ribeiro, S., Delerue-Matos, C. (2012) Molecular imprinted nanoelectrodes for ultra sensitive detection of ovarian cancer marker, *Biosens. Bioelectron.*, **33**, 179–183.
64. Chandra, P., Koh, W. C. A., Noh, H., Shim, Y. (2012) In vitro monitoring of i-NOS concentrations with an immunosensor: the inhibitory effect of endocrine disruptors on i-NOS release, *Biosens. Bioelectron.*, **32**, 278–282.
65. Chen, H., Tang, J., Su, B., Chen, G., Huang, J., Tang, D. (2010) Nanogold-actuated biomimetic peroxidase for sensitized electrochemical immunoassay of carcinoembryonic antigen in human serum, *Anal. Chim. Acta*, **678**, 169–175.
66. Chen, H., Tang, D., Zhang, B., Liu, B., Cui, Y., Chen, G. (2012) Electrochemical immunosensor for carcinoembryonic antigen based on nanosilver-coated magnetic beads and gold-graphene nanolabels, *Talanta*, **91**, 95–102.
67. Han, J., Zhuo, Y., Chai, Y., Mao, L., Yuan, Y., Yuan, R. (2011) Highly conducting gold nanoparticles–graphene nanohybrid films for ultrasensitive detection of carcinoembryonic antigen, *Talanta*, **85**, 130–135.
68. He, X., Yuan, R., Chai, Y., Shi, Y. (2008) A sensitive amperometric immunosensor for carcinoembryonic antigen detection with porous nanogold film and nano-Au/chitosan composite as immobilization matrix, *J. Biochem. Biophys. Methods*, **70**, 823–829.
69. Huang, K., Niu, D., Xie, W., Wang, W. (2010) A disposable electrochemical immunosensor for carcinoembryonic antigen based on

nano-Au/multi-walled carbon nanotubes–chitosans nanocomposite film modified glassy carbon electrode, *Anal. Chim. Acta*, **659**, 102–108.

70. Jiang, W., Yuan, R., Chai, Y., Mao, L., Su, H. (2011) A novel electrochemical immunoassay based on diazotization-coupled functionalized bioconjugates as trace labels for ultrasensitive detection of carcinoembryonic antigen, *Biosens. Bioelectron.*, **26**, 2786–2790.
71. Kong, F., Xu, M., Xu, J., Chen, H. (2011) A novel lable-free electrochemical immunosensor for carcinoembryonic antigen based on gold nanoparticles–thionine–reduced graphene oxide nanocomposite film modified glassy carbon electrode, *Talanta*, **85**, 2620–2625.
72. Li, J., Gao, H., Chen, Z., Wei, X., Yang, C. F. (2010) An electrochemical immunosensor for carcinoembryonic antigen enhanced by self-assembled nanogold coatings on magnetic particles, *Anal. Chim. Acta*, **665**, 98–104.
73. Li, Y., Yang, W., Fan, M., Liu, A. (2011) A sensitive label-free amperometric CEA immunosensor based on graphene–nafion nanocomposite film as an enhanced sensing platform, *Anal. Sci.*, **27**, 727–731.
74. Liao, Y., Yuan, R., Chai, Y., Zhuo, Y., Yang, X. (2010) Study on an amperometric immunosensor based on Nafion–cysteine composite membrane for detection of carcinoembryonic antigen, *Anal. Biochem.*, **402**, 47–53.
75. Liu, K., Yuan, R., Chai, Y., Tang, D., An, H. (2010) $[AuCl_4]^-$ and $Fe^{3+}/[Fe(CN)_6]^{3-}$ ions-derivated immunosensing interface for electrochemical immunoassay of carcinoembryonic antigen in human serum, *Bioprocess Biosyst. Eng.*, **33**, 179–185.
76. Liu, Y., Yuan, R., Chai, Y., Hong, C., Liu, K., Guan, S. (2009) Ultrasensitive amperometric immunosensor for the determination of carcinoembryonic antigen based on a porous chitosan and gold nanoparticles functionalized interface, *Microchim. Acta*, **167**, 217–224.
77. Liu, Z., Yuan, R., Chai, Y., Zhuo, Y., Hong, C., Yang, X. (2008) Highly sensitive, reagentless amperometric immunosensor based on a novel redox-active organic–inorganic composite film, *Sens. Actuators, B*, **134**, 625–631.
78. Lv, P., Min, L., Yuan, R., Chai, Y., Chen, S. (2010) A novel immunosensor for carcinoembryonic antigen based on poly(diallyldimethylammonium chloride) protected Prussian blue nanoparticles and double-layer nanometer-sized gold particles, *Microchim. Acta*, **171**, 297–304.

79. Shi, W., Ma, Z. (2011) A novel label-free amperometric immunosensor for carcinoembryonic antigen based on redox membrane, *Biosens. Bioelectron.*, **26**, 3068–3071.
80. Song, Z., Yuan, R., Chai, Y., Yin, B., Fu, P., Wang, J. (2010) Multilayer structured amperometric immunosensor based on gold nanoparticles and Prussian blue nanoparticles/nanocomposite functionalized interface, *Electrochim. Acta*, **55**, 1778–1784.
81. Su, B., Tang, J., Yang, H., Chen, G., Huang, J., Tang, D. (2011) A graphene platform for sensitive electrochemical immunoassay of carcinoembryoninc antigen based on gold-nanoflower biolabels, *Electroanalysis*, **23**, 832–841.
82. Tang, D., Ren, J. (2008) In situ amplified electrochemical immunoassay for carcinoembryonic antigen using horseradish peroxidase-encapsulated nanogold hollow microspheres as labels, *Anal. Chem.*, **80**, 8064–8070.
83. Tang, D., Yuan, R., Chai, Y. (2008) Ultrasensitive electrochemical immunosensor for clinical immunoassay using thionine-doped magnetic gold nanospheres as labels and horseradish peroxidase as enhancer, *Anal. Chem.*, **80**, 1582–1588.
84. Tang, D., Xia, B. (2008) Electrochemical immunosensor and biochemical analysis for carcinoembryonic antigen in clinical diagnosis, *Microchim. Acta*, **163**, 41–48.
85. Tang, H., Chen, J., Nie, L., Kuang, Y., Yao, S. (2007) A label-free electrochemical immunoassay for carcinoembryonic antigen (CEA) based on gold nanoparticles (AuNPs) and nonconductive polymer film, *Biosens. Bioelectron.*, **22**, 1061–1067.
86. Tang, J., Tang, D., Li, Q., Su, B., Qiu, B., Chen, G. (2011) Sensitive electrochemical immunoassay of carcinoembryonic antigen with signal dual-amplification using glucose oxidase and an artificial catalase, *Anal. Chim. Acta*, **697**, 16–22.
87. Wang, G., Zhang, G., Huang, H., Wang, L. (2011) Graphene-Prussian blue/gold nanoparticles based electrochemical immunoassay of carcinoembryonic antigen, *Anal. Methods*, **3**, 2082–2087.
88. Zhang, Y., Chen, H., Gao, X., Chen, Z., Lin, X. (2012) A novel immunosensor based on an alternate strategy of electrodeposition and self-assembly, *Biosens. Bioelectron.*, **35**, 277–283.
89. Chikkaveeraiah, B. V., Mani, V., Patel, V., Gutkind, J. S., Rusling, J. F. (2011) Microfluidic electrochemical immunoarray for ultrasensitive detection of two cancer biomarker proteins in serum, *Biosens. Bioelectron.*, **26**, 4477–4483.

90. Jensen, G. C., Krause, C. E., Sotzing, G. A., Rusling, J. F. (2011) Inkjet-printed gold nanoparticle electrochemical arrays on plastic. Application to immunodetection of a cancer biomarker protein, *Phys. Chem. Chem. Phys.*, **13**, 4888–4894.

91. Liao, J., Tang, D. (2009) High-throughput miniaturized immunoassay for human interleukin-6 using electrochemical sandwich-type enzyme immunosensors, *Curr. Pharm. Anal.*, **5**, 164–170.

92. Munge, B. S., Krause, C. E., Malhotra, R., Patel, V., Silvio Gutkind, J., Rusling, J. F. (2009) Electrochemical immunosensors for interleukin-6. comparison of carbon nanotube forest and gold nanoparticle platforms, *Electrochem. Commun.*, **11**, 1009–1012.

93. Wang, G., Huang, H., Zhang, G., Zhang, X., Fang, B., Wang, L. (2011) Dual amplification strategy for the fabrication of highly sensitive interleukin-6 amperometric immunosensor based on poly-dopamine, *Langmuir*, **27**, 1224–1231.

94. Zhao, B. Y., Wei, Q., Xu, C., et al. (2011) Label-free electrochemical immunosensor for sensitive detection of kanamycin, *Sens. Actuators, B*, **155**, 618–625.

95. Tong, P., Tang, S., He, Y., Shao, Y., Zhang, L., Chen, G. (2011) Label-free immunosensing of microcystin-LR using a gold electrode modified with gold nanoparticles, *Microchim. Acta*, **173**, 299–305.

96. Liu, B., Lu, L., Li, Q., Xie, G. (2011) Disposable electrochemical immunosensor for myeloperoxidase based on the indium tin oxide electrode modified with an ionic liquid composite film containing gold nanoparticles, poly(*o*-phenylenediamine) and carbon nanotubes, *Microchim. Acta*, **173**, 513–520.

97. Lu, L., Liu, B., Li, S., Zhang, W., Xie, G. (2011) Improved electrochemical immunosensor for myeloperoxidase in human serum based on nanogold/cerium dioxide-BMIMPF6/l-cysteine composite film, *Colloids Surf., B*, **86**, 339–344.

98. Suprun, E., Bulko, T., Lisitsa, A., et al. (2010) Electrochemical nanobiosensor for express diagnosis of acute myocardial infarction in undiluted plasma, *Biosens. Bioelectron.*, **25**, 1694–1698.

99. Suprun, E. V., Shilovskaya, A. L., Lisitsa, A. V., et al. (2011) Electrochemical immunosensor based on metal nanoparticles for cardiac myoglobin detection in human blood plasma, *Electroanalysis*, **23**, 1051–1057.

100. Szymanski, M., Turner, A. P. F., Porter, R. (2010) Electrochemical dissolution of silver nanoparticles and its application in metalloimmunoassay, *Electroanalysis*, **22**, 191–198.

101. Szymanski, M., Porter, R., Dep, G. V., Wang, Y., Haggett, B. G. D. (2011) Silver nanoparticles and magnetic beads with electrochemical measurement as a platform for immunosensing devices, *Phys. Chem. Chem. Phys.*, **13**, 5383–5387.

102. Zhou, F., Lu, M., Wang, W., Bian, Z., Zhang, J., Zhu, J. (2010) Electrochemical immunosensor for simultaneous detection of dual cardiac markers based on a poly(dimethylsiloxane)-gold nanoparticles composite microfluidic chip: a proof of principle, *Clin. Chem.*, **56**, 1701–1707.

103. Zhong, Z., Shan, J., Zhang, Z., Qing, Y., Wang, D. (2010) The signal-enhanced label-free immunosensor based on assembly of Prussian blue-SiO_2 nanocomposite for amperometric measurement of neuron-specific enolase, *Electroanalysis*, **22**, 2569–2575.

104. Cao, Q., Zhao, H., Yang, Y., et al. (2011) Electrochemical immunosensor for casein based on gold nanoparticles and poly(L-arginine)/multi-walled carbon nanotubes composite film functionalized interface, *Biosens. Bioelectron.*, **26**, 3469–3474.

105. Zhuo, Y., Yi, W., Lian, W., et al. (2011) Ultrasensitive electrochemical strategy for NT-proBNP detection with gold nanochains and horseradish peroxidase complex amplification, *Biosens. Bioelectron.*, **26**, 2188–2193.

106. Zhang, N., Xiao, F., Bai, J., et al. (2011) Label-free immunoassay for chloramphenicol based on hollow gold nanospheres/chitosan composite, *Talanta*, **87**, 100–105.

107. Sharma, P., Sablok, K., Bhalla, V., Suri, C. R. (2011) A novel disposable electrochemical immunosensor for phenyl urea herbicide diuron, *Biosens. Bioelectron.*, **26**, 4209–4212.

108. Bone, L., Vidal, J. C., Duato, P., Castillo, J. R. (2010) Ochratoxin A nanostructured electrochemical immunosensors based on polyclonal antibodies and gold nanoparticles coupled to the antigen, *Anal. Methods*, **2**, 335–341.

109. Vidal, J. C., Bonel, L., Duato, P., Castillo, J. R. (2011) Improved electrochemical competitive immunosensor for ochratoxin A with a biotinylated monoclonal antibody capture probe and colloidal gold nanostructuring, *Anal. Methods*, **3**, 977–984.

110. Zhang, X., Geng, P., Liu, H., et al. (2009) Development of an electrochemical immunoassay for rapid detection of *E. coli* using anodic stripping voltammetry based on Cu@Au nanoparticles as antibody labels, *Biosens. Bioelectron.*, **24**, 2155–2159.

111. Li, Y., Cheng, P., Gong, J., et al. (2012) Amperometric immunosensor for the detection of *Escherichia coli* O157:H7 in food specimens, *Anal. Biochem.*, **421**, 227–233.

112. Sharma, M. K., Agarwal, G. S., Rao, V. K., et al. (2010) Amperometric immunosensor based on gold nanoparticles/alumina sol–gel modified screen-printed electrodes for antibodies to *Plasmodium falciparum* histidine rich protein-2, *Analyst*, **135**, 608–614.

113. Liu, X., Wong, D. K. Y. (2009) Picogram-detection of estradiol at an electrochemical immunosensor with a gold nanoparticle|Protein G-(LC-SPDP)-scaffold, *Talanta*, **77**, 1437–1443.

114. Liu, X., Duckworth, P. A., Wong, D. K. Y. (2010) Square wave voltammetry versus electrochemical impedance spectroscopy as a rapid detection technique at electrochemical immunosensors, *Biosens. Bioelectron.*, **25**, 1467–1473.

115. Qu, F., Lu, H., Yang, M., Deng, C. (2011) Electrochemical immunosensor based on electron transfer mediated by graphene oxide initiated silver enhancement, *Biosens. Bioelectron.*, **26**, 4810–4814.

116. Wang, X., Tao, G., Meng, Y. (2009) Nanogold hollow microsphere-based electrochemical immunosensor for the detection of ferritin in human serum, *Microchim. Acta*, **167**, 147–152.

117. Chen, S., Yu, X., Xu, J., Chen, H. (2011) Gold nanoparticles-coated magnetic microspheres as affinity matrix for detection of hemoglobin A1c in blood by microfluidic immunoassay, *Biosens. Bioelectron.*, **26**, 4779–4784.

118. Zhuo, Y., Chai, Y., Yuan, R., Mao, L., Yuan, Y., Han, J. (2011) Glucose oxidase and ferrocene labels immobilized at Au/TiO_2 nanocomposites with high load amount and activity for sensitive immunoelectrochemical measurement of ProGRP biomarker, *Biosens. Bioelectron.*, **26**, 3838–3844.

119. de la Escosura-Muñiz, A., Maltez-da Costa, M., Sánchez-Espinel, C., et al. (2010) Gold nanoparticle-based electrochemical magnetoimmunosensor for rapid detection of anti-hepatitis B virus antibodies in human serum, *Biosens. Bioelectron.*, **26**, 1710–1714.

120. Ding, C., Li, H., Hu, K., Lin, J. (2010) Electrochemical immunoassay of hepatitis B surface antigen by the amplification of gold nanoparticles based on the nanoporous gold electrode, *Talanta*, **80**, 1385–1391.

121. Fu, X., Pu, L., Wang, J., Zhong, Z. (2010) Ionic liquids-doped organic-inorganic hybrid film for electrochemical immunoassay for hepatitis B surface antigen, *Ionics*, **16**, 51–56.

122. Liang, R., Chen, Y., Qiu, J. (2011) A sensitive amperometric immunosensor for hepatitis B surface antigen based on biocompatible redox-active chitosan-toluidine blue/gold nanoparticles composite film, *Anal. Methods*, **3**, 1338–1343.
123. Qiu, J., Liang, R., Wang, R., Fan, L., Chen, Y., Xia, X. (2009) A label-free amperometric immunosensor based on biocompatible conductive redox chitosan-ferrocene/gold nanoparticles matrix, *Biosens. Bioelectron.*, **25**, 852–857.
124. Qiu, J., Huang, H., Liang, R. (2011) Biocompatible and label-free amperometric immunosensor for hepatitis B surface antigen using a sensing film composed of poly(allylamine)-branched ferrocene and gold nanoparticles, *Microchim. Acta*, **174**, 97–105.
125. Shen, G., Zhang, Y. (2010) Highly sensitive electrochemical stripping detection of hepatitis B surface antigen based on copper-enhanced gold nanoparticle tags and magnetic nanoparticles, *Anal. Chim. Acta*, **674**, 27–31.
126. Tang, D., Li, H., Liao, J. (2009) Ionic liquid and nanogold-modified immunosensing interface for electrochemical immunoassay of hepatitis B surface antigen in human serum, *Microfluid. Nanofluid.*, **6**, 403–409.
127. Wu, S., Zhong, Z., Wang, D., et al. (2009) Gold nanoparticle-labeled detection antibodies for use in an enhanced electrochemical immunoassay of hepatitis B surface antigen in human serum, *Microchim. Acta*, **166**, 269–275.
128. Monerris, M. J., Arévalo, F. J., Fernández, H., Zon, M. A., Molina, P. G. (2012) Integrated electrochemical immunosensor with gold nanoparticles for the determination of progesterone, *Sens. Actuators, B*, **166–167**, 586–592.
129. Chen, H., Cui, Y., Zhang, B., Liu, B., Chen, G., Tang, D. (2012) Poly(*o*-phenylenediamine)-carried nanogold particles as signal tags for sensitive electrochemical immunoassay of prolactin, *Anal. Chim. Acta*, **728**, 18–25.
130. Lin, J., He, C., Pang, X., Hu, K. (2011) Amperometric immunosensor for prostate specific antigen based on gold nanoparticles/ionic liquid/chitosan hybrid film, *Anal. Lett.*, **44**, 908–921.
131. Mani, V., Chikkaveeraiah, B. V., Patel, V., Gutkind, J. S., Rusling, J. F. (2009) Ultrasensitive immunosensor for cancer biomarker proteins using gold nanoparticle film electrodes and multienzyme-particle amplification, *ACS Nano*, **3**, 585–594.

132. Wei, Q., Xiang, Z., He, J., et al. (2010) Dumbbell-like Au-Fe_3O_4 nanoparticles as label for the preparation of electrochemical immunosensors, *Biosens. Bioelectron.*, **26**, 627–631.

133. Gan, N., Hou, J., Hu, F., Zheng, L., Ni, M., Cao, Y. (2010) An amperometric immunosensor based on a polyelectrolyte/gold magnetic nanoparticle supramolecular assembly-modified electrode for the determination of HIV p24 in serum, *Molecules*, **15**, 5053–5065.

134. Chikae, M., Idegami, K., Nagatani, N., Tamiya, E., Takamura, Y. (2010) Highly sensitive method for electrochemical detection of silver nanoparticle labels in metalloimmunoassay with preoxidation/reduction signal enhancement, *Electrochemistry*, **78**, 748–753.

135. Yang, H., Yuan, R., Chai, Y., Zhuo, Y., Su, H. (2010) Electrochemical immunoassay for human chorionic gonadotrophin based on Pt hollow nanospheres and silver/titanium dioxide nanocomposite matrix, *J. Chem. Technol. Biotechnol.*, **85**, 577–582.

136. Li, F., Zhou, R., Zhao, K., Chen, H., Hu, Y. (2011) Magnetic beads-based electrochemical immunosensor for detection of pseudorabies virus antibody in swine serum, *Talanta*, **87**, 302–306.

137. Lu, L., Liu, B., Zhao, Z., et al. (2012) Ultrasensitive electrochemical immunosensor for HE4 based on rolling circle amplification, *Biosens. Bioelectron.*, **33**, 216–221.

138. An, Y., Jiang, X., Bi, W., et al. (2012) Sensitive electrochemical immunosensor for α-synuclein based on dual signal amplification using PAMAM dendrimer-encapsulated Au and enhanced gold nanoparticle labels, *Biosens. Bioelectron.*, **32**, 224–230.

139. Yin, Z., Liu, Y., Jiang, L., Zhu, J. (2011) Electrochemical immunosensor of tumor necrosis factor a based on alkaline phosphatase functionalized nanospheres, *Biosens. Bioelectron.*, **26**, 1890–1894.

140. Chen, C., Liu, D., Wu, Z., Luo, Q., Shen, G., Yu, R. (2009) Sensitive label-free electrochemical immunoassay by electrocatalytic amplification, *Electrochem. Commun.*, **11**, 1869–1872.

141. Kim, G., Kim, K., Oh, M., Sung, Y. (2010) Electrochemical detection of vascular endothelial growth factors (VEGFs) using VEGF antibody fragments modified Au NPs/ITO electrode, *Biosens. Bioelectron.*, **25**, 1717–1722.

142. Schefer, H., Mattmann, S., Joss, R. A. (1998) Hereditary persistence of α-fetoprotein, *Ann. Oncol.*, **9**, 667–672.

143. Wang, L., Gan, X. (2009) Antibody-functionalized magnetic nanoparticles for electrochemical immunoassay of α-1-fetoprotein in human serum, *Microchim. Acta*, **164**, 231–237.
144. Giannetto, M., Mori, L., Mori, G., Careri, M., Mangia, A. (2011) New amperometric immunosensor with response enhanced by PAMAM-dendrimers linked via self assembled monolayers for determination of alpha-fetoprotein in human serum, *Sens. Actuators, B*, **159**, 185–192.
145. Zhuo, Y., Yuan, P., Yuan, R., Chai, Y., Hong, C. (2009) Bienzyme functionalized three-layer composite magnetic nanoparticles for electrochemical immunosensors, *Biomaterials*, **30**, 2284–2290.
146. Tang, J., Tang, D., Niessner, R., Knopp, D. (2011) A novel strategy for ultra-sensitive electrochemical immunoassay of biomarkers by coupling multifunctional iridium oxide (IrO_x) nanospheres with catalytic recycling of self-produced reactants, *Anal. Bioanal. Chem.*, **400**, 2041–2051.
147. Huang, K., Sun, J., Xu, C., Niu, D., Xie, W. (2010) A disposable immunosensor based on gold colloid modified chitosan nanoparticles-entrapped carbon paste electrode, *Microchim. Acta*, **168**, 51–58.
148. Liao, K., Huang, H. (2005) Femtomolar immunoassay based on coupling gold nanoparticle enlargement with square wave stripping voltammetry, *Anal. Chim. Acta*, **538**, 159–164.
149. Chu, X., Fu, X., Chen, K., Shen, G., Yu, R. (2005) An electrochemical stripping metalloimmunoassay based on silver-enhanced gold nanoparticle label, *Biosens. Bioelectron.*, **20**, 1805–1812.

Chapter 7

Electrochemical DNA Sensors Based on Nanoparticles

Claudio Parolo, Alfredo de la Escosura-Muñiz, and Arben Merkoçi

Nanobioelectronics and Biosensors Group, Institut Català de Nanotecnologia, CIN2 (ICN–CSIC), Barcelona 08193, Spain
ICREA, Institució Catalana de Recerca i Estudis Avançats, Barcelona 08010, Spain
arben.merkoci@icn.cat

7.1 Introduction

The identification/quantification of specific DNA sequences is of crucial importance for many fields. For example, the detection of DNA mutations is already correlated to specific diseases. The most frequent of these mutations are single-nucleotide polymorphisms (SNPs), the change of just one base in a DNA sequence that can provoke serious damage to organisms. Furthermore specific DNA sequences can be indicators of pathogens that can be related to food and water contamination. Finally forensic applications need to be as fast and specific as possible to correlate DNA sequences to a person's identification. In optics the development of sensitive, specific,

Electrochemical Biosensors
Edited by Serge Cosnier

ISBN 978-981-4411-46-2 (Hardcover), 978-981-4411-47-9 (eBook)
www.panstanford.com

easy-to-use, fast, and cheap DNA biosensors is of extremely importance [1, 2].

In most cases, DNA analysis using nanomaterial (NM)-based biosensors consists of detection of the DNA sequence of interest using single-stranded DNA (ssDNA) as a bioreceptor and final optical or electrochemical transducing, while using NMs in any of the sensing technology steps. The specificity of the device depends on the hybridization between two ssDNA molecules (see more details of the main DNA biosensor components in Section 7.2.4). Given the relatively complicated technology or the employment of harmful reagents from other DNA sensor technologies, the electrochemical-based devices seem to take advantage of the inherent properties of electrochemical techniques. Electrochemical sensors have many advantages: they have a very fast response; are cheap, sensitive, and specific; and can be easily miniaturized. All these characteristics make them the perfect technology to study DNA with interest in different fields of research and diagnostics [3–10].

NMs, with their unique electrical and optical properties, thanks to their nanometer-size scale, are showing to be important building blocks in the design and fabrication of novel or improved biosensors. Between various NMs, nanoparticles (NPs) are extensively used in biosensors since they can act as labels, label carriers, and modifiers of transducers [11–13].

The purpose of this chapter is to show the latest trends in the use of NPs in the development of electrochemical DNA sensors.

7.2 General Aspects of DNA and Nanoparticles

7.2.1 *DNA Structure*

DNA is a linear polymer, whose each monomer unit consists of three components: a sugar (deoxyribose), a phosphate, and a nitrogenized base (Fig. 7.1a). The base sequence characterizes uniquely a DNA molecule. The deoxyribose molecules are linked to each other by phosphodiester bridges, creating the backbone of the polymer. Specifically, the 3-hydroxyl (3-OH) group of the sugar moiety of one nucleotide is esterified to a phosphate group, which is, in turn, joined to the 5-hydroxyl group of the adjacent sugar. The bases vary from one monomer to the other. Two of the bases are derivatives of purine, adenine (A) and guanine (G), and two of pyrimidine, cytosine (C) and thymine (T). Each purine base can form hydrogen bonds just with a pyrimidine base in particular: A binds with T, and C binds with G. The two types of base pairs form different numbers of hydrogen bonds, AT forming two hydrogen bonds and GC forming three hydrogen bonds. The connections within bases allow the formation of a helical structure consisting of two antiparallel strands. DNA with a high GC content is more stable than DNA with a low GC content, but contrary to popular belief, this is not due to the extra hydrogen bond of a GC base pair but rather the contribution of stacking interactions; hydrogen bonding merely provides specificity of the pairing, not stability. So all the information carried by a DNA molecule is contained in its sequence of bases. Any change in the sequence would produce a new molecule carrying new information. It is both the percentage of GC base pairs and the overall length of a DNA double helix that determine the strength of the association between the two strands of DNA. As hydrogen bonds are not covalent, they can be broken, for example, by heating, and rejoined relatively easily. Strands may also be separated by adding acid or alkali to ionize the nucleotide bases and disrupt base pairing. All these properties of the DNA molecule are essential for the design of a good DNA biosensor [14].

7.2.2 *DNA Amplification Techniques*

Although the true objective of DNA biosensing consists of the direct detection of DNA extracted from real samples, the small quantity of the specific DNA to be detected in the immense matrix of the total DNA makes a previous amplification often necessary [15].

In this context, a revolutionary milestone in scientific research was the development of the polymerase chain reaction (PCR), and for its invention Kary Mullis won the Nobel Prize in chemistry in 1993 [16]. PCR is a technique capable of generating an incredible amount of a specific sequence of the DNA molecule in a short period of time. PCR was essential to complete the sequencing of the human genome. It brought an enormous amount of data regarding DNA sequences, which are far from being completely understood. A wide knowledge of the human genome would open the way to personal medicine, increasing incredibly the quality of life of human beings [17].

PCR is based on the activity of DNA polymerase, an enzyme that synthesizes new strands of DNA in a 5′–3′ direction from a single strand template (Fig. 7.1b). PCR allows having billion copies of the target DNA, starting from just a few molecules. Essentially two primers (oligonucleotides of 20–30 bases) flank and define the target sequence to be amplified. These primers hybridize to opposite strands of the DNA to serve as initial points for the synthesis of the new DNA strand. DNA polymerase catalyzes this synthesis. The reagents to carry out PCR are generally a pair of primers

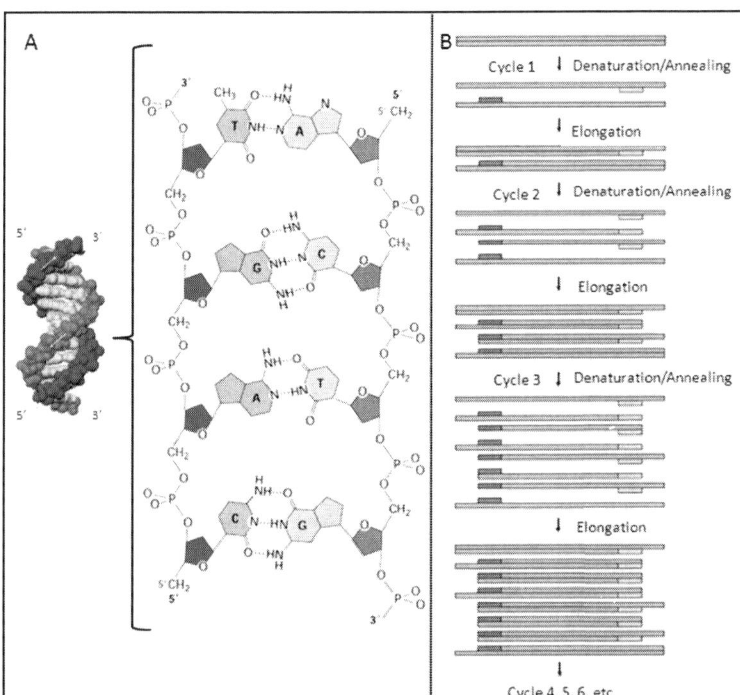

Figure 7.1 (a) Space-filling model and chemical structure of DNA double helix and (b) PCR scheme. Adapted with permission from Refs. [19–21].

that hybridize with the flanking sequences of the target, all four deoxyribonucleoside triphosphates, heat-stable DNA polymerase, an enzyme buffer, and an amount of magnesium chloride. The reaction is done in a thermal cycler, a machine that allows a rapid change in the temperature for several cycles. A general PCR cycle consists of three steps:

(1) Denaturation step: The solution is heated up to 94°C–95°C to separate the two strands of the target DNA; in fact at this temperature just the hydrogen bonds are broken, whereas the bonds of the backbone stay intact.
(2) Hybridization step: The temperature in this step depends on the primers' features, such as length and base composition, because the aim is the specific anneal of the primers to the target sequence. One primer hybridizes to the 3′ end of the target on one strand, and the other primer hybridizes to the 3′ end on the complementary target strand.
(3) Extension step: The solution temperature is raised up to 72°C, at which the DNA polymerase starts the amplification of the new double-stranded DNA (dsDNA) molecules, which are identical to the target DNA.

The detection of PCR products, called also amplicons, is generally done by gel electrophoresis, which involves the use of DNA intercalators, which are generally toxic. The main disadvantage of PCR for DNA sensing is the need for strict temperature control, which involves the use of a thermal cycler. Furthermore the integration of such a temperature controller in a portable device results in a very complicated process. For this reason many techniques that allow the amplification of DNA molecules at a constant temperature have been developed. Detailed information about these techniques has been extensively reported in the literature [18].

7.2.3 *What Nanoparticles Can Bring to DNA Sensors?*

NMs and especially NPs have been extensively used in various optical and electrochemical DNA sensors. NP properties such as large fraction of surface atoms, high surface energy, spatial confinement, and reduced imperfections do not exist in the corresponding bulk materials [22], making them of interest in the design of DNA sensors.

Gold nanoparticles (Au NPs) are the most used NPs in biosensing, not only of nucleic acids, but also of other biomolecules. Au NPs have some unique features such as easy synthesis, a surface plasmon resonance effect, catalytic activity, good biocompatibility, and easy functionalization. In particular functionalization is essential in the development of a biosensor, and in the case of Au NPs it happens generally through the Au–S bond using biomolecules modified with a thiol group. In the case of DNA, generally an ssDNA is modified with a thiol group on its 3′ or 5′ end. Besides Au NPs other types of NPs are also considered in this chapter: silver nanoparticles (Ag NPs), lead sulfide nanoparticles (PbS NPs), cadmium sulfide nanoparticles (CdS NPs), magnetic NPs, etc. In general all the mentioned NPs act as labels or carriers of other labels while being employed in DNA detection. Au NPs, besides few other metallic NPs, are also used as modifiers of electrotransducers in order to enhance their electroactive surface [23, 24].

7.2.4 *Designs of Electrochemical DNA Biosensors*

The first and most used DNA sensor design is based on the formation of a "sandwich" between a capture probe, the DNA target, and a detection probe (Fig. 7.2a). The three DNA molecules involved are ssDNA. The capture probe and the detection probe are complementary to the two different ends of the target molecule. The capture probe is generally attached directly to the electrode surface or to the surface of magnetic beads, and its function is to recognize the target sequence. In a second step the detection probe recognizes the captured target DNA, and this hybridization is followed by a second hybridization step, with the detection probe bearing an electroactive label (i.e., Au NPs) [9, 25].

It is also possible to design a DNA sensor starting from the dsDNA produced by PCR (Fig. 7.2b). In this case it is important to label the two primers with two different molecules, obtaining in this way at the end of PCR many dsDNA molecules modified at each end with a specific label. Generally one primer carries biotin, to perform in a second step the streptavidin/biotin bond, and the other primer a

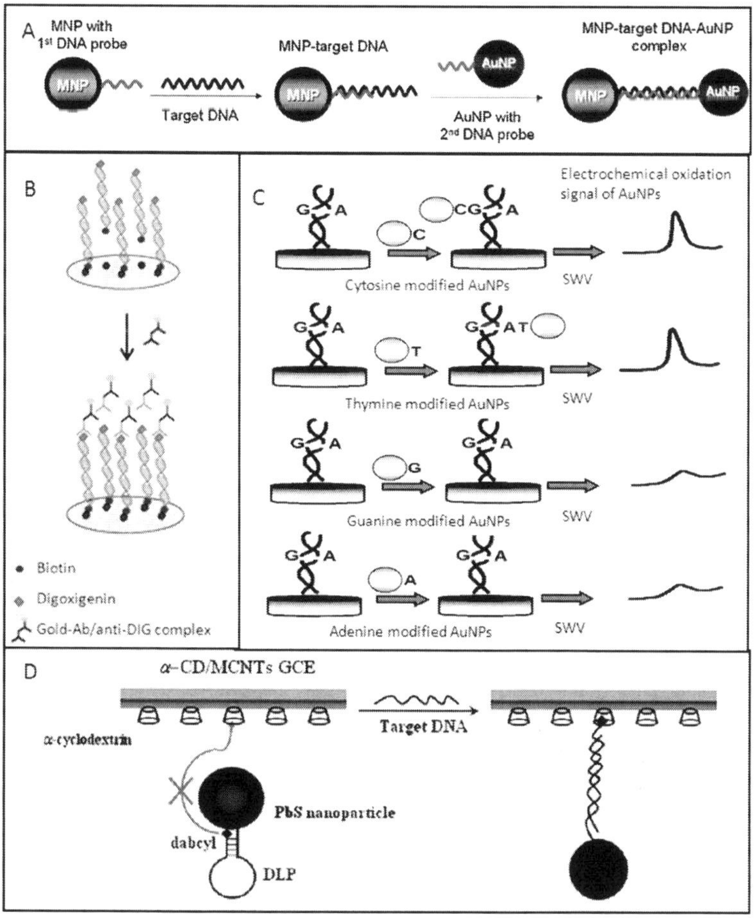

Figure 7.2 Designs of electrochemical DNA biosensors. (a) Example of a magnetosandwich assay. (b) Stem-loop-based DNA sensor. (c) Labeled monobase DNA sensor scheme. (d) DNA sensor based on a double-labeled PCR product scheme. Adapted with permission from Refs. [25–28].

thiol or a protein (i.e., digoxigenin), to be attached to NPs directly or through an antibody (i.e., antidigoxigenin) attached to the NPs [26].

Another way to detect DNA, in particular SNPs, is based on the use of monobases labeled with a marker (Fig. 7.2c). In fact when a mismatch is present in a dsDNA, there is an error in the hybridization and the conformation of the dsDNA is not so closed, making the base of the mismatch available to interact with a free base. In this way using a solution of a known monobase modified with a label, such as Au NPs, it is possible to electrochemically detect if an SNP is present in the dsDNA target [27].

A quite new approach consists of the use of stem-loop ssDNA (Fig. 7.2d). These molecules are a particular kind of ssDNA able to hybridize in two regions of their own sequence. This makes them obtain a conformation where the two ends of the ssDNA are closed. If the target ssDNA is present, it hybridizes with the stem loop, opening the conformation and making the two ends free [28].

Finally it has to be mentioned that DNA can interact not only with other DNA molecules but also with DNA-binding proteins, peptide nucleic acid, and RNA [29].

7.3 Applied Technologies

7.3.1 *Gold Nanoparticle–Based Technologies*

The excellent electroactivity of metallic NPs, together with their easy bioconjugation, has given rise to their extensive use as labels in DNA sensors. Several electrochemical routes have been exploited for the sensitive detection of these NP tags in bioassays.

In this section several examples of electrochemical DNA biosensors based on the use of Au NPs as labels/carriers first or as electrodes modifiers are considered.

7.3.1.1 Gold nanoparticles as electroactive and catalytic labels

The excellent electroactivity of Au NPs, together with their catalytic activity toward other reactions, has allowed their application in various DNA-sensing strategies.

For example, Yeung et al. [30] developed DNA biosensors integrated in a microchip for the multiplexed detection of *Escherichia coli* and *Bacillus subtilis* cells. The developed device contains a thin-film heater and a temperature sensor patterned on the silicon substrate, while indium tin oxide electrodes integrated inside the reaction microchamber are used as transduction elements modified with DNA strands. The chip performed all the steps required: the thermal lysis of pathogens, the magnetic particle–based isolation of the target genomes, asymmetric PCR, and electrochemical sequence-specific detection using Au NP labels. In this case, the catalytic activity of the Au NPs on silver reduction is approached to selectively deposit silver on their surface, followed by the electrochemical oxidative dissolution of the silver ions, which is measured and related with the concentration of cells at levels of 10^2 cells/sample.

A similar approach based on the use of streptavidin-modified Au NPs and biotinylated probes was performed by Pinijsuwan et al. [31], taking advantage also of the use of latex microspheres as carriers of the Au NPs labels (Fig. 7.3a). The authors immobilized the target DNA onto a screen-printed carbon electrode (SPCE), and then hybridization with a complementary sequence modified with biotin was performed. Finally the Au NP–streptavidin complexes loaded onto the latex microspheres were captured by the streptavidin–biotin bond. The Au NPs were detected by anodic stripping voltammetry after the NP chemical dissolution/oxidation, obtaining a limit of detection of 0.5 fM.

Castañeda et al. [32] took advantage of easy isolation of specific DNA sequences by magnetic particles (Fig. 7.3b). In fact, the authors used Au NPs as labels in a magneto-sandwich assay. In detail they detected two sequences, one related to BRCA1 and one to cystic fibrosis. They used magnetic microparticles modified with streptavidin, to easily attach the capture probe modified with biotin, to concentrate the target sequences. Then the use of a detection sequence functionalized with Au NPs allowed them to detect the analyte, performing in this case direct voltammetric detection of the Au NPs without previous dissolving, by simply electrochemical oxidizing their surface. Then the detection was done measuring the electrochemical reduction of the generated Au(III) ions by differential pulse voltammetry on graphite–epoxy electrodes (modified with a magnet), reaching a limit of detection of 0.198 μg/mL.

A similar approach was performed by Torres-Chavolla et al. [25]. They detected *Mycobacterium tuberculosis*, combining thermophilic helicase-dependent isothermal amplification, amine-terminated magnetic particles, and dextrin-coated Au NPs. In particular, the authors first amplified the DNA of interest by thermophilic helicase-

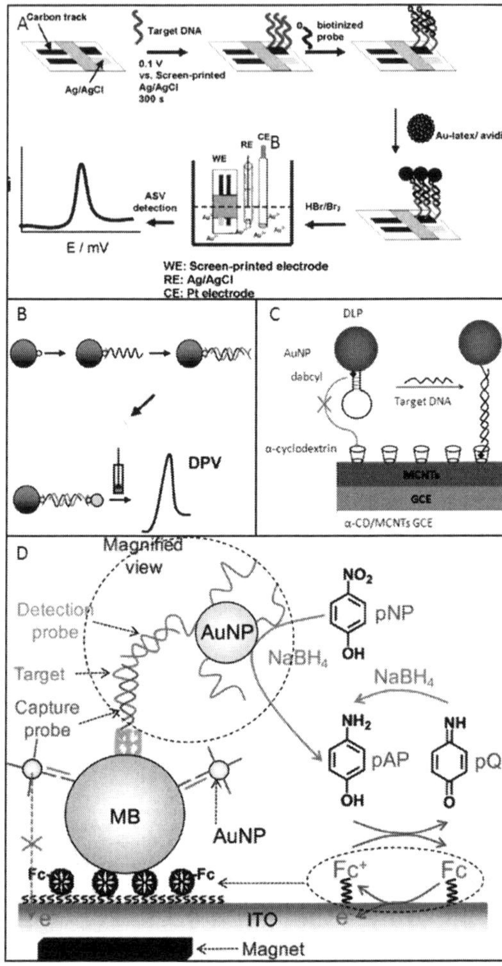

Figure 7.3 Au NPs as electroactive and catalytic labels. (a) Magneto-sandwich assay using Au NPs. (b) Au NPs used in a stem loop–based DNA biosensor. (c) Au NPs carried by latex microspheres. (d) Au NPs as catalysts. Adapted from Refs. [31, 32, 34, 35].

dependent isothermal amplification, and then the amplicons produced were hybridized in a sandwich-like assay. The Au NPs were finally detected by the same direct voltammetric approach explained above using SPCEs and obtaining a detection limit of 0.01 ng/μL of the isothermally amplified target.

A different platform was used by Liao et al. [33]. The authors developed a very sensitive DNA sensor for the detection of the mutated *BRAF* gene associated with papillary thyroid carcinomas on the basis of a 96-well streptavidin modified microplate. The authors first attached to the wells the biotinylated capture probes; then the target sequence modified also with biotin was added and allowed to hybridize with the capture probe. Streptavidin-modified Au NPs were then added and finally dissolved in bromide, the generated gold ions being detected by square-wave stripping voltammetry technique using a glassy carbon electrode. The authors obtained a limit of detection of 0.35 aM.

A DNA biosensor based on the use of the stem loop instead of a sandwich was proposed by Fan et al. (Fig. 7.3c) [34]. They developed a device based on a stem-loop ssDNA, which was modified with Au NPs on one end and with dabycil on the other end. In ssDNA conformation the two ends of the ssDNA are very close, making it impossible for the dabycil to interact with the cyclodextrin immobilized on glassy carbon electrodes modified with multiwalled carbon nanotubes. On the other hand, in the presence of the complementary sequence the dabycil was free to interact with the cyclodextrin attaching the dsDNA, and consequentially the Au NPs, to the electrode. Detecting the Au NPs by the direct voltammetric approach explained before, a limit of detection of 2.6×10^{-10} M of the DNA target was reached, differentiating also single mismatches.

An example of the approaching of catalytic properties on the reduction of chemical compounds is the work of Selvaraju et al. [35], where Au NPs were used as catalyzers of the generation of *p*-aminophenol starting from *p*-nitrophenol and $NaBH_4$ (Fig. 7.3d). The *p*-aminophenol is then electrooxidized to *p*-quinoneimine at the electrode. The *p*-aminophenol redox cycling by $NaBH_4$ offers large signal amplification. The authors used a magneto-sandwich format to concentrate the Au NPs onto an indium tin oxide electrode. In this way, a limit of detection of 1 fM of the target ssDNA was reached.

The different approach reported by Kerman et al. also deserves to be mentioned [27]. The authors detected SNPs using monobase-modified Au NPs. In fact, the monobases can hybridize the dsDNA just in the presence of an SNP using DNA polymerase. In this way, it was possible to detect not only the presence of SNPs but also which bases were involved, detecting in this case the oxidation signal of Au by square-wave voltammetry.

7.3.1.2 Gold nanoparticles as signal amplifiers and carriers of other labels

The presence of Au NPs attached to the electrotransducer surface through the DNA hybridization reaction produces an increase in the conductivity of the surface, exerting an amplification effect in detection strategies based on both conductimetric and impedimetric measurements.

Park et al. [36] used catalytic silver deposition onto Au NPs to detect DNA, taking advantage in this case of the changes in the conductivity of a microelectrode (60 nm Au on 5 nm Ti) produced by the presence of the silver catalytically deposited on the Au NPs. First they performed a classical sandwich assay with the Au NPs attached to the detection probe, and then they measured the difference in conductivity after silver deposition, which is enhanced by the presence of Au NPs. In this way a limit of detection of 500 fM was achieved.

The changes in the electron transfer resistance on the electrode surface by the presence of Au NPs without any amplification were approached by Gao et al. [37] for DNA detection (Fig. 7.4a). The authors first immobilized onto a gold electrode ssDNA, and then Au NPs were attached to the ssDNA. In this way, the electron transfer resistance decreased. In the presence of the complementary sequence, the Au NPs were displaced, resulting in an increase of the impedance, indicating hybridization events.

Bonanni et al. [26] took advantage of the strength of PCR in order to amplify specifically the target sequence and, at the same time, used modified primers, obtaining a dsDNA modified with biotin on one end and with digoxin on the other end of the dsDNA. In this way they could concentrate the PCR amplicons to the streptavidin-modified electrode surface. In the second step the digoxin was recognized by an antibody specific for the digoxin. Finally using a secondary antibody functionalized with Au NPs the authors could increase the impedimetric response of an avidin bulk-modified graphite–epoxy biocomposite electrode from 4 fM to 0.04 fmol.

A special case of the use of Au NPs as signal amplifiers was recently proposed by De la Escosura-Muñiz et al. [38]. The authors developed an innovative DNA hybridization biosensor based on the use of nanoporous alumina filter membranes attached onto the working area of an SPCE (Fig. 7.4b). The membrane contained nanochannels (200 nm in diameter and thickness of 60 μm), which were functionalized by a capture ssDNA probe, and the presence of the complementary ssDNA target was detected through a decrease

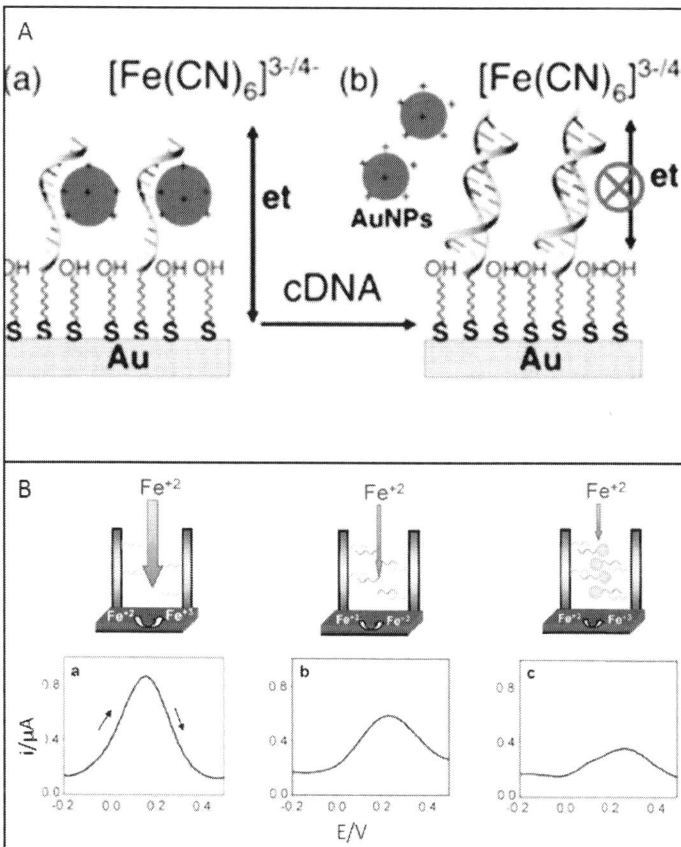

Figure 7.4 Au NPs used as amplifiers. (a) Displacement of Au NPs by hybridization affects electron transfer. (b) Au NPs as amplifiers in a nanochannel-based DNA sensor. Adapted from Refs. [37, 38].

in the voltammetric signal of the $[Fe(CN)_6]^{4-/3-}$ redox system due to both steric and electrostatic effects. Using Au NPs labels as additional blockage agents, they detected ssDNA at levels of 42 ng/mL.

Finally, an example of the use of Au NPs as carriers of other electrochemical labels was the work of Thiruppathiraja et al. [39]. The authors could detect the genomic DNA of *Mycobacterium sp.* in a clinical specimen using Au NPs loaded with alkaline phosphatase enzyme as a signal amplifier. The authors first modified the indium tin oxide electrode surface with Au NPs in order to attach easily a capture probe. Then, they incubated the electrode with the sample and finally with dual-label Au NPs, which carried a DNA detection sequence and the enzyme. The alkaline posphatase hydrolyzed the *p*-nitrophenol phosphate, and the reaction was characterized by voltammetric and impedimetric measurements. The authors studied both genomic DNA and sputum samples, reaching a detection limit of 1.25 ng/mL of genomic DNA.

Au NPs have also been used as carriers of other NPs [40] that are finally detected, taking advantage of their electroactive properties, as will be explained in Section 7.3.2.

7.3.1.3 Gold nanoparticles as modifiers of electrotransducers

Attempts to develop DNA hybridization assays using nanostructurated surfaces have been reported in the last few years. The introduction of NPs into the transducing platform is generally achieved by their integration within conventional electrodes in various forms, including that of a composite [41].

Li et al. [42] generated a mercatpophenyl film onto a glassy carbon electrode by electrografting. Then the modified electrode surface was dipped in a Au NP solution for six hours in order to obtain a Au NP–modified electrode. In the second step, a thiolated DNA probe was attached to the Au NPs. Finally hybridization was detected by differential pulse voltammetry of $Co(phen)_3{}^{3+}$ as the electrochemical indicator, with a limit of detection of 7.2 $\times$ 10^{-11} M. In a previous work [43] the same authors proposed a similar approach but one based on the use of a mercapto-diazoaminobenzene monolayer, obtaining a detection limit of 9.10 $\times 10^{-11}$ M.

Instead of using classical Au NPs, Li et al. [44] produced a very sensitive electrode for DNA sensing based on the formation of dendritic gold nanostructures on a planar gold electrode by applying a potential of -1.5 V and a solution of 2.8 mM $HAuCl_4$ and 0.1 M H_2SO_4 (Fig. 7.5a). With such an electrode the authors studied the hybridization events of two DNA sequences, reaching a detection limit of 1 fM performing cyclic voltammetry and differential pulse voltammetry of methylene blue used as the electrochemical hybridization indicator.

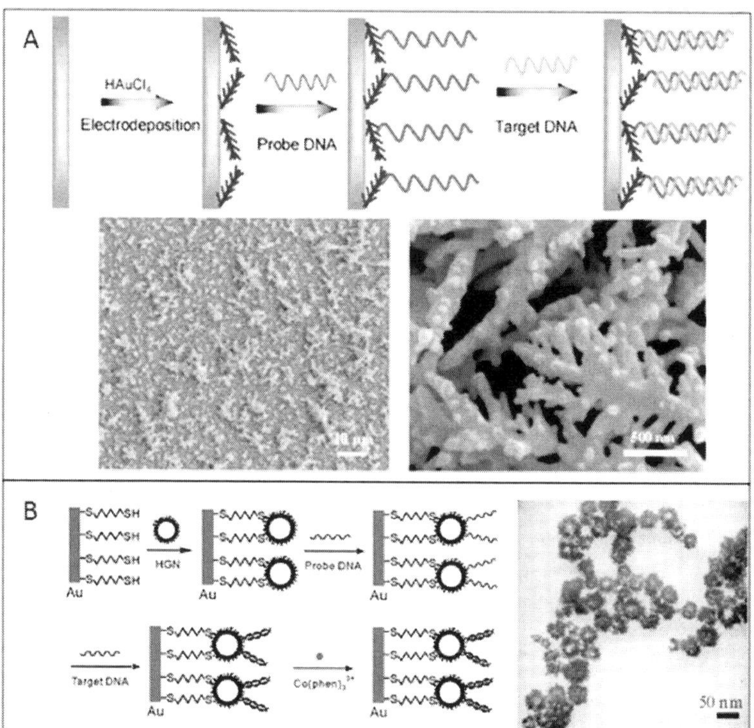

Figure 7.5 Au NPs used as transducer modifiers. (a) Dendritic gold nanosctructure used in a planar gold electrode. (b) Hollow gold nanospheres immobilized onto a gold electrode. Adapted from Refs. [44, 45].

Also Liu et al. [45] developed a DNA hybridization biosensor based on hollow gold nanospheres prepared using Co NPs as sacrificial templates (Fig. 7.5b). After the immobilization of the hollow gold nanospheres onto the gold electrode, the authors functionalized them with capture ssDNA to detect the hybridization events. A limit of detection of 1 pM was achieved with a range of detection between 1 pM and 10 nM, measuring cyclic voltammetry and differential pulse voltammetry of $Co(phen)_3{}^{3+}$, which has a high affinity for dsDNA and not for ssDNA.

A different approach was proposed by Spain et al. [46]. They developed a DNA biosensor based on the growth of Au NPs onto polyaniline nanofibres. The nanocomposite material produced was deposited onto a gold electrode. Then a sandwich assay, which detected a DNA sequence specific of *Staphylococcus aureus*, was performed with a DNA strand modified with horseradish peroxidase (HRP) as a detection probe. The signal measured was the reduction of a hydroquinone mediator in solution. A limit of detection of a pM level was obtained without the need for amplification. Also Hu et al. [47] used a nanocomposite material combining graphene sheets and Au NPs to obtain a DNA sensor. First they modified the graphene sheets with 3,4,9,10-perylene tetracarboxylic acid to obtain a good separation of them and increase the number of negatively charged –COOH sites. In this way the efficiency of decoration of the graphene sheets with Au NPs was increased. The Au NPs were synthetized by reduction of $HAuCl_4$ by amine-terminated ionic liquid (NH2-IL). This green synthesis produced 3 nm homogenously dispersed NH2-IL-protected Au NPs, which, in combination with the highly negative charged graphene sheets, allow the electrostatic interaction and adsorption of DNA without the need for its modification. Finally, impedimetric measurements allow them to detect the hybridization event, obtaining a limit of detection of 3.4×10^{-14} M.

7.3.2 *Other Nanoparticles*

7.3.2.1 Silver nanoparticles

Silver nanoparticles (Ag NPs) are other NPs with high interest in electrochemical biosensors due to their high electroactivity, although some drawbacks related to their synthesis and toxicity have minimized their extensive use. Kong et al. [48] took advantage of the use of Ag NPs for the detection of DNA by measuring the differences in conductance in interdigited microelectrodes. First they immobilized peptide nucleic acid in the gaps of the electrodes, and then the complementary DNA sequence was hybridized with the peptide nucleic acid molecules. In the next step hematin molecules were introduced into the DNA strand via zirconium phosphate and zirconium carbonate chemistries. The hematin catalyzed the reduction of silver ions into Ag NPs, which span the gap between electrodes. A limit of detection of 1 fM was achieved.

A different approach was proposed by Jian et al. [49]. They developed a DNA biosensor based on the aggregation of Ag NPs. First they modified a gold electrode with a capture ssDNA, which recognizes the short target ssDNA. Then biotinilated Ag NPs functionalized with signalling ssDNA were associated to the target strands. In a further step streptavidin was added to bind the biotin of the Ag NPs. Finally a solution of unlabeled biotinilated Ag NPs was added, inducing the aggregation of the NPs. The aggregates were detected, taking advantage of impedimetric measurements, allowing a detection limit of 10 fM.

7.3.2.2 Semiconductor nanoparticles

The use of semiconductor NPs with different compositions, the so-called quantum dots, has enormous potential in multidetection designs, since each of these materials can give rise to a specific electrochemical signal characteristic of the contained metal (i.e., Pb, Cd, Zn, etc.). Furthermore, the conditions for electrochemical detection are generally less aggressive (less acidic conditions, lower potentials), which can be advantageous for some biosensing designs.

Fan et al. [28] developed a biosensor to study hybridization events based on the use of a double-tagged stem loop: on one end with dabycil and on the other end with Pb NPs, a similar conformation as in Section 7.3.1.1. In the closed conformation the dabycil cannot interact with the α-cyclodextrin present on the surface of the multiwalled carbon nanotube–modified/glassy carbon electrode, but when the stem loop is opened with the complementary DNA sequence the dabycil interacts with the α-cyclodextrin, making possible the Pb NPs' detection. In this way, it was possible to obtain a limit of detection of 7.1×10^{-10} M by measuring the voltammetric signal of Pb in a mercury film electrode. The same authors proposed a similar approach based on the use of CdS NPs [50]. In this work the electrode was a β-CD-modified poly(*N*-acetylaniline) glassy carbon electrode [51]. The stem-loop probes were used also by Kjällman et al. in combination with CdTe NPs to detect up to 4,7 fM DNA molecules by impedance spectroscopy in a gold electrode.

Different uses of Pb NPs and CdS NPs were reported by Hu et al. and Zhu et al. In the first work [40] the authors reported a DNA hybridization biosensor based on a biobarcode using Pb NPs and a nanoporous gold glassy carbon electrode. The device is based on the formation of a sandwich between the capture probe attached to the glassy carbon electrode and the detection probe of the label. In this case the label consists of Au NPs carrying the Pb NPs and the detection sequence. The authors reached a limit of detection of 2.6 $\times 10^{-16}$ M of the DNA target, detecting the Pb released from the NPs by anodic stripping voltammetry. In the work of Zhu et al. [52], the hybridization of DNA was detected by the use of a CdS nanocluster and a mercury-modified glassy carbon electrode. First they modified the electrode surface with pyrrole in order to attach the target DNA, and then they hybridized it with CdS NP–modified DNA. They used anodic stripping voltammetry to detect the released Cd ions and obtained a limit of detection of 0.2 pmol/L.

Dong et al. created a new label based on the functionalization of poly(styrene-co-acrylic acid) microbeads with CdTe quantum dots. The quantum dots were carried by the microbeads and functionalized with streptavidin in order to bind the biotin of a detection probe of a sandwich assay, where the capture probe was attached to a glassy carbon electrode. In this way up to 0.52 fM of the DNA sequence related to breast cancer was detected [53]. A different type of DNA hybridization biosensor was developed by Chen et al. [54], taking advantage of the specificity of BfuCI nuclease for a particular four-nucleotide-long dsDNA. In detail, a stem loop modified with CdSe/ZnS core–shell NPs having the enzyme recognition sequence can be cleaved, washing away the quantum dots. On the other hand if the complementary sequence was present the nuclease was not able to cut the strand (in fact it would hybridize in a different region, leaving the nuclease recognition site in an ssDNA conformation). In this work two different electrodes were used: first a gold electrode to perform all the hybridization events and then, after the dissolution of Cd into nitric acid solution, a mercury-modified electrode to detect the Cd ions. In this way it was possible to achieve a detection limit of 3.3×10^{-14} M.

Finally an interesting multidetection DNA hybridization biosensor design was developed by Hansen et al. [55]. The authors used three different sulfide NPs to detect electrochemically three different target sequences: CdS, ZnS, and PbS (Fig. 7.6a). Each type of NP was functionalized with a specific sequence complementary to one of the three sequences immobilized onto the electrode. Then using stripping analysis after the NPs' dissolving/oxidation, it was possible to detect the three different NP labels down to 100 aM [55].

7.3.2.3 Other nanoparticles

The advantageous properties of NPs with different compositions from those detailed in previous sections have also approached, in a minor extent, for the electrochemical detection of DNA.

A classical sandwich assay was proposed by Cai et al. [56] using Cu@Au alloy NPs as labels. The electrochemical signal was recorded by using a glassy carbon electrode and performing the anodic stripping voltammetry of released Cu ions, obtaining a detection limit of 5.0 pM.

Dong et al. developed a DNA biosensor based on the use of an electrochemically reduced graphene oxide–modified electrode, where they adsorbed thiolated DNA strands, which hybridize with DNA strands attached to Au NPs (Fig. 7.6b) [57]. In a further step the target DNA modified with biotin hybridizes with the DNA of the Au NPs. Finally carbon NPs modified with streptavidin and HRP were used as labels. The limit of detection of 5 aM was reached measuring the differential pulse voltammetry signal of the HRP. The range of detection was between 1×10^{-17} M and 1×10^{-13} M. Another work based on the use of HRP as a label was reported by Li et al. [58]. They could detect up to 0.01 pM of the DNA target or

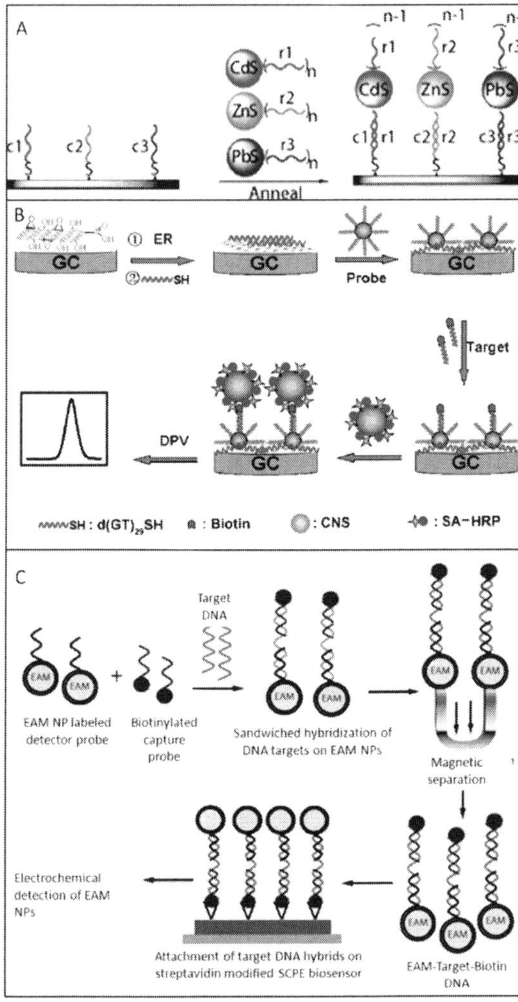

Figure 7.6 Other NPs. (a) Sulfide-based NPs for multiplex DNA detection. (b) EAM NP–based DNA sensor. (c) Triplex amplification based on graphene, Au NPs, carbon NPs, and HRP. Adapted with permission from Refs. [55, 57, 61].

500 cfu/mL of *E. coli* cells, without the nucleic acid amplification step, using Fe_2O_3@Au core–shell NPs and a probe modified with HRP in a sandwich assay. The Fe_2O_3@Au core–shell NPs have two functions, functionalization with the capture sequence through gold and magnetic separation, thanks to the Fe_2O_3 paramagnetic properties. The amperometric detection of the enzymatic reaction of reduction of 2,2′-diaminoazobenzene catalyzed by HRP was finally detected.

Another enzyme was used by Chen et al. [59]. The authors developed a biosensor based on a sandwich assay and the use of glucose oxidase as a label. The detection probe was modified with streptavidin in order to bind the avidin-labeled glucose oxidase. The capture probe was attached to the gold electrode. The signal was produced by enzyme-catalyzed deposition of cupric hexacyanoferrate (CuHCF) NPs in the presence of glucose, cupric ions, and ferricyanide. The limit of detection was 1 fM, obtained with differential pulse voltammetry measurements of the oxidation of the deposited NPs.

Similar to Fe_2O_3@Au core–shell NPs, also alginic acid–coated cobalt magnetic beads used by Geng et al. [60] can be used both as carrier and as magnetic platforms. The beads capped with a 5-(NH_2) oligonucleotide were used to detect a specific gene of *E. coli.* The electrochemical detection was based on the use of daunomycin. It was observed that if the target sequence hybridizes with the target sequence the reduction peak of daunomycin decreases proportionally to the concentration of the hybridized target. With this method a limit of detection of 10 cells/mL in a real water sample, after an enrichment process, was obtained. In the work of Pal et al. [61] electrically active magnetic (EAM) polymer NPs were also used as magnetic platforms and as labels in a sandwich-type assay (Fig. 7.6c). In particular the capture probe was modified with EAM NPs, whereas the detection probes were modified with biotin. The complexes were washed using the magnetic properties of EAM NPs and captured to the surface of a streptavidin-modified SPCE and cyclic voltammetry registered, reaching a detection limit of 0.01 ng/μL.

A different approach was followed by Kerman et al. [62], who developed a DNA biosensor based on ferrocene-conjugated chitosan

NPs and nuclease S1. On a gold electrode surface a peptide nucleic acid sequence complementary to the target was fixed, and then the sample was added to the surface. If the hybridization was perfect the nuclease would not cut the DNA target, but if a mismatch was present the DNA target would be degraded. The ferrocene-conjugated chitosan NPs were sensitive to the DNA target only in the way that when it was present the label got accumulated, and consequently an increase of the voltammetric signal of the ferrocene appears, reaching a detection limit of up to 1 fM.

Liao et al. [63] used liposomes to carry $[Ru(NH_3)_6]^{3+}$ as a redox indicator in a DNA sensor. First they modified an SPCE with Au NPs in order to attach the capture probe. Then they incubated with the target sequence and liposomes, functionalized with a competitive DNA sequence. The limit of detection obtained measuring the voltammetric signal of the redox indicator was 0.75 amol (equivalent to the amount present in 5 μL of a 0.15 pM solution) for a DNA sequence specific of *E. coli*.

7.4 Conclusion and Future Perspective

Due to their sensitivity and specificity electrochemical DNA biosensors seem to be excellent devices for the detection of DNA. Furthermore electrochemical biosensors are easy to be integrated in various platforms (i.e., microfluidics) where the various assay steps can be included, making them a good alternative for DNA detection in real samples. The synergy between electrochemical detection and the use of NPs, as labels or as carriers, is giving to these devices higher sensitivity besides higher stability if compared to other technologies where enzymes or other labels are being used. Given the advances in nanotechnology in general and particularly in NMs, the cost of NP-based DNA biosensors is expected to go in line with mass production, making them ideal tools for future point-of-care and easy-to-use devices for diagnostics and environmental and security applications.

An increase in electrochemical DNA biosensors based on microchip devices can be noticed. Microfluidics platforms (lab-on-a-chip or lateral-flow devices) are excellent avenues for an integrated

DNA technology. These platforms are already integrating in a single device all the steps required for the detection of DNA: sample pretreatment, amplification reaction, labeling, preconcentration, and detection. Such an integration level is expected to better approach the DNA technology to point-of-care applications in many places with less resources and a lack of specialized personnel to perform such analysis.

To further increase the sensitivity of DNA detection the integration of isothermal amplification techniques with electrochemical detection may be expected to bring significant advantages. This integration may simplify much more the design and operation of DNA biosensors. The possibility to perform multidetection of different DNA sequences or different SNPs in the same run (sample) can be simply reached by using various NPs (with various electrochemical/catalytic properties). The parallel study of different DNA sequences with a cheap, fast, and easy-to-use biosensor would significantly simplify the screening of mutations besides other applications.

References

1. Sassolas, A., Leca-Bouvier, B. D., Blum, L. J. (2008) DNA biosensors and microarrays, *Chem. Rev.*, **108**, 109–139.
2. Pérez-López, B., Merkoçi, A. (2011) Nanomaterials based biosensors for food analysis applications, *Trends Food Sci. Technol.*, **22**, 625–639.
3. Bonanni, A, del Valle, M. (2010) Use of nanomaterials for impedimetric DNA sensors: a review, *Anal. Chim. Acta*, **678**, 7–17.
4. Wei, F., Lillehoj, P. B., Ho, C. (2010) DNA Diagnostics: nanotechnology-enhanced electrochemical detection of nucleic acids, *Pediatr. Res.*, **67**, 458–468.
5. de-los-Santos-Alvarez, P., Lobo-Castañón, M. J., Miranda-Ordieres, A. J., Tuñón-Blanco, P. (2004) Current strategies for electrochemical detection of DNA with solid electrodes, *Anal. Bioanal. Chem.*, **378**, 104–118.
6. Erdem, A. (2007) Nanomaterial-based electrochemical DNA sensing strategies, *Talanta*, **74**, 318–325.

7. Hvastkovs, E. G., Buttry, D. (2010) Recent advances in electrochemical DNA hybridization sensors, *Analyst*, **135**, 1817–1829.
8. Tosar, J. P., Brañas, G., Laíz, J. (2010) Electrochemical DNA hybridization sensors applied to real and complex biological samples, *Biosens. Bioelectron.*, **26**, 1205–1217.
9. Odenthal, K. J., Gooding, J. J. (2007) An introduction to electrochemical DNA biosensors, *Analyst*, **132**, 603–610.
10. Mir, M., Homs, A., Samitier, J. (2009) Integrated electrochemical DNA biosensors for lab-on-a-chip devices, *Electrophoresis*, **30**, 3386–3397.
11. de la Escosura-Muñiz, A., Parolo, C., Merkoçi, A. (2010) Immunosensing using nanoparticles, *Mater. Today*, **13**, 24–34.
12. Perfézou, M., Turner, A., Merkoçi, A. (2011) Cancer detection using nanoparticle-based sensors, *Chem. Soc. Rev.*, **41**, 2606–22.
13. Medina-Sánchez, M., Miserere, S., Merkoçi, A. (2012) Nanomaterials and lab-on-a-chip technologies, *Lab Chip*, **12**, 1932–43.
14. Krishnan, Y., Simmel, F. C. (2011) Nucleic acid based molecular devices, *Angew. Chem., Int. Ed.*, **50**, 3124–3156.
15. Caliendo, A. M. (2011) Multiplex PCR and emerging technologies for the detection of respiratory pathogens, *Clin. Infect. Dis.*, **52**(Suppl 4), S326–S330.
16. Mullis, K. B., Faloona, F. A. (1987) Specific synthesis of DNA in vitro via a polymerase-catalyzed reaction, *Methods Enzymol.*, **155**, 335–350.
17. Beaudet, A. L., Belmont, J. W. (2008) Array-based DNA diagnostics: let the revolution begin, *Annu. Rev. Med.*, **59**, 113–129.
18. Yang, S., Rothman, R. E. (2004) Review PCR-based diagnostics for infectious diseases: uses, limitations, and future applications in acute-care settings, *Lancet Infect. Dis.*, **4**, 337–348.
19. Wing, R., Drew, H., Takano, T., et al. (1980) Crystal structure analysis of a complete turn of B-DNA, *Nature*, **287**, 755–758.
20. Dickerson, R. E. (1983) The DNA helix and how it is read, *Sci. Am.*, **249**, 94–111.
21. Lodish, H., Berk, A., Zipursky, S. L., et al. (2003) *Molecular Cell Biology*, 5th ed. (W. H. Freeman).
22. Merkoçi, A. (2010) Nanoparticles-based strategies for DNA, protein and cell sensors, *Biosens. Bioelectron.*, **26**, 1164–1177.
23. Lord, H., Kelley, S. O. (2009) Nanomaterials for ultrasensitive electrochemical nucleic acids biosensing, *J. Mater. Chem.*, **19**, 3127.
24. Willner, I., Willner, B. (2010) Biomolecule-based nanomaterials and nanostructures, *Nano Lett.*, **10**, 3805–3815.

25. Torres-Chavolla, E., Alocilja, E. C. (2011) Nanoparticle based DNA biosensor for tuberculosis detection using thermophilic helicase-dependent isothermal amplification, *Biosens. Bioelectron.*, **26**, 4614–4618.

26. Bonanni, A., Pividori, M. I., Campoy, S., Barbé, J., del Valle, M. (2009) Impedimetric detection of double-tagged PCR products using novel amplification procedures based on gold nanoparticles and protein G, *Analyst*, **134**, 602–608.

27. Kerman, K., Saito, M., Morita, Y., et al. (2004) Electrochemical coding of single-nucleotide polymorphisms by monobase-modified gold nanoparticles, *Anal. Chem.*, **76**, 1877–1884.

28. Fan, H., Zhao, K., Lin, Y., et al. (2011) A new electrochemical biosensor for DNA detection based on molecular recognition and lead sulfide nanoparticles, *Anal. Biochem.*, **419**, 168–172.

29. Brandt, O., Hoheisel, J. D. (2004) Peptide nucleic acids on microarrays and other biosensors, *Trends Biotechnol.*, **22**, 617–622.

30. Yeung, S.-W., Lee, T. M.-H., Cai, H., Hsing, I.-M. (2006) A DNA biochip for on-the-spot multiplexed pathogen identification, *Nucleic Acids Res.*, **34**, e118.

31. Pinijsuwan, S., Rijiravanich, P., Somasundrum, M., Surareungchai, W. (2008) Sub-femtomolar electrochemical detection of DNA hybridization based on latex/gold nanoparticle-assisted signal amplification, *Anal. Chem.*, **80**, 6779–6784.

32. Castañeda, M. T., Merkoçi, a, Pumera, M., Alegret, S. (2007) Electrochemical genosensors for biomedical applications based on gold nanoparticles, *Biosens. Bioelectron.*, **22**, 1961–1967.

33. Liao, K.-T., Cheng, J.-T., Li, C.-L., Liu, R.-T., Huang, H.-J. (2009) Ultrasensitive detection of mutated papillary thyroid carcinoma DNA using square wave stripping voltammetry method and amplified gold nanoparticle biomarkers, *Biosens. Bioelectron.*, **24**, 1899–1904.

34. Fan, H., Xu, Y., Chang, Z., et al. (2011) A non-immobilizing electrochemical DNA sensing strategy with homogenous hybridization based on the host-guest recognition technique, *Biosens. Bioelectron.*, **26**, 2655–2659.

35. Selvaraju, T., Das, J., Jo, K., et al. (2008) Nanocatalyst-based assay using DNA-conjugated Au nanoparticles for electrochemical DNA detection, *Langmuir*, **24**, 9883–9888.

36. Park, S.-J., Taton, T. A., Mirkin, C. (2002) Array-based electrical detection of DNA with nanoparticle probes, *Science*, **295**, 1503–1506.

37. Gao, Q., Zhang, W., Guo, Y., Qi, H., Zhang, C. (2011) Highly sensitive impedimetric sensing of DNA hybridization based on the target DNA-

induced displacement of gold nanoparticles attached to ssDNA probe, *Electrochem. Commun.*, **13**, 335–337.

38. de la Escosura-Muñiz, A., Mekoçi, A. (2010) Nanoparticle based enhancement of electrochemical DNA hybridization signal using nanoporous electrodes, *Chem. Commun.*, **46**, 9007–9009.
39. Thiruppathiraja, C., Kamatchiammal, S., Adaikkappan, P., Santhosh, D. J., Alagar, M. (2011) Specific detection of Mycobacterium sp. genomic DNA using dual labeled gold nanoparticle based electrochemical biosensor, *Anal. Biochem.*, **417**, 73–79.
40. Hu, K., Liu, P., Ye, S., Zhang, S. (2009) Ultrasensitive electrochemical detection of DNA based on PbS nanoparticle tags and nanoporous gold electrode, *Biosens. Bioelectron.*, **24**, 3113–3119.
41. Liu, S., Liu, J., Wang, L., Zhao, F. (2010) Development of electrochemical DNA biosensor based on gold nanoparticle modified electrode by electroless deposition, *Bioelectrochemistry*, **79**, 37–42.
42. Li, F., Feng, Y., Dong, P., Yang, L., Tang, B. (2011) Gold nanoparticles modified electrode via simple electrografting of in situ generated mercaptophenyl diazonium cations for development of DNA electrochemical biosensor, *Biosens. Bioelectron.*, **26**, 1947–1952.
43. Li, F., Feng, Y., Dong, P., Tang, B. (2010) Gold nanoparticles modified electrode via a mercapto-diazoaminobenzene monolayer and its development in DNA electrochemical biosensor, *Biosens. Bioelectron.*, **25**, 2084–2088.
44. Li, F., Han, X., Liu, S. (2011) Development of an electrochemical DNA biosensor with a high sensitivity of fM by dendritic gold nanostructure modified electrode, *Biosens. Bioelectron.*, **26**, 2619–2625.
45. Liu, S., Liu, J., Han, X., Cui, Y., Wang, W. (2010) Electrochemical DNA biosensor fabrication with hollow gold nanospheres modified electrode and its enhancement in DNA immobilization and hybridization, *Biosens. Bioelectron.*, **25**, 1640–1645.
46. Spain, E., Kojima, R., Kaner, R. B., et al. (2011) High sensitivity DNA detection using gold nanoparticle functionalised polyaniline nanofibres, *Biosens. Bioelectron.*, **26**, 2613–2618.
47. Hu, Y., Hua, S., Li, F., et al. (2011) Green-synthesized gold nanoparticles decorated graphene sheets for label-free electrochemical impedance DNA hybridization biosensing, *Biosens. Bioelectron.*, **26**, 4355–4361.
48. Kong, J. M., Zhang, H., Chen, X. T., Balasubramanian, N., Kwong, D. L. (2008) Ultrasensitive electrical detection of nucleic acids by hematin

catalysed silver nanoparticle formation in sub-microgapped biosensors, *Biosens. Bioelectron.*, **24**, 793–797.

49. Jiang, X., Chen, K., Han, H. (2011) Ultrasensitive electrochemical detection of Bacillus thuringiensis transgenic sequence based on in situ Ag nanoparticles aggregates induced by biotin-streptavidin system, *Biosens. Bioelectron.*, **28**, 464–468.
50. Kjällman, T. H. M., Peng, H., Soeller, C., Travas-Sejdic, J. (2010) A CdTe nanoparticle-modified hairpin probe for direct and sensitive electrochemical detection of DNA, *Analyst*, **135**, 488–494.
51. Fan, H., Xing, R., Xu, Y., Wang, Q., He, P., Fang, Y. (2010) A new electrochemical method for DNA sequence detection with homogeneous hybridization based on host–guest recognition technology, *Electrochem. Commun.*, **12**, 501–504.
52. Zhu, N., Zhang, A., He, P., Fang, Y. (2003) Cadmium sulfide nanocluster-based electrochemical stripping detection of DNA hybridization, *Analyst*, **128**, 260–264.
53. Dong, H., Yan, F., Ji, H., Wong, D. K. Y., Ju, H. (2010) Quantum-dot-functionalized poly(styrene-co-acrylic acid) microbeads: step-wise self-assembly, characterization, and applications for sub-femtomolar electrochemical detection of DNA hybridization, *Adv. Funct. Mater.*, **20**, 1173–1179.
54. Chen, J., Zhang, J., Yang, H., Fu, F., Chen, G. (2010) A strategy for development of electrochemical DNA biosensor based on site-specific DNA cleavage of restriction endonuclease, *Biosens. Bioelectron.*, **26**, 144–148.
55. Hansen, J. A., Mukhopadhyay, R., Hansen, J. Ø., Gothelf, K. V. (2006) Femtomolar electrochemical detection of DNA targets using metal sulfide nanoparticles, *J. Am. Chem. Soc.*, **128**, 3860–3861.
56. Cai, H., Zhu, N., Jiang, Y., He, P., Fang, Y. (2003) Cu@Au alloy nanoparticle as oligonucleotides labels for electrochemical stripping detection of DNA hybridization, *Biosens. Bioelectron.*, **18**, 1311–1319.
57. Dong, H., Zhu, Z., Ju, H., Yan, F. (2012) Triplex signal amplification for electrochemical DNA biosensing by coupling probe-gold nanoparticles–graphene modified electrode with enzyme functionalized carbon sphere as tracer, *Biosens. Bioelectron.*, **33**, 228–232.
58. Li, K., Lai, Y., Zhang, W., Jin, L. (2011) Fe_2O_3@Au core/shell nanoparticle-based electrochemical DNA biosensor for Escherichia coli detection, *Talanta*, **84**, 607–613.

59. Chen, X., Xie, H., Seow, Z. Y., Gao, Z. (2010) An ultrasensitive DNA biosensor based on enzyme-catalyzed deposition of cupric hexacyanoferrate nanoparticles, *Biosens. Bioelectron.*, **25**, 1420–1426.

60. Geng, P., Zhang, X., Teng, Y., et al. (2011) A DNA sequence-specific electrochemical biosensor based on alginic acid-coated cobalt magnetic beads for the detection of E. coli, *Biosens. Bioelectron.*, **26**, 3325–3330.

61. Pal, S., Alocilja, E. C. (2010) Electrically active magnetic nanoparticles as novel concentrator and electrochemical redox transducer in Bacillus anthracis DNA detection, *Biosens. Bioelectron.*, **26**, 1624–1630.

62. Kerman, K., Saito, M., Tamiya, E. (2008) Electroactive chitosan nanoparticles for the detection of single-nucleotide polymorphisms using peptide nucleic acids, *Anal. Bioanal. Chem.*, **391**, 2759–2767.

63. Liao, W.-C., Ho, J.-A. A. (2009) Attomole DNA electrochemical sensor for the detection of Escherichia coli O157, *Anal. Chem.*, **81**, 2470–2476.

Chapter 8

Electroenzymatic Labeling for Immunosensors and DNA Sensors

Dianping Tang

Department of Chemistry, Fuzhou University, Fuzhou 350108, China
dianping.tang@fzu.edu.cn

This chapter describes recent advances, by coupling with bioactive enzyme labels, in electrochemical immunosensors and DNA sensors. And some sophisticated analytical devices for electrochemical immunosensors and DNA sensors, including potentiometry, amperometry, impedimetry, and conductometry, based on various signal generation principles from the antigen–antibody interaction and DNA hybridization are introduced. Signal amplification associates with the use of electroactive enzyme labels and electroenzymatic nanolabels. It covers the basic principles and biomedical and clinic applications of immunosensors and DNA sensors and indicates the future prospects in this field.

8.1 Introduction

Biosensors are defined as analytical devices incorporating a biological material (e.g., tissue, microorganisms, organelles, cell

Electrochemical Biosensors
Edited by Serge Cosnier

ISBN 978-981-4411-46-2 (Hardcover), 978-981-4411-47-9 (eBook)
www.panstanford.com

receptors, enzymes, antibodies, nucleic acids, natural products), a biologically derived material (e.g., recombinant antibodies, engineered proteins, aptamers), or a biomimic (e.g., synthetic receptors, biomimetic catalysts, combinatorial ligands, imprinted polymers) intimately associated with or integrated within a physicochemical transducer or transducing microsystem, which may be optical, electrochemical, thermometric, piezoelectric, magnetic, or micromechanical [1]. Electrochemical biosensors combine the sensitivity of electroanalytical methods with the inherent bioselectivity of the biological component. These devices contain a biological recognition element (enzymes, proteins, antibodies, nucleic acids, cells, tissue or receptors) that selectively reacts with the target analyte and produces an electrical signal that is related to the concentration of the analyte being studied [2]. They have been applied to a wide variety of analytical problems, including uses in medicine, biomedical research, drug discovery, the environment, food, process industries, security, and defense [3].

Electrochemical immunosensors have recently attracted considerable interest because of their high sensitivity, low cost, and inherent miniaturization [4]. The measurement principle is based on a specific reaction between antibodies and antigens with a sandwich-type or competitive-type binding protocol. To acquire high sensitivity, the online amplification of the electrochemical signal is extremely important. Although the antigen–antibody reaction by itself can cause a change in the electrochemical signal to some extent, it is very minor. Therefore, labeling of the signal antibody is essential. A common resort is to exploit signal transduction labels, including ligand-conjugated enzymes and nanolabels [5]. These labels can be usually transformed to readily detectable electroactive species through enzymatic conversion of certain substrates or chemical decomposition of the metal or its insoluble salt. Electrochemical DNA sensors combine the specificity of biological recognitions with the sensitivity of physicochemical transducers. These sensors commonly rely on the immobilization of single-stranded oligodeoxyribonucleic acid on different electrodes (or chips) to recognize its complementary target sequence. The hybridization event can be detected via monitoring changes in the current response of a redox indicator, or electronic and interfacial properties resulting from the duplex formation [6].

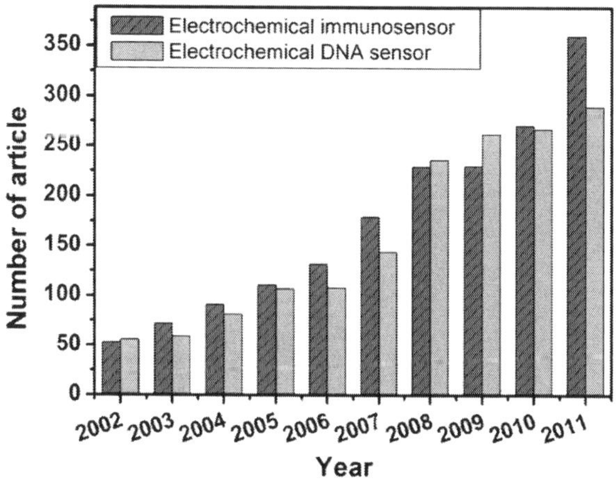

Figure 8.1 Number of published articles referring to electrochemical immunosensors and DNA sensors during the period from 2002 to 2011.

Figure 8.1 shows the number of published articles referring to electrochemical immunosensors and DNA sensors during the period from 2002 to 2011. Accordingly to their detection working principle, electrochemical immunosensors and DNA sensors can be classified as potentiometric, amperometric, impedimetric, and conductometric sensors. This chapter mainly focuses on recent developments in electrochemical immunosensors and DNA sensors by using bioactive enzyme as labels with signal amplification.

8.2 Electroenzymatic Labeling for Immunosensors

Immunosensors, miniaturized analytical devices that combine high specificity of immunological reactions with sensitivity and convenience of the detection technique, have been developed rapidly as alternatives. Compared with conventional immunoassays, immunosensors are specific, simple, and convenient and can offer multitarget analytes and miniaturization. Electrochemical immunosensors combine antigen–antibody reactions with electrochemical measurements. For electrochemical immunosensors, the immunologic materials are immobilized on an electrochemical transducer, for example, a gold electrode, a glassy carbon

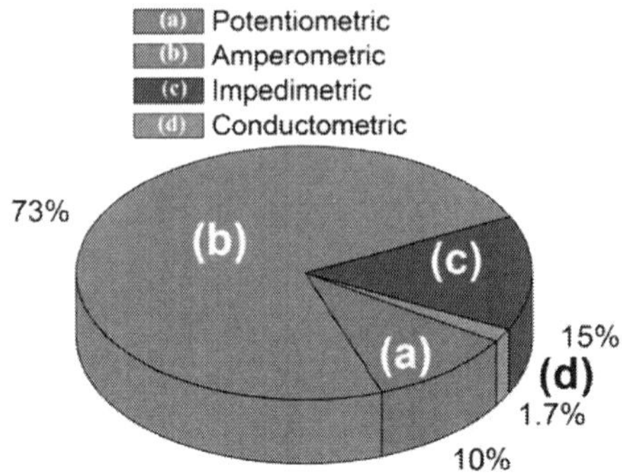

Figure 8.2 Main classes of electrochemical immunosensors identified. Pie chart showing the percentage of published articles referring to electrochemical immunosensors during the period from 2002 to 2011.

electrode, or a platinum electrode. The method usually involves the following steps: immobilization of the antibody/or antigen on the electrode, incubation with the analyte, separation, and measurement. Typically, sensitive and specific determination of disease-related biomarkers is essential in modern medicine and clinic diagnosis. Efficient labeling of biomolecules is of fundamental importance for various biodiagnostic assays. A bioactive enzyme is one of protein-based enzymes that act as catalysts to facilitate a variety of biological processes. Specifically, electroenzymatic activity involves donating electrons to bind to other substrate substances. Hence, enzyme labels are usually employed for detectable signal amplification. Assay protocols have been developed using different techniques, including potentiometry, amperometry, impedimetry, and conductometry. Due to the overwhelming amount of literature available, it is our choice to underline the most recent trends in this field (Fig. 8.2).

8.2.1 *Electroenzymatic Labeling for Potentiometric Immunosensors*

Potentiometric biosensors are based on the ion-selective electrode (ISE) and the ion-sensitive field-effect transistor (ISFET). The primary outputting signal is possibly due to ions accumulated at the ion-selective membrane interface. Current flowing through the electrode is equal to or near zero. Potentiometric immunosensors are based on the surface charge or potential change upon immunoreaction on the interface of the detection device, usually using a nonlabeling method. Recently, there are mainly two opinions about the potentiometric response mechanism of the membrane of immunosensors—(i) the first-degree kinetic response mechanism that the transformation of potentiometric value exhibits a linear dependence on the concentration of analytes and (ii) the mechanism of charge and electricity membrane that the transformation of potentiometric value exhibits a linear dependence on the logarithm of the concentration of analytes [7]. So the membrane potentiometric value mainly consists of the surface potentiometric value (i.e., Donnan potentiometric value) and the diffusion potentiometric value. When the concentration of the electrolyte on both sides of the membrane is a fixed value, the transformation of the membrane potentiometric value depends on the density of the electron charge on the membrane and the transference quantity of ions. Meanwhile, either antibodies or antigens in aqueous solution have a net electrical charge polarity, which is correlated to the isoelectric points of the species and the ionic composition of the solution. If the antibody complex combines with the antigen, the electrical charge of the resulting complex will be different from that of the antibody alone. This change can be measured potentiometrically against the reference electrode immersed in the same solution [8]. The merit of this methodology shortens the assay time. Unfortunately, the change in the potential is very small, and it is not conducive to determination of the low-concentration analyte. One preferable strategy is to utilize enzyme-labeled antibodies for amplification of the electrochemical signal.

To achieve high sensitivity for the successful development of potentiometric immunosensors, enzyme-labeled antibodies can be utilized. Typically, the most convenient labeling enzymes usually include urease, peroxidase, glucose oxidase, and alkaline phosphatase (ALP), which can change either pH or ionic strength during the measurement. In the early 1990s, Brown and Meyerhoff designed a potentiometric immunosensor for the detection of human IgG on an ammonium ISE with a competitive-type immunoassay format by coupling with the bioelectrocatalytic process of adenosine deaminase and ALP [9]. Adenosine deaminase and IgG were initially immobilized on the ammonium ISE, and then the signal antibodies were employed by using anti-IgG–protein A–ALP conjugates. With the competitive-type immunoassay mode, the immobilized IgG on the electrode and the analyte (IgG) competed with ALP-conjugated anti-IgG antibodies. Therefore, the conjugation amount of ALP-anti-IgG on the electrode decreased with the increase of IgG concentration. Upon addition of adenosine monophosphate (AMP) in the assay solution, AMP was initially transferred to adenosine and phosphate by ALP, and the formed adenosine could be then inosine and ammonia by the adenosine deaminase. Instead, the intermediate in the two-step reaction diffused to the membrane surface, reducing the rate of product (ammonium) formation within the diffusion layer detected by the membrane. ALP and adenosine deaminase sequentially facilitate the production of ammonium ions from AMP. Following that, the Wilkins group reported another potentiometric immunosensor for determination of IgG by using peroxidase as labels [10]. The ability of peroxidase is to catalyze the electrode reaction of hydrogen peroxide electroreduction by a direct (mediatorless) mechanism. Light-addressable potentiometric immunosensors (LAPSs) are capable of signal amplification due to the incorporated field-effect transistor. Usually, an LAPS consists of a silicon base covered with silicon oxide and silicon nitride layers. The backward illumination with a light-emitting diode (LED) induces a change of conductivity in the sensing area, and after formation of the immunocomplex labeled with urease, the change of pH is registered. Thompson and Lee used an LAPS for determination of *Francisella tularensis*, with a limit of detection (LOD) of 3.4×10^3 cells/mL [11].

8.2.2 *Electroenzymatic Labeling for Amperometric Immunosensors*

Amperometric immunosensors, as an analytical method combination of the electrochemical reaction with the antigen–antibody immune response, are among the most promising and most interesting immunosensors. They are usually designed to measure current generated by an electrochemical reaction. However, since most analytes (e.g., antigen, hapten) cannot intrinsically act as redox partners in the electrochemical reaction, an electrochemically active label is needed for the electrochemical reaction of the analyte at the sensing electrode. Enzymes such as horseradish peroxidase (HRP) and ALP are the active labels most commonly used to catalyze the reaction of substrates to form electroactive products. In electrochemical immunoassays, sandwich-type and competitive-type binding protocols are the most popular modes for antigen/antibody detection. Owing to the use of two matched antibodies in the sandwich assay, that is, a sandwich with an antigen and a conjugate with an antibody (which combines with another epitope of the antigen and the label), it has the advantages of high specificity and high sensitivity. The Ju group constructed an electrochemical enzyme immunosensor for detection of carcinoembryonic antigen (CEA) by using HRP-labeled anti-CEA antibodies as signal antibodies on a thionine monolayer–modified gold electrode [12]. During the electrochemical measurement, the electrochemical signal was mainly derived from the labeled enzymes toward the bioelectrocatalytic reaction of the substrate. Usually, the measurable signal is difficultly achieved from a direct immunoreaction; thus an appropriate label is indispensable. A classic example is based on an enzyme-labeled secondary antibody. However, the association constant for small analyte–antibody complexes may be as high as 10^{10} to 10^{12} M^{-1}, and creating a further increase is almost impossible. There is, because of sterical reasons, usually a ratio of 1:1 for the enzyme and the secondary antibody used in the traditional enzyme immunoassay. In contrast, bionanolabels usually depend on the use of direct nanoparticle–ligand conjugates, wherein affinity ligands (e.g., antibodies) are covalently attached to the nanoparticle surface. Such a conjugation strategy often requires time-consuming optimization processes to maximize the affinity, the ligand-to-nanoparticle ratio, and the colloidal stability of each new design.

A gold nanoparticle label is an ideal one in biotechnological systems due to its inherent advantages, such as easy preparation and good biocompatibility. The electrocatalytic properties of nanoparticle labels have been used for signal amplification in clinical immunoassays. One major merit of using nanoparticles is that the nanoparticles can provide unique chemical and physical properties to enable new and advanced functions, such as high surface-to-volume ratio and surfacefree energy, in comparison to bulk materials. The Tang group reported an ultrasensitive electrochemical immunosensor for clinical immunoassays using thionine-doped magnetic gold nanospheres as labels and HRP as the enhancer [13]. The high surface-to-volume ratio of bionanospheres might greatly enhance the immobilization density of HRP–anti-CEA bound. Meanwhile, thionine doped into the bionanospheres acts as a good mediator, which could effectively shuttle electrons from the base electrode surface to the redox center of HRP. The bioelectrocatalytic reaction of the immobilized HRP could amplify the amperometric signal output. Compared with classical enzyme-linked immunosorbent assays (ELISAs), the double-codified, label-based immunoassay exhibited better analytical properties. Following that, an improved work was developed on the basis of an in situ amplified electrochemical immunoassay for CEA using HRP-encapsulated nanogold hollow microspheres (GHSs) as labels [5]. In this design, thousands of HRP molecules were entrapped into the nanoGHS. When one antibody molecule bound to the HRP–GHS surface was reacted with the corresponding antigen (analyte), the carried HRP molecules exhibited higher catalytic efficiency relative to the H_2O_2 system than that of directly using a HRP-labeled secondary antibody. Meanwhile, the high surface-to-volume ratio of nanoGHSs greatly enhanced the immobilization amount of HRP and anti-CEA.

Nanotechnology is multidisciplinary and interdisciplinary and covers diverse fields, including chemistry, physics, materials science, engineering, biology, and even medicine. It provides excitingly new possibilities for advanced development of new analytical methods and instruments for bioanalytical and biotechnological applications. Currently, a vast library of nanostructures has been synthesized and documented, with a wide variety of properties and applications. The nanolabels mainly consisted of nanogold, nanosilica, semiconductor nanoparticles, carbon nanotubes (CNTs), etc. [14]. In the past, we also synthesized magnetic nanogold microspheres, nanoGHSs, and enzyme-doped silica nanoparticles for labels of biomolecules [15]. When one antibody on the nanoparticles reacted with

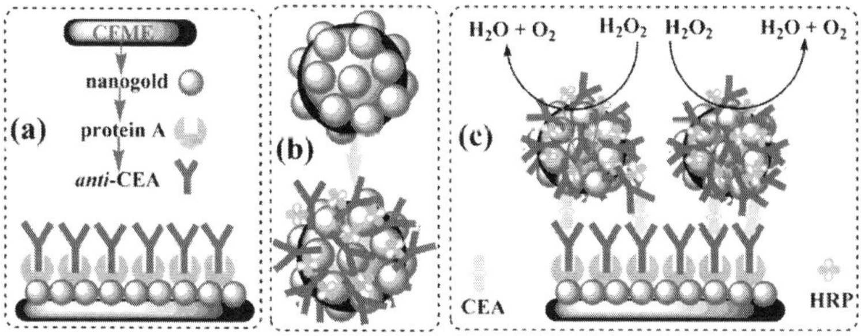

Figure 8.3 (a) Schematic representation of the preparation of an immunosensing layer, (b) nanosphere (top) and bionanosphere (bottom), and (c) schematic view of sandwich-type electrochemical detection of CEA. From Ref. [15] with permission.

the corresponding antigen, other HRP-labeled antibodies on the nanostructures would be carried over and thus participated in the electrochemical reaction. In this case, the nanocarriers and enzyme molecules would exhibit higher catalytic efficiency relative to the substrate system than those of the enzyme-labeled antibody (Fig. 8.3).

8.2.3 *Electroenzymatic Labeling for Impedimetric Immunosensors*

Electrochemical impedance spectroscopy (EIS) analysis is an effective tool used to detect biological characteristics of the biomolecule-modified electrode interface. Frequency dependence of the impedance of the electrode double layer yields useful information about the adsorption kinetics, and the dynamics of charge transfer at electrode interface are strongly influenced by the nature of the electrode surface and the structure of the electrical double layer. The adsorption or desorption of insulating materials on conductive supports is anticipated to alter the interfacial electron transfer features (capacitance and resistance) at the electrode surface. EIS contains electrolyte resistance (R_s), lipid bilayer capacitance (C_{dl}), charge transfer resistance (R_{et}) and the Warburg element (Z_w). The complex impedance can be presented as the sum of the real, Z_{re}, and imaginary, Z_{im}, components that originate mainly from the resistance and capacitance of the cell. The two components of the scheme, R_s and Z_w, represent bulk properties of the electrolyte solution and diffusion of the applied redox probe in solution, respectively. Thus, they are not affected by chemical transformations occurring at the electrode interface. The other two components of the circuit, C_{dl} and R_{et}, depend on the dielectric and insulating features at the electrode/electrolyte interface. In EIS, the semicircle diameter of EIS equals the electron transfer resistance, R_{et}. This resistance controls the electron transfer kinetics of the redox probe at the electrode interface. Its value varies when different substances are adsorbed onto the electrode surface.

In electrochemical immunosensors, the redox couple $Fe(CN)_6{}^{4-}/{}^{3-}$ often serves as a probe for the insulating properties and density of the adsorbed layer. In the presence of $Fe(CN)_6{}^{4-}/{}^{3-}$, electron transfer is observed and faradic impedance is measured. The formation of an antigen–antibody complex will change the electrochemical impedance because the electrode is coated with a blocking layer. When antigens bind to the surface-immobilized antibodies, access to the redox couple is hindered to a higher degree than in the absence of antigens. As the faradic reaction of a redox couple becomes increasingly hindered, the electron transfer resistance will increase and the capacitance will decrease accordingly. Dijksma et al. developed self-assembled monolayers of cysteine or acetylcysteine formed on electropolished polycrystalline Au electrodes, which structured an electrochemical immunosensor for direct detection of interferon [16].

Certainly a bioactive enzyme can be also used for detection of biomolecules in the impedimetric immunosensor. Vermeeren et al. designed an impedimetric, diamond-based immunosensor for the detection of C-reaction protein on the basis of an ELISA reference method [17]. The Rusling group compared several immunosensor-like methodologies for sensitive detection of antibodies to a peptide sequence from the major peanut allergen *Arachis hypogaea* 2 (Ara h2) [18]. The sensors feature a synthetic peptide layer of the major IgE-binding epitope from Ara h2 attached to a dense gold nanoparticle film on a pyrolytic graphite (PG) electrode. The gold nanoparticle–peptide sensor was used to determine model chicken antipeanut antibodies (IgY) in serum. Faradic and nonfaradic impedance strategies were compared to amperometric detection. Measurements employed goat antichicken secondary antibodies (Ab_2) labeled with HRP to bind to IgY on the sensor and provide amplified signals. The best impedimetric sensor configuration featured HPR-catalyzed precipitation of the enzyme product onto the sensor measured by nonfaradic impedance. The HRP-labeled secondary antibodies were also employed for construction of an impedance immunosensor based on a receptor protein adsorbed directly on a porous gold film [19]. Using a sandwich-type immunoassay format, the conjugated HRP-labeled secondary antibodies on the gold film could catalyze 3,3′-diaminobenzidine tetrahydrochloride dihydrate (DAB) in the solution. For the biocatalytic precipitation of an insoluble product on the electrode surface, which resulted in a remarkable R_{ct} increase, the impedance signals could be amplified by precipitation of an insoluble product on the electrode.

8.2.4 *Electroenzymatic Labeling for Conductometric Immunosensors*

Conductometric sensors for biosensing devices, which consist of a planar glass support with interdigitated gold electrode pairs on one surface in a planar configuration, have been introduced by Watson et al. [20]. The principle of detection is based on changes of the electrical resistance between two parallel electrodes by many biochemical reactions in solution. Conductometric enzyme immunosensors can detect products of enzymatic reactions due to increasing conductivity of the enzyme membrane. When enzyme-labeled antibodies were immobilized on the electrode and conjugated with antigens in the sample solution, the antigen–antibody complex coating on the surface of the electrode inhibited the biocatalytic efficiency of the immobilized enzyme and the conductivity of the supporting electrolyte was changed. The antigen–antibody reaction can change the enzyme activity and hence the immunosensor signal due to the hindrance of access to a substrate or the electron transduction between the electrode and the enzyme active site. Furthermore, conductometric immunosensors have a number of merits, such as suitability for miniaturization and large-scale production, without a reference electrode and with a low driving voltage. In addition, the conductometric detection mode has a major advantage, that is, a large number of enzymatic reactions involve either consumption or production of the charged species and, therefore, lead to a change in the ionic composition of the reacting solution.

Simple conductivity sensors are constructed of an insulating material embedded with platinum, graphite, stainless steel, or other metallic pieces. These metal contacts serve as sensing elements and are placed apart at a fixed distance to make contact with a solution whose conductivity is to be determined. The detection principle of a conductometric enzyme immunoassay is based on the change in conductivity between two parallel electrodes through detecting the products of enzymatic reactions due to increasing conductivity of the enzyme membrane. The Liu group developed a microcomb electrode-based conductometric immunosensor for the detection of aflatoxin B_1 using nanogold particles as an immobilized matrix, and the detection was based on the immobilized HRP as a trace and H_2O_2 and KI as enzyme substrates [21]. In this configuration, the active site of the immobilized enzyme was shielded and the access of target molecules to the enzyme was either partially or completely blocked after the immobilized antibodies interacted with the antigens to form an immunocomplex during the incubation. An enzyme immunobioassay format was employed to detect the analyte in solution, since most enzymatic reactions were accompanied by changes in the free iodine concentration when KI was used as a supporting electrolyte. Thus, the formation of the antibody–antigen complex introduced a local change of conductivity inside the membrane. Recently, our group also developed a simple and sensitive conductometric immunosensor for detection of alpha-fetoprotein using carbon nanoparticles as labels [22]. With a sandwich-type immunoassay format, the carbon nanosphere-conjugated HRP-labeled secondary antibodies on the transducer were increased with the increase of the target analyte in the sample, and the conductivity of the immunosensor was decreased in the H_2O_2–KI system. Combination of carbon nanospheres with enzyme labels enhanced the sensitivity of the conductometric immunosensors.

8.3 Electroenzymatic Labeling for DNA Sensors

DNA detection is an important research area in modern life sciences and is relevant for applications ranging from drug discovery, rapid pathogen detection, and single-nucleotide polymorphism detection to the assessment of water and food quality. Various techniques, including enzymatic, fluorescent, electrochemical, and surface plasmon resonance spectroscopy and quartz crystal microbalance, have been proposed for DNA hybridization detection. DNA hybridization biosensors, also known as genosensors, are analytical devices for the determination of specific DNA target sequences in solution upon hybridization of the targets with complementary probes immobilized on a solid substrate. DNA electrochemical sensors hold great promise to serve as devices suitable for point-of-care diagnostics and multiplexed platforms for fast, simple, and inexpensive nucleic acid analysis. This system usually relies on the immobilization of a single-stranded DNA (ssDNA) probe onto a surface to recognize its complementary DNA target sequence by hybridization. An electrical signal is produced when target DNA binds to the complementary sequence of the capture or probe DNA in a process called hybridization. Although a lot of progress has been made in the past few years, the analytical properties of DNA sensors in complex biological samples has been assayed in only a small fraction of published research articles. Usually, the electrochemical signal derives from the electroactive indicator that binds preferentially to the DNA duplexes instead of ssDNA probes [23]. However, the signal is always limited. Enzyme labels have long been used in electrochemical immunosensors as a means of amplifying the electrochemical signal for bioaffinity interactions. Similarly, enzyme labels can be applied to electrochemical DNA sensors for hybridization detection by labeling the target DNA sequence (simple assay) or reporter DNA probe (sandwich assay) with a redox-active enzyme [24]. The hybridization is indicated by the enzyme-catalyzed electro-oxidation/electroreduction of a substrate to an electrochemically detectable product. The enzymatic amplification of the binding event allows measurement down to 3000 copies or zmols of target DNA. Nanoparticle labels such as colloidal gold have been used for quantitate binding [25].

8.3.1 *Electroenzymatic Labeling for Potentiometric DNA Sensors*

In genetic molecular diagnostics there is a large need for instruments for the detection of different DNA variants and mutations for human genetic diseases. Current methods are usually based on fluorescently labeled optical systems. They are often expensive, time consuming, and not always suitable for extensive multiplex analyses. Potentiometric DNA sensors are essentially passive electrochemical devices in which changes in the electromotive force are monitored under virtually zero current conditions. Potentiometric ion sensors or ISEs are an important subgroup of electrochemical sensors. ISEs based on polymeric membranes containing neutral or charged carriers (ionophores) are available for the determination of a large number of inorganic and organic ions. Such devices can be applied for bioelectronic detection of DNA hybridization, due to the low cost and simplicity of potentiometric ISEs and the inherent miniaturization and protability of the supporting instrumentation. Bakker et al. demonstrated for the first time the use the potentiometric microsensors for monitoring DNA hybridization [26]. The new potentiometric nucleic acid measurements rely on sandwich DNA hybridization for capturing a secondary oligonucleotide bearing CdS nanocrystal tags. The high sensitive of the Cd^{2+}-selective microelectrodes can reach a femtomolar detection limit.

Enzyme labels such as ALP or HRP have been widely used for bioelectronic detection of DNA hybridization. The Wang group has described the use of an ISE transducer for highly sensitive potentiometric detection of enzyme-linked DNA hybridization connected to biocatalytic metallization [27]. The assay uses a low-volume solid-contact silver ISE to monitor the depletion of silver ions induced by the biocatalytic reaction of the ALP enzyme tag (Fig. 8.4). The resulting potentiometric hybridization assay relies on a thiolated DNA probe, which is confined to the gold surface. Then the capture probe with the target/biotinylated detection–probe conjugate results in a sandwich structure. Subsequently, ALP binding and addition of the *p*-aminophenylphosphate (pAPP) monosodium salt hydrate substrate leads to the biocatalytic generation of *p*-aminophenol (pAP). The new potentiometric protocol is comparable

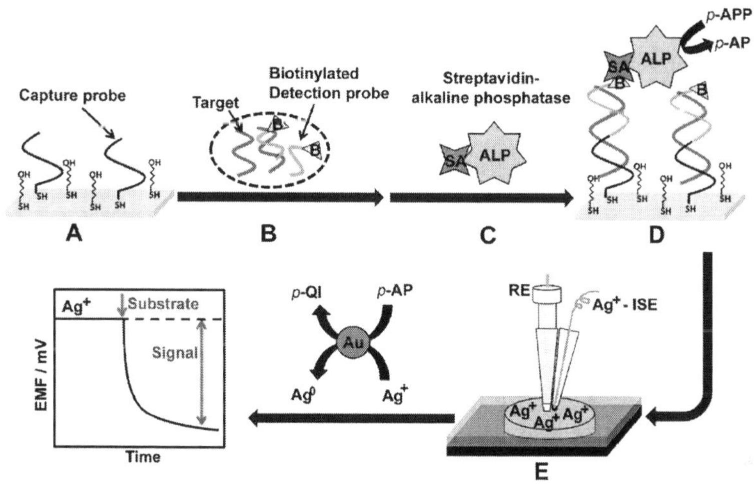

Figure 8.4 Representation of the potentiometric detection of DNA hybridization. (a) Formation of the mixed thiol monolayer (thiolated DNA capture probe and mercaptohexanol (MCH)) on the gold substrate (b) hybridization of the target DNA/biotinylated detection-probe mixture with the surface capture probe (c) binding of the streptavidin-labeled alkaline phosphatase (SA-ALP) enzyme (d) addition of the ALP substrate to initiate the enzymatic reaction, and (e) potentiometric detection of changes in the level of silver ions upon adding an aliquot of the enzymatic reaction mixture to the Ag^+-ion-selective electrode (ISE) cell. From Ref. [27] with permission.

with advanced fluorescence-based enzyme-linked DNA assays that yield detection limits in the picomolar to femtomolar ranges.

8.3.2 *Electroenzymatic Labeling for Amperometric DNA Sensors*

Amperometric DNA sensors with high sensitivity have become one of the most widely used analytical tools in biochemical analyses. Usually, amperometric DNA sensing is performed by using the labeling method due to the irreversible redox behavior of the nucleobases and the use of high overpotentials, resulting in significant background currents and limited sensitivity. Signal amplification of the amperometric detection of DNA can be obtained by labeling the target DNA sequence or reporter DNA probe with a redox-active enzyme, which catalyzes the electro-oxidation/electroreduction of a substrate to an electrochemically detectable product. This mode was firstly demonstrated by the Heller group [28]. In the system, bonding of biotinylated HRP to avidin covalently attached to an electron-conducting redox hydrogel resulted in the "wiring" of the enzyme, that is, its electrical connection to the electrode. The currents flow as a result of continuous electroreduction of H_2O_2, electrocatalyzed by the HRP label of an oligonucleotide strand when the complementary strand is covalently bound to a hydrogel that electrically "wires" the HRP. Just as the merit of electroenzymes, a one-step enzyme-amplified amperometric sandwich hybridization test for RNA and DNA was designed by the same group [29]. Further, the enzyme-amplified amperometric immunosensor was used for detection of 3000 copies of DNA in a 10 μL droplet at 0.5 fM concentration [30].

Neugebauer et al. used biotinylated target DNA to subsequently attach an avidin–ALP enzyme label at electrode locations where only double-stranded DNA (dsDNA) was present [31]. Enzymatic conversion of *p*-aminophenol 2-phosphate sesquimagnesium salt hydrate (pAPP) by the ALP supplied reduced *p*-aminophenol (pAP). DNA hybridization detection was accomplished by oxidizing pAP to *p*-quinone imine (pQI). Bionanotechnology, the emerging research field of manipulating matter at the molecular or atomic level, has provided excitingly new possibilities for successful development of new labels for signal amplification of amperometric DNA detection. The Ju group developed a novel biosensing strategy for selective electrochemical detection of DNA by using streptavidin–HRP-functionalized CNTs as labels for signal amplification [32]. In the paper, a multiwalled CNT was used as a carrier for loading numerous enzyme tags and accelerating electron transfer after an enzymatic catalytic reaction of HRP. This novel method can detect target DNA down to the attomolar level with a linear range of 5 orders of magnitude.

8.3.3 *Electroenzymatic Labeling for Impedimetric DNA Sensors*

EIS is the electrochemical technique where the electrode impedance is monitored as a function of the frequency of an applied alternative voltage [33]. Besides, EIS is a rapidly developing electrochemical technique for the characterization of biomaterial-functionalized electrodes and biocatalytic transformations at electrode surfaces, specifically for the transduction of biosensing events at electrodes or field-effect transistor devices [34]. However, EIS has been utilized for DNA detection according to the change of electron transfer on the surface of electrodes [35], while in recent years, it is slight interesting for the investigation of enzyme-labeled impedimetric DNA sensors, which achieve a strong signal by the enzyme. Fausto designed a disposable oligonucleotide-modified screen-printed gold electrode used for enzyme-based impedimetric detection of a biotinylated target sequence [36]. The thiol-tethered probes were immobilized on the gold electrode, and then sandwith hybridization with the surface-tethered probe and a biotinylated signaling probe was done under the addition of the biotinylated target and unmodified polymerase chain reaction (PCR) products complemented with a surface probe and a biotintlated probe. The biotinylated hybrid was then coupled with a streptavidin–ALP conjugate and finally exposed to the substrate solution. Electron transfer resistance was finally achieved by the enzyme-catalyzed precipitation of an insulating product, blocking the electrical communication between the gold surface and the $Fe(CN)_6^{3-/4-}$ redox probe. Therefore, electronic transfer resistance increased with the increasing of the biotinylated target. Recently, the Kaatz group reported another impedimetric DNA sensor on a gold electrode with the chain hybrid format by coupling with the bioelectrocatalytic process of ALP [37]. The probe ssDNA immobilized on the gold electrode initially captured target ssDNA-labeled biotin, and then streptavidin-labeled ALP was employed by interaction between biotin and streptavidin. The additional soluble substrates were converted to insoluble dye products, which precipitate and cause blocking of the electrode surface, by an enzymatic reaction. Certainly, the electronic transfer resistance increased with the increase of dye products, and then a strong signal on faradic EIS could be obtained. The above two articles investigate the target DNA quantitatively based ALP-catalyzed precipitation of an insulating product on the electrode surface blocking the electron transfer.

8.3.4 *Electroenzymatic Labeling for Conductometric DNA Sensors*

The conductometric measuring method can be used in enzyme catalysis to determine substance concentration and enzyme activity, selectivity in this case being provided just by the enzymes that catalyze only defined reactions. The behavior of DNA attached onto metallic and nonmetcallic surfaces via self-assembly with various chemistries (e.g., Au-S) may have applications in biomedical devices. For example, ssDNA self-assembled on a metallic interface such as gold or nonmetals such as CNTs and diamond has potential of use in DNA microarrays. Previously, faradic EIS has been used to detect the hybridization of target DNA strands with DNA probes tethered onto diamond and silicon surfaces [38]. Label-free electrochemical measurement of hybridization can induce changes in capacitance or conductivity at the transducer surface. Rivera-Gandia et al. constructed electrochemical capacitance DNA sensors at a hairpin-modified gold electrode [39]. The double-layer capacitance (C_{dl}) was determined with electrochemical perturbations from 0.2 V to 0.5 V versus Ag/AgCl. The results demonstrated that nonfaradic EIS can be used to monitor the conformational changes of ssDNA under potential perturbation at the electrode–liquid interface. Moreover, the double-layer capacitance behavior of the attached DNA strands on the gold surface can be precisely measured by using the change in C_{dl} in EIS.

Recently, a good variety of electrochemical DNA biosensors have been developed by using nanoparticles [40] and enzymes [41] as reporters for the transduction and amplification of DNA hybridization events. Weizmann et al. reported a conductivity-based DNA detection method utilizing CNT–DNA nanowire devices and oligonucleotide-functionalized enzyme probes [41]. The sensing surface consists of ssDNA bridging a gap between two singlewalled carbon nanotubes (SWCNTs) via covalent attachment at their termini, leading to the formation of a network of ssDNAlinked CNT wires fixed between two gold electrodes. In the presence of the ss-DNA analyte, selective binding occurs at the ssDNA junction between contiguous nanotubes resulting in a dsDNA assembly (Fig. 8.5). The ssDNA analyte has adjacent recognition sequences that are

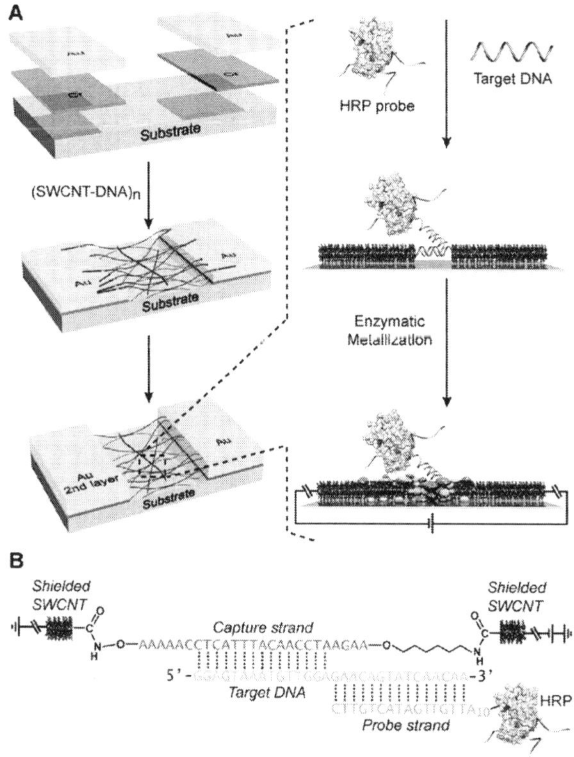

Figure 8.5 Sensing strategy and device assay. (a) Device assembly process and sensing scheme. (b) DNA–CNT nanowire and HRP probe hybridization for DNA junction visualization. From Ref. [41] with permission.

complementary to the nanotubebridging capture strand at one end and to an oligonucleotide-functionalized enzyme (HRP) probe at the other end. When the device is immersed in a solution consisting of the probe and the analyte, the enzyme probe will hybridize with the analyte recognition domain and occupy the sensing gap between contiguous nanotubes. Exposure of the device to the appropriate reagents initiates enzymatic metallization, resulting in conductive connection between the interrupted nanotube wires upon analyte detection and provides significant signal amplification. A highlight of this work is a DNA-linked CNT wire motif, which forms a network of interrupted CNT wires connecting two electrodes. Sensing occurs at the DNA junctions linking CNTs, followed by amplification using enzymatic metallization leading to a conductimetric response. The DNA analyte detection limit is 10 fM with the ability to discriminate single, double, and triple base pair mismatches. This method represents a straightforward approach to highly sensitive and selective detection of oligonucleotide analytes.

8.4 Conclusion

Inherent sensitivity, simplicity, speed, and cost benefits continue to be strong driving forces for the development of electrochemical immunosensors and DNA sensors. Despite historic achievements in the field of label-free bioassays, labeling techniques will continue to play a leading role in this field. Enzyme labels offer very elegant ways of interfacing biomolecule recognition events with inherent signal amplification. Coupling enzyme labels and ultrasensitive electrochemical detection is one of the most exciting and challenging aspects of this field. Success will play a vital role in advancing numerous scientific disciplines, including biomedicine, biology, chemistry, environmental science, toxicology, and materials science. To achieve a better understanding of electroenzymatic labeling, great efforts will need to be made worldwide to design and develop new strategies that improve the properties of electroenzymatic labeling. In addition, developing uses for a variety of nanomaterials coupled with enzyme labels and micro-/nanofluidic devices will offer advanced miniaturized, high-throughput, and cost-effective multiplex assays for tagging of a wide variety of important chemical and biological targets in a single assay.

References

1. Ronkainen, N. J., Halsall, H. B., Heineman, W. (2010) Electrochemical biosensor, *Chem. Soc. Rev.*, **39**, 1747–1763.
2. Palchetti, I., Mascini, M. (2012) Electrochemical nanomaterial-based nucleic acid aptasensors, *Anal. Bioanal. Chem.*, **402**, 3103–3114.
3. Dahlin, A., Bielacher, B., Rajendran, P., et al. (2012) Electrochemical plasmonic sensors, *Anal. Bioanal. Chem.*, **402**, 1773–1784.

4. Tang, J., Tang, D., Niessner, R., Chen, G., Knopp, D. (2011) Magneto-controlled graphene immunosensing platform for simultaneous multiplexed electrochemical immunoassay using distinguishable signal tags, *Anal. Chem.*, **83**, 5407–5414.
5. Tang, D., Ren, J. (2008) In situ amplified electrochemical immunoassay for carcinoembryonic antigen using horseradish peroxidase-encapsulated nanogold hollow microspheres as labels, *Anal. Chem.*, **80**, 8064–8070.
6. Drummond, T., Hill, M., Barton, J. (2003) Electrochemical DNA sensor, *Nat. Biotechnol.* **21**, 1192–1199.
7. Tang, D., Yuan, R., Chai, Y., et al. (2005) Preparation and application on a kind of immobilization method of anti-diphtheria for potentiometric immunosensor modified colloidal Au and polyvinyl butyral as matrixes, *Sens. Actuators, B*, **104**, 199–206.
8. Yuan, R., Tang, D., Chai, Y., Zhong, X., Liu, Y., Dai, J. (2004) Ultrasensitive potentiometric immunosensor based on SA and OCA techniques for immobilization of HBsAb with colloidal Au and polyvinyl butyral as matrixes, *Langmuir*, **20**, 7240–7245.
9. Brown, D. V. (1991) Potentiometric enzyme channeling immunosensor for proteins, *Biosens. Bioelectron.*, **6**, 615–622.
10. Ghindilis, A. L., Atanasov, P., Wilkins, E. (1996) Potentiometric immunoelectrode for fast assay based on direct electron transfer catalyzed by peroxidase, *Sens. Actuators, B*, **34**, 528–532.
11. Thompson, H. C., Lee, W. E. (1992) *Rapid Immunofiltration Assay of Francisella tularensis*. Defence Research Establishment Suffield, Canada. Suffield Memorandum No. 1376, 1–17.
12. Dai, Z., Chen, J., Yan, F., Ju, H. (2005) Electrochemical sensor for immunoassay of carcinoembryonic antigen based on thionine monolayer modified gold electrode, *Cancer Detect. Prevent.*, **29**, 233–240.
13. Tang, D., Yuan, R., Chai, Y. (2008) Ultrasensitive electrochemical immunosensor for clinical immunoassay using thionine-doped magnetic gold nanospheres as labels and horseradish peroxidase as enhancer, *Anal. Chem.*, **80**, 1582–1588.
14. Su, B., Tang, J., Chen, H., Huang, J., Chen, G., Tang, D. (2010) Thionine/nanogold multiplayer film for electrochemical immunoassay of alpha-fetoprotein in human serum using biofunctional double-codified gold nanoparticles, *Anal. Methods*, **2**, 1702–1709.
15. Tang, D., Su, B., Tang, J., Ren, J., Chen, G. (2010) Nanoparticle-based sandwich electrochemical immunoassay for carbohydrate antigen

125 with signal enhancement using enzyme-coated nanometer-sized enzyme-doped silica beads, *Anal. Chem.*, **82**, 1527–1534.

16. Dijksma, M., Kamp, B., Hoogvliet, J., van Bennekom, W. (2001) Development of an electrochemical immunosensor for direct detection of interferon-gamma at the attomolar level, *Anal. Chem.*, **73**, 901–907.
17. Vermeeren, V., Grieten, L., Vanden Bon, N., et al. (2011) Impedimetric, diamond-based immunosensor for the detection of C-reaction protein, *Sens. Actuators, B*, **157**, 130–138.
18. Liu, H., Malhotra, R., Peczuh, M., Rusling, J. (2010) Electrochemical immunosensors for antibodies to peanut allergen Ara H2 using gold nanoparticle-peptide films, *Anal. Chem.*, **82**, 5865–5871.
19. Chen, Z., Jiang, J., Shen, G., Yu, R. (2005) Impedance immunosensor based on receptor protein adsorbed directly on porous gold film, *Anal. Chim. Acta*, **553**, 190–195.
20. Watson, L. D., Maynard, P., Cullen, D. C., et al. (1987) A microelectronic conductimetric biosensor, *Biosensors*, **3**, 101–105.
21. Liu, H. M., Yang, Y. D., Chen, P., Zhong, Z. Y. (2009) Enhanced conductometric immunoassay for hepatitis B surface antigen using double-codified nanogold particles as labels, *Biochem. Eng. J.*, **49**, 107–112.
22. Tang, J., Huang, J., Su. B., Chen. H., Tang, D. (2011) Sanwich-type conductometric immunoassay of alpha-fetoprotein in human serum using carbon nanoparticles as label, *Biochem. Eng. J.*, **53**, 223–228.
23. Takenaka, S., Yamashita, K., Takagi, M., Uto, Y., Kondo, H. (2000) DNA sensing on a DNA probe-modified electrode using ferrocenylnaphthalene diimide as the electrochemically active ligand, *Anal. Chem.*, **72**, 1334–1341.
24. Zhang, Y., Pothukuchy, A., Shin, W., Kim, H., Heller, A. (2004) Detection of approximately 10^3 copies of DNA by an electrochemical enzyme-amplified sandwich assay with ambient O_2 as the substrate, *Anal. Chem.*, **76**, 4093–4097.
25. Wang, J. (2006) *Analytical Electrochemistry* (John Wiley and Sons VCH, Hoboken, New Jersey).
26. Numnuam, A., Chumbimuni-Torres, K. Y., Xiang, Y., et al. (2007) Potentiomtric detection of DNA hybridization, *J. Am. Chem. Soc.*, **130**, 410–411.
27. Wu. J., Chumbimuni-Torres, K. Y., Galik, M., Thammakhet, C., Haake, D. A., Wang, J. (2009) Potentilmetric detection of DNA hybridization using enzyme-induced metallization and a silver ion selective electrode, *Anal. Chem.*, **81**, 10007–10012.

28. Lumley-Woodyear, T., Campbell, C. N., Heller, A. (1996) Direct enzyme-amplified electrical recognition of a 30-base model oligonucleotide, *Anal. Chem.*, **118**, 5504–5505.

29. Campbell, C., Gai, D., Cristler, N., Banditrat, C., Heller, A. (2002) Enzyme-amplified amperometric sandwich test for RNA and DNA, *Anal. Chem.*, **74**, 158–162.

30. Zhang, Y., Kim, H., Heller, A. (2003) Enzyme-amplified amperometric detection of 3000 copies of DNA in a 10-microL droplet at 0.5 fM concentration, *Anal. Chem.*, **75**, 3267–3269.

31. Neugebauer, S., Zimdars, A., Liepold, P., Gebala, M., Schuhmann, W., Hartwich, G. (2009) Optimization of an electrochemical DNA assay by using a 48-electrode array and redox amplification studies by means of scanning electrochemical microscopy, *Chem. Bio. Chem*, **10**, 1193–1199.

32. Cao, C., Dong, H., Lei, J. Ji, H., Ju, H. (2011) Signal amplification of streptavidin-horseradish peroxidase functionalized carbon nanotubes for amperometric detection of attomolar DNA, *Chem. Commun.*, **47**, 5220–5222.

33. Bard, A., Faulkner, L. (2001) *Electrochemical Methods: Fundamentals and Applications*, John Wiley & Sons, INC., 2nd Edition.

34. Katz, E., Willner, I. (2003) Probing biomolecular interactions at conductive and semiconductive surfaces by impedance spectroscopy: routes to impedimetric immunosensors, DNA-sensors, and enzyme biosensors, *Electroanalysis*, **15**, 913–947.

35. Tang, J., Tang, D., Zhou, J., Yang, H., Chen, G. (2012) Nuclease cleavage-assisted target recycling for signal amplification of free-label impedimetric aptasensors, *Chem. Commun.*, **48**, 2627–2629.

36. Lucarelli, F., Marrazza, G., Mascini, M. (2005) Enzyme-based impedimetric detection of PCR products using oligonucleotide-modified screen-printed gold electrodes, *Biosens. Bioelectron.*, **20**, 2001–2009.

37. Kaatz, M., Schulze, H., Ciani, I., Lisdat, F., Mount, A., Bachmann, T. (2012) Alkaline phosphatase enzymatic signal amplification for fast, sensitive impedimetric DNA detection, *Analyst*, **137**, 59–63.

38. Yang, W., Butler, J., Russell, J., Jr., Hamers, R. (2004) Interfacial electrical properties of DNA-modified diamond thin films: intrinsic response and hybridization-induced field effects, *Langmuir*, **20**, 6778–6787.

39. Rivera-Gandia, J., Maldonado, M., Torre-Melendez, Y., et al. (2011) Electrochemical capacitance DNA sensing at hairpin-modified Au electrode, *J. Sensor*, **2011**, article ID 735279.

40. Du, P., Li, H., Cao, W. (2009) Construction of DNA sandwich electrochemical biosensor with nanoPbS and nanoAu tags on magnetic microbeads, *Biosens. Bioelectron.*, **24**, 3223–3228.
41. Weizmann, Y., Chenoweth, D., Swager, T. (2011) DNA-CNT nanowire networks for DNA detection, *J. Am. Chem. Soc.*, **133**, 3238–3241.

Chapter 9

Conductometric Enzyme Biosensors

Sergei Dzyadevych[a] and Nicole Jaffrezic-Renault[b]

[a] *Laboratory of Biomolecular Electronics, Institute of Molecular Biology and Genetics, National Academy of Sciences of Ukraine, Kiev 03143, Ukraine*

[b] *Institute of Analytical Sciences, Claude Bernard University Lyon 1, 69622 Villeurbanne Cedex, France*

dzyad@yahoo.com, nicole.jaffrezic@univ-lyon1.fr

Biosensors could be useful tools to compete with often tedious, complex, and expensive standard methods. Conductometric enzyme biosensors have important advantages as they do not need the use of a reference electrode; they operate at low-amplitude alternating voltages, thus preventing Faraday processes on electrodes; they are insensitive to light; and they can be miniaturized and integrated easily by using cheap standard thin-film technology. In this chapter some examples of enzyme conductometric biosensors with their advantages and disadvantages and perspectives of their application for different human domains such as biomedicine, environmental monitoring, agriculture, and biotechnology have been described and summarized.

Electrochemical Biosensors
Edited by Serge Cosnier

ISBN 978-981-4411-46-2 (Hardcover), 978-981-4411-47-9 (eBook)
www.panstanford.com

9.1 Introduction

The requirements and regulations in the fields of environmental protection, control of biotechnological processes, and certification of food and water quality are becoming more and more drastic. At the same time stricter requirements regarding human and animal health have led to a rising number of clinical and veterinary tests. This means that highly sensitive, fast, and commercial methods of analysis need to be developed. Analytical devices of the new generation—biosensors—can be useful tools to compete with often tedious, complex, and expensive standard methods [1–3].

A biosensor converts the modification of the physical or chemical properties of a biomatrix, which occurs as a result of biochemical interactions, into an electric or an optic signal whose amplitude depends on the concentration of defined analytes in the measuring medium. Functionally, the device consists of two parts, a biomatrix, that is, a detecting layer of immobilized material (enzymes [4, 5], antibodies [6, 7], receptors [8], organelles [9], microorganisms [10, 11]), and a transducer (potentiometric [12, 13], impedimetric [14, 15], amperometric [4, 16], conductometric [17, 18], acoustic [19], optic [20, 21], or colorimetric [22, 23]. Numerous reviews [24–26], books [27–29], and a lot of experimental research concerning various types of biosensors have been published. However, only some works are devoted to the development of conductometric biosensors, including those used for chromatography and chemical sensors for the determination of air moisture and the concentration of certain gases [30–32].

Conductometric biosensors have important advantages as they do not need the use of a reference electrode; they operate at low-amplitude alternating voltages, thus preventing Faraday processes on electrodes; they are insensitive to light; and they can be miniaturized and integrated easily by using cheap standard thin-film technology.

The liquids analyzed are mostly considered to have significant background conductivity, which is easily modified by different factors; therefore the selectivity of this method is presumed to be low and consequently its potential use for different applications rather doubtful [33]. However, in the case of an integral microbiosensor, most of these difficulties can be overcome using a differential measuring scheme, which compensates for changes in background conductivity, the influence of temperature variations, and other factors [34, 35].

9.2 Conductometry in Enzyme Catalysis

The conductometric measuring method can be used in enzyme catalysis to determine substance concentration and enzyme activity, selectivity in this case being provided by the enzymes that catalyze only certain reactions. As a matter of fact, the subject under consideration is not a biosensor as such but an application of this method in enzymology.

In 1961 one of the first studies in this field was published showing how it might be possible to determine urea concentration in solutions [36]. This method is based on the difference between electric conductivity of a urea solution and that of a solution of ammonium carbonate formed as a result of urea hydrolysis by urease. In the experiments a bridge-measuring scheme was used. The urease activity was shown to decrease in the presence of heavy metal (Ag, Hg, etc.) ions in the solution. Electrolytes such as NaCl or KCl do not influence urease activity, but if their concentration in the solution is high it can lead to a wrong result, especially at low urea concentrations. At low electrolyte concentrations in experiments without a buffer solution, during urea hydrolysis the medium pH gradually changed from 7.0 to 9.0. However, this causes only an insignificant change in the urease activity, while the solution conductivity during the reaction varied substantially. The urea concentration was determined within the 0.1 μM–2 mM range, at an optimal pH of 7.0. A comparison of the conductometric method with other methods of urea analysis carried out in that work has shown that the former is characterized by high accuracy, speed, and simplicity. Besides, in contrast to optical methods, the measurement accuracy of conductometry does not depend on the solution color.

However, conductometric methods have some limitations. The ratio between the signal and noise levels should not be lower than 2%. For this reason, the concentrations of the buffer and some other ingredients, which can be added to the reaction mixture, are important. The method sensitivity is reduced in the presence of nonreacting ions in solution. Buffers with low ionic strength can be used, though, to measure low concentration until the signal-to-noise ratio is of proper value. A disadvantage of conductometry is also its low specificity—it is incapable of distinguishing between simultaneous reactions that can cause an artefact. The capacity of the double layer and the electrode polarization during the reaction can also be sources of the method error.

In conductometric enzyme biosensors, the enzymatic reaction is confined close to the interdigitated electrode (IDE) surface because the enzyme is crosslinked in contact with this surface. The IDEs allow the measurement of the change of conductivity in the region defined by field lines. The involved thickness is of the order of the interdigit distance (few tens of micrometers) [37]. As it has been modeled [38], the observed steady-state response of a conductometric enzyme biosensor is the result of the reaction rate–limited kinetics of the enzymatic reaction and the diffusive flux of urea hydrolysis products away from the transducer surface, in the boundary layer.

9.2.1 *Conductometric Enzyme Biosensors Based on Direct Analysis*

9.2.1.1 Glucose biosensors

Glucose sensors dominate undoubtedly among the biosensors developed by scientific laboratories since glucose, a vital metabolite of living organisms, is the most frequently determined in biological media [39]. Besides, glucose is widely used in the food industry and biotechnology since it is the actual source of carbon in numerous microbial fermentation processes and in the growth of cell culture. Moreover, it is the glucose sensor that has been first described [40], its development has attracted the greatest attention, and glucose together with enzyme glucose oxidase has been used in further development of novel biosensor prototypes as a model system.

The principle of operation of conductometric biosensors for glucose determination is based on the following enzymatic reaction:

Glucose oxidase

$$\begin{array}{c}\beta\text{-D-glucose} + O_2 \rightarrow \text{D-gluconolactone} + H_2O_2 \\ \Downarrow \\ \text{D-gluconic acid} + H_2O \leftrightarrows \text{Acid residue} + H^+\end{array} \quad (9.1)$$

The first conductometric biosensor for glucose determination has been described in Refs. [41, 42]. Conductometric transducers based on thin-film interdigitated metal (Au, Cr, Cu, Ni) electrodes were studied, and enzyme glucose oxidase was immobilized on the gold electrode–based chips. The influence of ionic strength and buffer capacity of the samples on the biosensors' response in kinetic and steady-state modes of measurements was thoroughly tested.

The typical calibration curves for a conductometric glucose biosensor are presented in Fig. 9.1. It is distinctly seen that the dynamic ranges of the biosensor toward glucose are quite narrow (up to 1.5–2 mM) and do not depend on the solution buffer capacity. The biosensor sensitivity depends highly on buffer concentration. The detection limit of glucose concentration was 0.01 mM.

Limitation of the enzymatic reaction by oxygen, a cosubstrate of this reaction, is the reason of a narrow dynamic range of glucose biosensors. The use of potassium ferrocyanide in biocatalytic

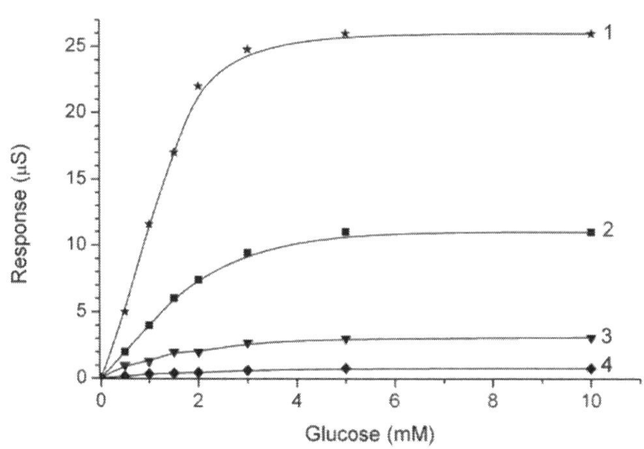

Figure 9.1 Calibration curves of conductometric glucose biosensors obtained in 2 mM (1), 5 mM (2), 10 mM (3), and 20 mM (4) phosphate buffer, pH 7.4.

glucose oxidation as the oxidizing agent results in a higher level of solution acidity inside the enzymatic layer because three protons per glucose molecule are generated compared to one proton per molecule in the case of the natural oxidizing cosubstrate, oxygen. Depending on the potassium ferrocyanide concentration, a 10–100 times higher biosensor response and a remarkably wider dynamic working range were revealed as well [43].

A further study was mainly focused on the use of different additional membranes deposited on top of the enzymatic membrane for the sensor modification. In this case the diffusion barrier is formed for membrane penetration by glucose molecules, while oxygen diffuses to the membrane easily [44]. Utilization of additional membranes allowed getting of a wider (up to 10 mM) operation dynamic range at a minor decrease in sensitivity. Similar results were received for polytetrahydrofuran membranes, facilitating the increase in the dynamic range up to 15 mM without any decrease in sensitivity. Lower sensitivity of conductometric biosensors toward changes in buffer capacity and medium ion strength was shown to be the result of usage of the additional membranes as well.

9.2.1.2 Urea biosensors

Urea determination in biological liquids is an important diagnostic test as the increase in urea concentration in blood and its decrease in urine actually evidence renal dysfunction of the organism [45].

Fundamentally, urease biosensors are based on the following enzymatic reaction:

$$(H_2N)_2C{=}O + 2H_2O + H^+ \xrightarrow{\text{Urease}} 2NH_4^+ + HCO_3^- \quad (9.2)$$

This reaction results in proton consumption as well as in the generation of additional ions (NH_4^+ and HCO_3^-); a joint effect causes changes in conductivity inside the membrane.

The first conductometric biosensor for urea determination has been described in Ref. [46]. It was a device consisting of a silicon substrate with a pair of gold interdigitated and serpentine electrodes. The experiments were carried out in both a laboratory and clinics; the biosensor response to urea was in the range of 0.1–10 mM in imidasole buffer, pH 7.5. The K_M of the immobilized enzyme was higher than that of the native one; the authors explained it as a result of diffusion limitation. A comparison of the data obtained by the biosensor in the laboratory with the results of conventional clinical tests showed good agreement (the correlation coefficient was higher than 0.99).

Similar conductometric biosensors have also been used as a multisensor [18]. Urease was immobilized on the surface of the first electrode pair in a gel layer; on the second pair there was L-asparaginase; and on the third pair, a three-enzyme system "urease–creatinase–creatininase." This sensor was used for the determination of urea, L-asparagine, and creatinine, respectively. The sensor was tested separately with each of the substrates, and in multisubstance mode the kinetic and calibration curves were determined.

The multisensor described in Ref. [47] consisted of a conductometric biosensor for urea analysis, combined with an amperometric biosensor for glucose determination. It was highly selective and simple to operate and was used in clinics.

Mikkelsen and Rechnitz [48] have characterized conductometric biosensors for urea and D-amino acids determination. The enzymes urease and D-amino acid oxidase were then used. The minimum detection limit for urea concentration was 5 μM; the linear dynamic range was of 3 orders. The dependence of the response on buffer capacity was studied. While the sensor for D-amino acid analysis was being developed, the D-amino acid oxidase was co-immobilized with catalase since hydrogen peroxide, being the product of the enzymatic reaction, is the inhibitor of D-amino acid oxidase. A comparative analysis of using copper and platinum electrodes, as well as different buffer solutions, showed that the platinum electrodes and glycine buffer were preferable. An optimal pH of the sensor for D-amino acids and its selectivity toward various amino acids were determined. The sensor showed stable results during 33-days' operation.

Bilitewski et al. [49] described two types of thick-film conductometric biosensors for urea determination. The first type was manufactured by printing two IDEs onto an Al_2O_3 substrate using

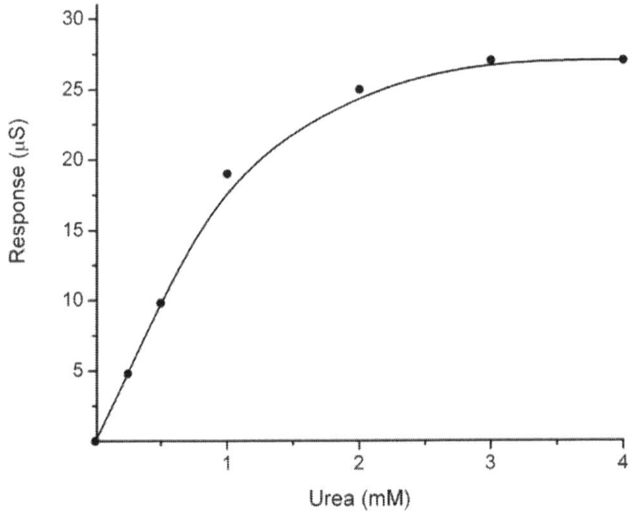

Figure 9.2 Calibration curve of conductometric urea biosensors obtained in 5 mM potassium–phosphate buffer, pH 7.4.

platinum paste, while the second one, consisting of four silver–palladium electrodes in parallel, was manufactured by the "green tape" technology. Urease was immobilized by covalent binding in albumin gel. The response time for both biosensors was about 10 minutes. The dynamic ranges for the first biosensor were 0.1–50 mM urea, and the linear part was 0.1–4 mM; for the second 10 μM–5 mM urea, and the linear part was 10–350 μM. These biosensors were shown to suit medical analysis.

A typical calibration curve of conductometric urease sensors for urea is presented in Fig. 9.2. The response value was shown to depend on medium conditions, that is, buffer capacity, pH, and ionic strength [50], which is a disadvantage since it requires taking into account these parameters in the analysis protocol.

The screen-printed thick-film IDE has proved to be an excellent conductometric transducer in which the admittance signal is dominated by the conductance signal and the resulting dynamic range is wide enough to be employed for the construction of a conductometric urea biosensor [51]. The sol–gel-derived urea biosensor exhibited a lower detection limit of 0.03 mM and a

2 orders of magnitude wide dynamic range, thus suggesting that the biosensor could be used in routine urea assays in real samples such as urine and serum. Measurements in serum were satisfactory. Although the current urea biosensor is far from completely optimized, it is comparable to other urea sensor systems, and it exhibits the advantages of a simple fabrication procedure, good sensor-to-sensor reproducibility, relatively good storage stability, a wide dynamic range, easy control of biosensor performance by changing the alkoxide–water ratio in the stock sol–gel solution in the construction of the biosensor, and no chemical modification of the substrate or enzyme.

It has been shown recently [52] that the use of gold nanoparticles for the immobilization of urease allows the increase of the sensitivity of detection (10–107 μS/mM) due to the decrease of the thickness of probed zone, the gold nanoparticles behaving as nanoelectrodes. A detection limit of 100 μM of urea is obtained when crosslinked urease is directly immobilized on top of the IDEs (interdigitated distance: 20 microns), whereas a detection limit of 2 μM is obtained when urease-functionalized gold nanoparticles are deposited on top of the IDEs.

9.2.1.3 Biosensors for arginine determination

Nowadays there is a considerable demand for the determination of L-arginine in clinical practices and the pharmaceutical industry. The levels of L-arginine, the most basic natural amino acid, are of great interest in life sciences due to the fact that L-arginine and its metabolic derivatives, such as urea, ornithine, creatine, nitric oxide, and citrulline, are involved in a wide range of biological activities, including energy metabolism, vasodilation, the urea cycle, immune response, and neurotransmission in mammals.

The conductometric biosensor for L-arginine detection is based on the arginase–urease system, in which excess of urease is added to ensure complete and prompt conversion of all the urea formed to ammoniacal nitrogen [53].

Arginase

$$\text{L-arginine} \rightarrow \text{L-ornithine} + \text{Urea} \qquad (9.3)$$

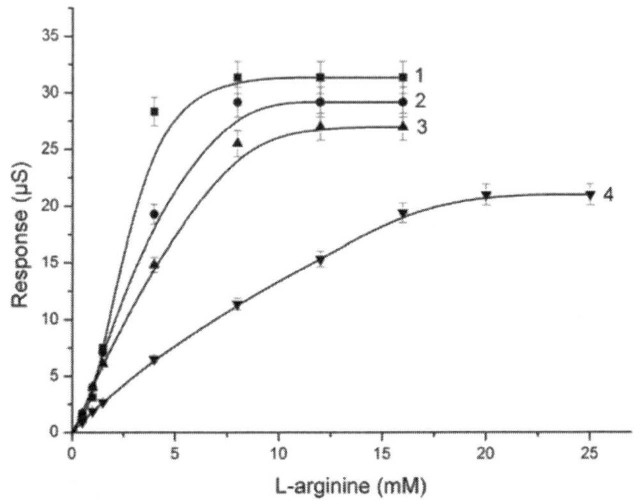

Figure 9.3 Calibration curves for L-arginine detection, obtained in 2.5 mM, (1), 5 mM, (2), 10 mM (3), and 20 mM (4) phosphate buffer, pH 6.0.

Urease

$$(NH_2)_2CO + 2H_2O + H^+ \rightarrow 2NH_4{}^+ + HCO_3{}^- \tag{9.4}$$

The dependences of the responses of L-arginine biosensors on buffer capacity were obtained in 2.5 mM, 5 mM, 10 mM, and 20 mM phosphate buffer (pH 6.0). According to the results, the sensitivity of the biosensors for L-arginine determination had an inversely proportional dependence on the concentration of phosphate buffer (Fig. 9.3).

To estimate the operational stability of an L-arginine biosensor, continuous measurements of its sensitivity were carried out for 10 hours. The relative standard deviation of the measurements was 4%–5%. The dependence "Response–Number of analysis" showed that L-arginine biosensors had high operational stability. An increase in the biosensor response observed with time might be related to the variation of the local pH due to $NH_4{}^+$ accumulation within the bioselective membrane.

9.2.1.4 Biosensor for testing protein denaturation

Proteomics technologies are a useful tool for the identification of disease biomarkers. Many cellular changes in proteins expression, as a response to an external stimulus or mutation, can be expressed directly on the proteins' sequence and structure. The amino acid sequence constituting proteins is the exact expression of the DNA sequence; a mutation on the encoding gene can induce the appearance of new protein properties or a change in its activity. One of these changes can be the appearance or disappearance of a protease cleavage site. The number of cleavage site characteristics of a known protein can be used in these cases as a mutation marker.

Conductometric microelectrodes were used [54] as a transducer to develop an enzyme biosensor based on proteinase K to quantify the number of the hydrolysis sites of different peptides and proteins. Asparatam was used for a negative test: there is no specific cleavage site for proteinase K (aspartame presents a short peptide sequence $\geq$4 amino acids). Angiotensin with only one cleavage site allows estimating the conductivity induced by the cleavage of only one site. Bovine serum albumin (BSA) with many cleavage sites allows us to show the cumulative induced conductivity as a response to the enzyme action for many specific sites.

Aspartame injection does not induce any signal variation (Fig. 9.4). This peptide is resistant because it does not have any proteinase K cleaving site. This result allows us to conclude about the efficiency of the sensor to screen the presence or absence of a cleaving site on peptide sequences. Figure 9.4 shows also the significant difference between the angiotensin and aspartame response for an equal concentration 6 mg/mL. At steady-state equilibrium, conductometric responses were 0.18 μS in the case of aspartame and 0.6 μS in the case of angiotensin. The aspartame response does not follow the shape of a kinetic response. The quick increase in the conductometric signal is related to the injection but not to proteinase K action.

The results obtained for native and denaturated BSA show the potentiality of this proteinase K conductometric biosensor to be able to detect a change in protein conformation. A sample of BSA was denaturized by heat at 100°C for 20 minutes. The monitoring of the

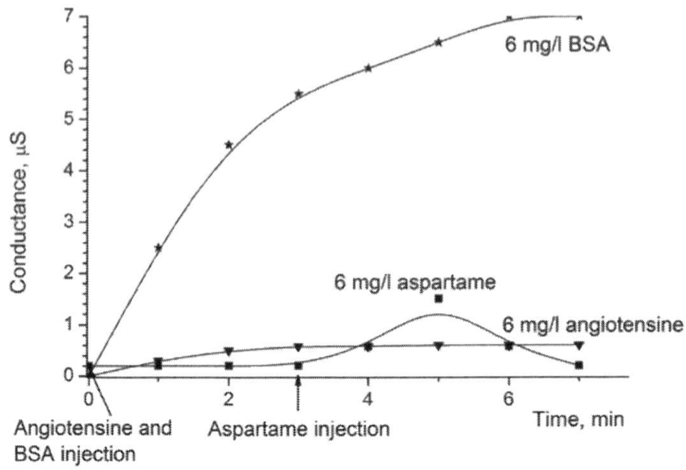

Figure 9.4 Comparison of response curves of a conductometric biosensor to aspartame and BSA additions: equal concentration of BSA and aspartam (6 mg/mL), buffer pH 7.5, and temperature 23 ± 2°C.

induced conductivity as a response to non-native BSA demonstrates that for a same concentration of native and non-native BSA the obtained conductivity is three times more important for non-native BSA (for a concentration of 8 mg/mL we obtained 8.72 μS in the case of native BSA and 28 μS in the case of non-native BSA).

Such a system could be applied for the study of some therapeutic proteins presenting the limit of their short half-time. Microencapsulation of proteins and peptides improves the therapeutic efficiency of these bioactive materials, protects them against degradation, and enhances the sustained drug release. This method can modify the fragile nature of proteins. The control of the proteins' properties constitutes an important objective. This can be controlled by conductometric biosensors; it is possible to monitor the properties and release kinetics as a function of time [55].

9.2.1.5 Biosensors for formaldehyde determination

Formaldehyde determination in food, the environment, and biological liquids is essential as well since it is one of the most widespread commercial products due to its chemical activity, technological prerequisites of high purity, and rather low cost [56]. The challenge has become stronger recently by reason of comprehension of formaldehyde high toxicity [57], mutagenicity, and carcinogenicity [58]. This has been ascertained by experiments with microorganisms (mutagenic effect), mice, and rats (cancer development) [59].

A conductometric enzyme biosensor for determination of formaldehyde in aqueous solutions has been developed using interdigitated thin-film planar electrodes, and immobilized alcohol oxidase from *Hansenula polymorpha* was presented in Ref. [60]. The enzymatic reaction was the following one:

$$CH_2O + O_2 + H_2O \xrightarrow{AOX} HCOOH + H_2O_2$$
$$HCOOH \downarrow HCOO^- + H^+ \quad (9.5)$$

The calibration curves obtained in buffer solutions of various buffer capacities (Fig. 9.5) demonstrate identical upper, but different lower, measurement limits.

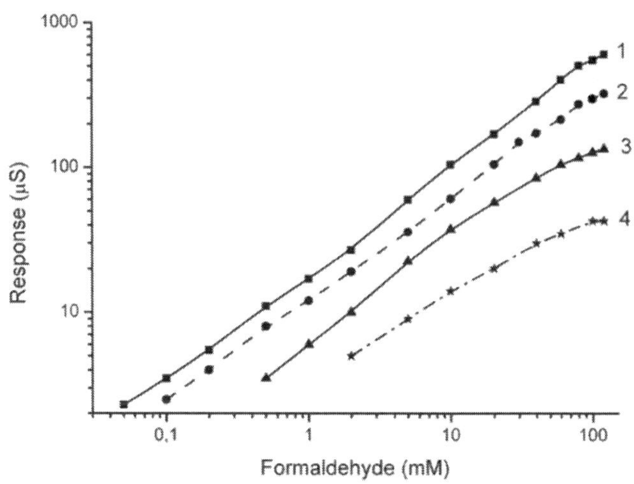

Figure 9.5 Calibration curves of a formaldehyde conductometric biosensor obtained in 1 mM (1), 2 mM (2), 5 mM (3), and 10 mM (4) phosphate buffer, pH 7.5.

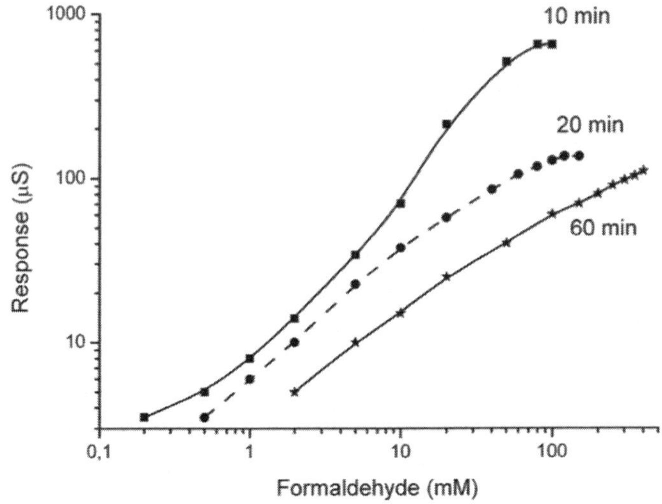

Figure 9.6 Calibration curves of a formaldehyde conductometric biosensor at various durations of alcohol oxidase immobilization in glutaraldehyde vapor. Measurements were performed in 5 mM phosphate buffer, pH 7.5.

The biosensor steady-state response was reached after about one minute. The lower measurement limit in 1 mM phosphate buffer was 0.05 mM (0.15 ppm) formaldehyde concentration, which is adequate as per Occupational Safety and Health Administration, USA (OSHA) standards [61].

For different durations of alcohol oxidase immobilization (Fig. 9.6), a shift of calibration curves toward both low and high formaldehyde concentrations was observed.

In the case of long immobilization, a decrease in the response value as well as a wider operation range for high substrate concentrations can be the result of the formation of numerous covalent bonds between glutaraldehyde and enzyme molecules, which causes partial blocking of the enzyme-active centers. Besides, such a dense membrane is capable of diminishing diffusion of the substrate and of products of the biochemical reaction and thus obtaining a wider range of measurable formaldehyde concentrations—up to 500 mM. On the other hand, in the case of short time of immobilization, the sensor demonstrated higher sensitivity and the ability to measure low concentrations.

Therefore, the linear range of alcohol oxidase–based conductometric biosensors covers a formaldehyde concentration of 0.05 mM to 500 mM and can be adjusted to actual practical needs by modifying both options of enzyme immobilization and measurement protocols.

The biosensor developed was not absolutely specific and selective. It demonstrated no response to primary alcohols and other substrates alone. Unfortunately, the response of this biosensor in a mixture of formaldehyde and methanol was decreased in comparison to the one observed for pure formaldehyde, even if no response was obtained with the interfering species alone. The operational stability was not less than 20 hours, and the relative standard deviation appeared to be about 3%. Moreover, the storage stability was more than one month.

9.2.1.6 Biosensors for proteins determination

As proteins constitute more than 30% of the total chemical oxygen demand (COD) in effluents, they seem to constitute a good indicator of organic matter content. However, the most widely used methods for protein analyses are based on classical laboratory methods based on colorimetric procedures that are not convenient for in situ monitoring of organic matter content.

Conductometric biosensors for proteins determination were developed on the basis of enzyme proteinase K [62, 63]. Proteinase K hydrolyzes proteins into different ionic amino acids, which results in local conductivity changes. In this work, the biosensor response using BSA as a standard protein was optimized.

The dependence of the steady-state responses of the proteinase K–based conductometric biosensor on the concentration of BSA is shown in Fig. 9.7.

The linear range for BSA determination was from 0.8 μg/mL to 6 μg/mL with good sensitivity. After 6 μg/mL, we have a saturation phenomenon. This range of response obtained with the biosensor is in adequacy with the values of the real concentrations of proteins in river water. Then, the response biosensor was tested with samples of river water. Good correlations between conductance changes and values given by standard methods (COD and protein

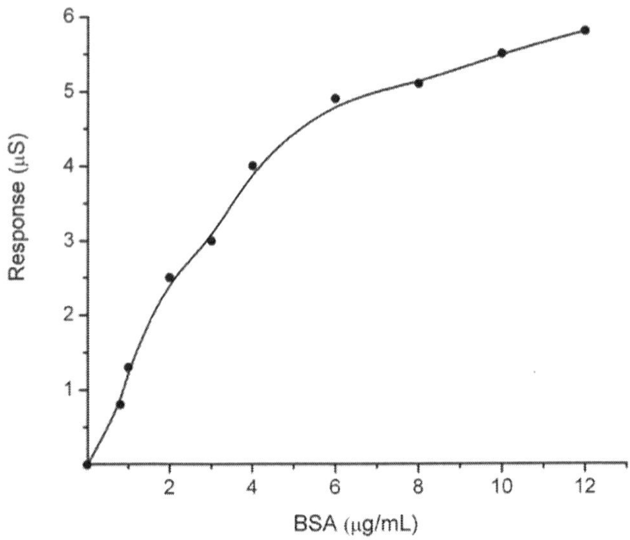

Figure 9.7 Dependency of biosensor response (μS) on BSA concentration.

concentration evaluated by a micro-BCA protein assay) have been shown. Correlation coefficients of 0.89 and 0.92 were, respectively, obtained.

In the next work [64], two enzymes, proteinase K and pronase, were used for the creation of conductometric biosensors for protein detection. The biosensor with proteinase K and pronase gives a higher signal and a response in close agreement with the variations in total organic carbon and organic nitrogen concentrations; this biosensor was then selected compared to the first conductometric biosensor for protein detection [62, 63]. In fact, the pronase composition (endo- and exopeptidases and in particular chymotrypsine, trypsin, carboxypeptidase, and aminopeptidase) and the association with proteinase K confer to the biosensor the capability to catalyze the hydrolysis of all the different proteins contained in the water samples.

The optimum conditions for the preparation of the biosensor increased the sensitivity and gave a limit of quantification of 0.583 μg/L for total organic carbon and of 0.218 μg/L for organic nitrogen in water samples. This sensor shows good reproducibility (2.28%), a capacity to be used at temperatures varying between 10°C and 30°C (temperature depending on the season), and a long lifetime (5 weeks).

9.2.1.7 Biosensors for phosphate determination

Phosphate is an essential nutrient for the growth of aquatic plants. Because of its widespread presence in detergents and fertilizers, increased phosphate concentration can lead to eutrophication of lakes and rivers. In clinical diagnosis, the determination of phosphate levels in body fluids can provide useful information about several diseases, the energetic state of cells, and bone function. Additionally, an excess diet of phosphate in food products will affect human health. Thus, phosphate determination is also relevant to food quality control.

A conductometric biosensor for phosphate determination was reported using maltose phosphorylase with the following reaction sequence [65]:

Maltose phosphorylase

$$\text{Maltose} + \text{Phosphate} \rightarrow \beta\text{-D-glucose-1-phosphate} + \alpha\text{-D-glucose} \quad (9.6)$$

The conductometric biosensor has two linear ranges: one is from 1.0 μM to 20 μM phosphate with a detection limit of 1.0 μM, and the other one is from 20 μM to 400 μM phosphate (Fig. 9.8). For the first linear section, the regression equation is ΔS (μS) = 0.182 + 0.298 [phosphate] (μM), $R^2 = 0.994$, and for the second section, the equation is ΔS (μS) = 3.820 + 0.133 [phosphate] (μM), $R^2 = 0.999$.

No interference from other anionic species was detected. Moreover, the conductometric biosensor presented here has long-term storage and operational stability as well as good thermal stability. The practical use of this biosensor in determination of phosphate in real samples (wastewater, river water, etc.) is being investigated.

9.2.1.8 Biosensors for nitrite determination

Water containing a high concentration of nitrite can create serious problems, such as eutrophication and potential hazards to human

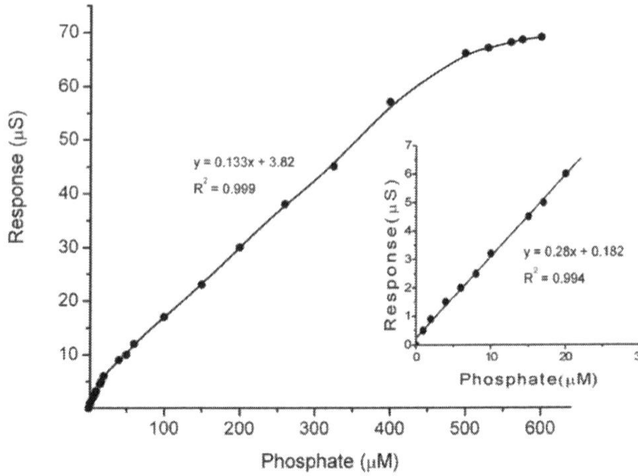

Figure 9.8 Calibration curve of a conductometric phosphate biosensor. Measurements were conducted in 0.1 M citrate buffer, pH 6.0 at room temperature.

health. Eutrophication in rivers, lakes, and coastal waters has become one of the most prevalent environmental problems. On the other hand, with many potential hazards to human health, nitrite has been considered to be a main toxic agent. Nitrite promotes the irreversible oxidization of hemoglobin to methemoglobin and reduces the blood capacity to transport oxygen. In addition, nitrite in the body can be converted into carcinogenic *N*-nitrosoamine compounds. Consequently, there is a growing demand to detect nitrite in food, drinking water, and environmental samples.

In Ref. [66], authors presented a conductometric nitrite biosensor based on the co-immobilization of cytochrome c nitrite reductase (ccNiR), BSA, Nafion®, methyl viologen (MV), and glycerol with saturated glutaraldehyde vapor on an IDE. ccNiR is usually purified and stored in oxidized form. However, this enzyme will only be active in a reduced state, which can be achieved with an artificial electron donor such as MV in the reduced state ($MV^{\bullet+}$). MV's electrochemical behavior involves reduction of the oxidized state (MV^{2+}). With sodium dithionite, MV^{2+} can be reduced to $MV^{\bullet+}$, which further reduces ccNiR to the active state.

In addition, since viologens are highly water soluble, any practical device containing the electron mediator should be based on immobilized viologens. The structure of MV^{2+} contains a hydrophobic part, which is capable of hydrophobic–hydrophobic interactions with Nafion® and two cationic pyridinium groups that undergo ion exchange with the sulphonate sites of Nafion® polymer chains according to the following reaction:

$$MV_{aq}{}^{2+} + 2(SO_3{}^-Na^+)_{film} \rightarrow [(SO_3{}^-)_2MV^{2+}]_{film} + 2Na^+_{aq} \qquad (9.7)$$

This interaction results in MV^{2+} accumulation in the membrane. In the presence of sodium dithionite as the electron donor, the biocatalyzed reduction of $NO_2{}^-$ to $NH_4{}^+$ is stimulated. The reaction is as follows:

$$NO_2{}^- + 6MV^{ffl+} + 8H^+ \rightarrow NH_4{}^+ + 6MV^{2+} + 2H_2O \qquad (9.8)$$

The calibration curve of an optimized nitrite conductometric biosensor is shown in Fig. 9.9.

From Fig. 9.9, the linear regression equation is: ΔS (μS) = 0.1250 + 0.1972 [$NO_2{}^-$] (μM), R^2 = 0.9993. The optimized biosensor exhibited a fast response to nitrite (about 10 seconds), a linear range of 0.2–120 μM nitrite, a sensitivity of 0.194 μS/μM [$NO_2{}^-$], and a low detection limit of 0.05 μM.

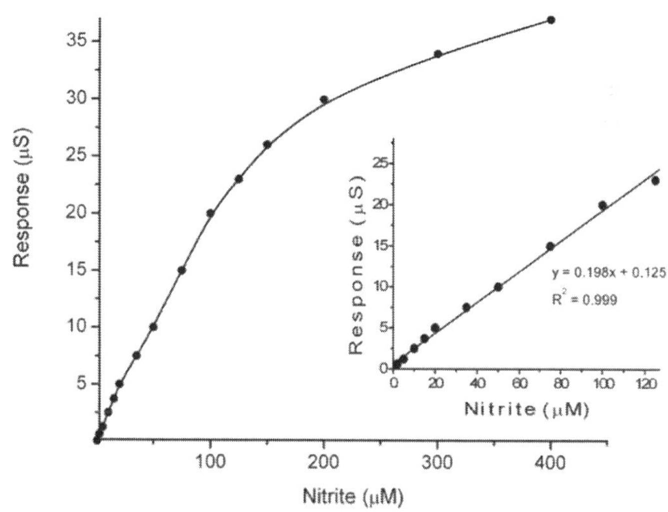

Figure 9.9 Calibration curve of an optimized conductometric nitrite biosensor. Measurements were performed in PBS (5.0 mM, pH 6.5) at room temperature.

The conductance response of the sensor was found to be fairly stable during the first week and then to drop gradually with time. After about three weeks, it still maintained over 50% of the initial response. Other experiments showed that the conductometric biosensor could be maintained at about 75% of the initial response for about one month, when the checking was reduced to less than five times within that period.

No obvious interference from other ionic species familiar in natural waters was detected. The application experiments show that the biosensor is suitable to be used for real water samples.

9.2.1.9 Biosensors for nitrate determination

A highly sensitive, fast, and stable conductometric enzyme biosensor for determination of nitrate in waters was described in Refs. [67, 68]. The conductometric electrodes were modified by an MV mediator mixed with nitrate reductase from *Aspergillus niger* by crosslinking with glutaraldehyde in the presence of BSA and Nafion® cation exchange polymer, allowing retention of the viologen mediator according to the following ion exchange reaction:

$$MV_{aq}{}^{2+} + 2(SO_3{}^{-}Na^{+})_{film} \rightarrow [(SO_3{}^{-})_2MV^{2+}]_{film} + 2Na_{aq}{}^{+} \quad (9.9)$$

The enzymatic reaction for the reduction of nitrate is the following one:

$$NO_3{}^{-} + 2MV^{\cdot+} + 2H \rightarrow NO_2{}^{-} + 2MV^{2+} + H_2O \quad (9.10)$$

Using the optimum conditions obtained in the above studies, a calibration curve of the conductometric nitrate biosensor was obtained over a nitrate concentration range of 0.02–0.5 mM using the biosensor in 5.0 mM phosphate buffer at pH 7.5. As shown in Fig. 9.10, the MV-/Nafion®-modified electrode showed a linear response in a concentration range of 0.02–0.25 mM, and the linear regression equation was ΔS (μS) $= 1.1058 + 30.469$ [$NO_3{}^{-}$] (mM), $R = 0.9975$. The detection limit of 0.005 mM of nitrate was obtained with a signal-to-noise ratio of 3. It has been reported that the oxygen interference is not negligible during the detection, because the reduced viologens ($MV^{\bullet+}$) can be auto-oxidized by oxygen. A calibration curve of the nitrate conductometric biosensor was also obtained in nondeoxygenated buffer (Fig. 9.10), which was

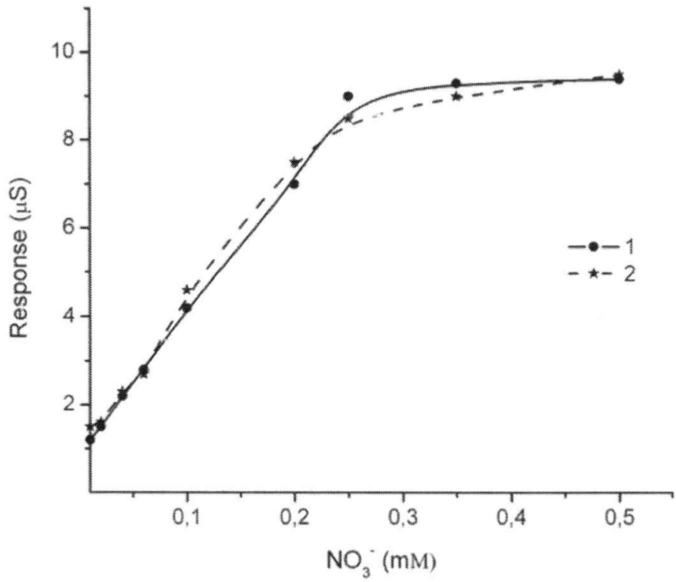

Figure 9.10 Typical calibration curves of MV/Nafion®/nitrate reductase (NR) electrode for increasing nitrate concentrations (1) in nitrogen-saturated buffer and (2) in nondeoxygenated buffer. Measurements were conducted in 5.0 mM phosphate buffer, pH 7.5 at 25°C.

comparable to that in nitrogen-saturated buffer. At the same time, we find that although the relative standard deviation obtained is higher in air-saturated buffer (9%) than in nitrogen-saturated buffer (6%) ($n = 10$), the difference is not very significant. These results confirm that in the presence of oxygen the interferential response is not significant, which is due to the excessive sodium dithionite that depletes the oxygen in the buffer solution.

Linear calibration in the range of 0.02–0.25 mM with a detection limit of 0.005 mM nitrate was obtained with a signal-to-noise ratio of 3. When stored in 5 mM phosphate buffer (pH 7.5) at 4°C, the sensor showed good stability over two weeks.

The developed nitrate conductometric biosensor was used to determine nitrate in some water samples using the method of standard addition. The results indicate that the nitrate conductometric biosensor can be successfully applied to the determination of nitrate at concentrations normally presented in some water samples.

Further practical use of the nitrate conductometric biosensor in determination of nitrate in real samples is being investigated.

9.2.1.10 Biosensors for sucrose determination

Sucrose, a key component of molasses, is used in the food industry as liquid sugar; some special sugars are consumed in pharmaceutics and the cosmetic industry. Since sucrose is a component of foodstuffs and beverages, precise information on sucrose presence and concentration is very important for assessment of its quality [69, 70].

The basic cascade of enzymatic reactions for sucrose detection by conductometric biosensors is as follows:

$$\text{Sucrose} + \text{H}_2\text{O} \xrightarrow{\text{Invertase}} \beta\text{-D-fructose} + \alpha\text{-D-glucose} \tag{9.11}$$

$$\alpha\text{-D-glucose} \xrightarrow{\text{Mutarotase}} \beta\text{-D-glucose} \tag{9.12}$$

$$\begin{gathered} \text{Glucose oxidase} \\ \beta\text{-D-glucose} + \text{O}_2 \rightarrow \text{D-gluconolactone} + \text{H}_2\text{O}_2 \\ \Downarrow \\ \text{D-gluconic acid} + \text{H}_2\text{O} \leftrightarrows \text{Acid residuum} + \text{H}^+ \end{gathered} \tag{9.13}$$

Calibration curves of the dependence of biosensor responses on the sucrose concentration in buffer solution are plotted for various buffer capacities of the solution (Fig. 9.11). The biosensor responses and linear range of measurements are seen to vary to some extent when the concentration of buffer solution changes. The highest sucrose sensitivity of the conductometric biosensor was in 2.5 mM phosphate buffer, pH 7.2; however, the linear range in this case shifts toward the region of low concentration, that is, 0.001–2.5 mM (Fig. 9.11), while in 5 mM phosphate buffer the linear range was slightly wider (0.002–5 mM). In 10 mM and 20 mM phosphate buffers, the biosensor sensitivity toward sucrose slightly decreased. Therefore, the biosensor for measurement of sucrose concentration in given ranges with required sensitivity can be attained by varying the buffer concentration, thus being adapted to actual practical needs.

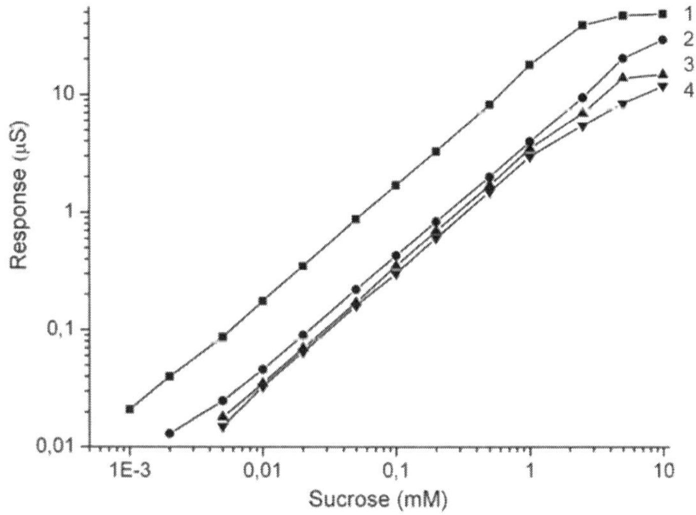

Figure 9.11 Calibration curves of a sucrose biosensor at different buffer concentrations. Measurements were done in 2.5 mM (1), 5 mM (2), 10 mM (3), and 20 mM (4) phosphate buffer, pH 7.2.

Operational stability, signal reproducibility, and essential characteristics of biosensors were tested. The responses to the same sucrose concentration (0.5 mM) were measured every 30 minutes for 4 days; during intervals the sensors were kept in a continuously stirred buffer. Biosensors were stored dry at room temperature. The chosen sucrose concentration was on the linear region of the sensor calibration curve. The measurement data was highly reproducible every day of the experiment, while sufficient operational stability was revealed for a week.

9.2.1.11 Biosensors for lactose determination

Lactose is of great significance in physiology due to its stimulating action on the nervous system and prophylactic and therapeutic effect upon cardiovascular diseases. In humans, lactose promotes calcium, manganese, and magnesium absorption, sustains *Bifidobacteria* growth; besides, it inhibits intestinal pathogen flora because of the generation of lactic acid resulting from lactose disintegration. For lactic acid bacteria, lactose is a main energetic source of lactic acid fermentation, which is a base for the production of numerous cultured milk foods. In the milk-processing industry, lactose is used as a crystallization inoculant upon concentrated milk production.

A conductometric biosensor for lactose determination is based on a cascade of the following enzymatic reactions [70]:

$$\text{Lactose} + H_2O \xrightarrow{\beta\text{-Galactosidase}} \text{Galactose} + \alpha\text{-D-glucose} \tag{9.14}$$

$$\alpha\text{-D-glucose} \xrightarrow{\text{Mutarotase}} \beta\text{-D-glucose} \tag{9.15}$$

$$\begin{aligned} \beta\text{-D-glucose} + O_2 &\xrightarrow{\text{Glucose oxidase}} \text{D-gluconolacton} + H_2O_2 \\ &\Downarrow \\ \text{D-gluconic acid} + H_2O &\leftrightarrows \text{Acid residue} + H^+ \end{aligned} \tag{9.16}$$

Dependence of the biosensor response on lactose concentration at different buffer concentrations is presented in Fig. 9.12. The change in buffer solution concentration is seen to result in varying biosensor responses and measurement linear ranges. In 20 mM and 30 mM phosphate buffers, the biosensor sensitivity to lactose and the linear range dropped considerably.

Storage stability was studied to evaluate the potential of the developed biosensor to be commercialized. The biosensors were stored dry at 4°C. The response to insertion of 0.1 mM lactose into the model solution, obtained on the first day, was taken as 100%. Further measurements were carried out at certain intervals (three to eight days). The biosensor activity dropped by 22% during three months.

9.2.1.12 Biosensors for maltose determination

Due to unique characteristics, maltose syrups are used in the production of a wide range of confectionery (caramel, frozen dairy produce, jellies, canned fruits, etc.). Besides, saccharose is frequently constituted of maltose in food manufacture, in particular in the production of child foodstuffs, due to lower allergenicity

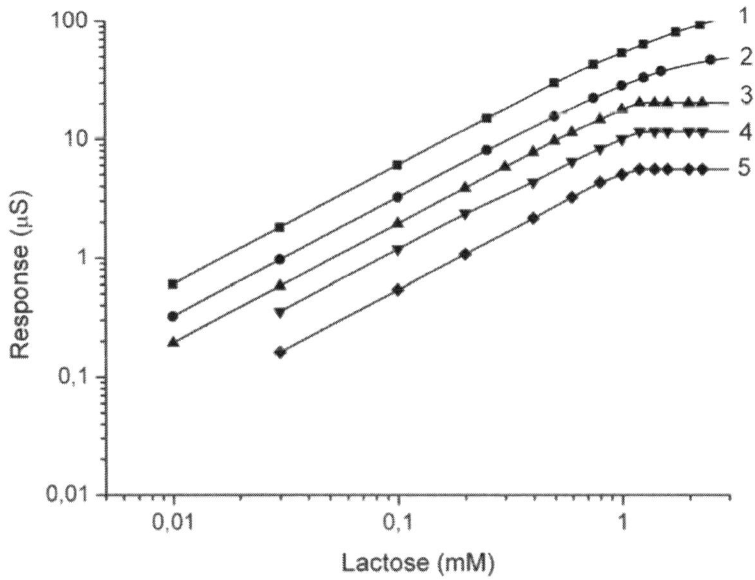

Figure 9.12 Calibration curves of a lactose biosensor at different buffer concentrations. Measurements were done in 5 mM (1), 10 mM (2), 15 mM (3), 20 mM (4), and 30 mM (5) phosphate buffer, pH 6.5.

of the latter. Maltose is used in the manufacture of dietetic and sports foodstuffs, breads, wines, preserves, and beverages. Besides, maltose is utilized in microbiology and pharmacology. In organisms, maltose is decomposed into two glucose molecules being exposed to the enzyme maltase (α-glucosidase) present in digestive juices of animals and humans. Genetically determined absence of maltase in the mucous coat of the human intestine causes congenital intolerance to maltose, which is a severe disease and requires the elimination of maltose, starch, and glycogen from the ration and supplementation with maltase.

A conductometric biosensor with a three-enzyme membrane as a sensitive element was developed for maltose determination, and its analytical characteristics for model samples (response dependence on pH, ion strength, buffer capacity of the working solution) were investigated [71]. The basic cascade of enzymatic reactions for maltose detection by a conductometric biosensor is as follows:

$$\text{Maltose} + H_2O \xrightarrow{\alpha\text{-Glucosidase}} \alpha\text{-D-glucose} + \alpha\text{-D-glucose} \quad (9.17)$$

$$\alpha\text{-D-glucose} \xrightarrow{\text{Mutarotase}} \beta\text{-D-glucose} \quad (9.18)$$

$$\beta\text{-D-glucose} + O_2 \xrightarrow{\text{Glucose oxidase}} \text{D-gluconolacton} + H_2O_2$$
$$\Downarrow \quad (9.19)$$
$$\text{D-gluconic acid} + H_2O \leftrightarrows \text{Acid residue} + H^+$$

The calibration curve of a maltose conductometric biosensor is shown in Fig. 9.13. The linear range of biosensor operation was up to 1 mM, and the minimum measurable concentration was 0.002 mM for maltose.

Selectivity as an essential characteristic of the conductometric maltose biosensor was studied by determination of its reaction to

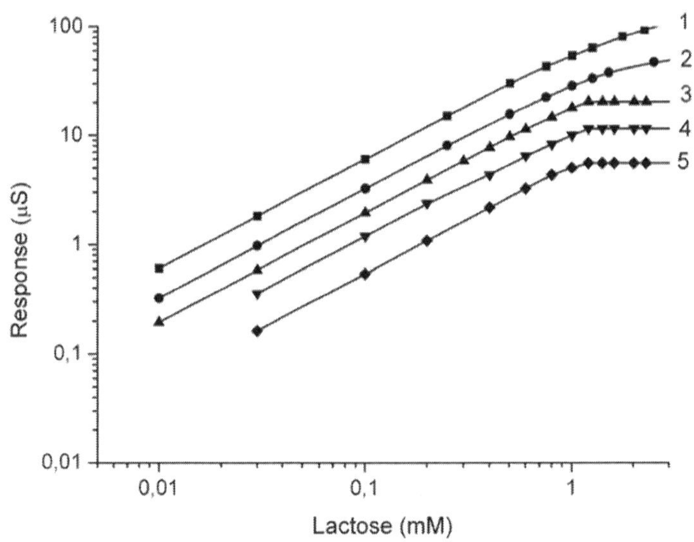

Figure 9.13 Calibration curve of a maltose conductometric biosensor. Measurements were done in 5 mM phosphate buffer, pH 6.5.

Table 9.1 Selectivity of the biosensor for maltose determination

0.5 mM substance	Relative response of maltose biosensor (%)
Maltose	100
Glucose	139
Sucrose	4
Fructose	0
α-Lactose	0
β-Lactose	0
Mannose	0

the content of interfering substances. Measurements were carried out in 5 mM phosphate buffer solution, pH 6.5. The interfering substances at 0.5 mM concentration were introduced into the cell; the response to 0.5 mM maltose was taken as 100% (Table 9.1).

Basically, the tested conductometric system was shown to be selective and, thus, can be suggested for further application in the analysis of real samples. The maltose biosensor response to glucose is quite comprehensible since glucose oxidase is a component of the enzyme membrane. This is why to measure maltose, in case glucose is present in the samples tested, another sensor, sensitive only to glucose, should be used along with the maltose one.

9.2.2 *Conductometric Enzyme Biosensors Based on Inhibition Analysis*

9.2.2.1 Biosensor for organophosphorous pesticides determination

Tons of pesticides annually used in agriculture and horticulture can degrade in the environment by microbial degradation and photodegradation and chemical hydrolysis. During these transformation processes, intermediate products, which may be more toxic than the initial products, are generated. These toxic compounds contaminate air, soil, and water over large areas. Consequently, it is necessary to develop tests for toxicity assessment of environmental samples.

A conductometric biosensor based on inhibition analysis, first described in Ref. [72], was intended for the determination of organophosphorous pesticides. As a sensitive element, the enzymes acetyl- and butyrylcholinesterase were used.

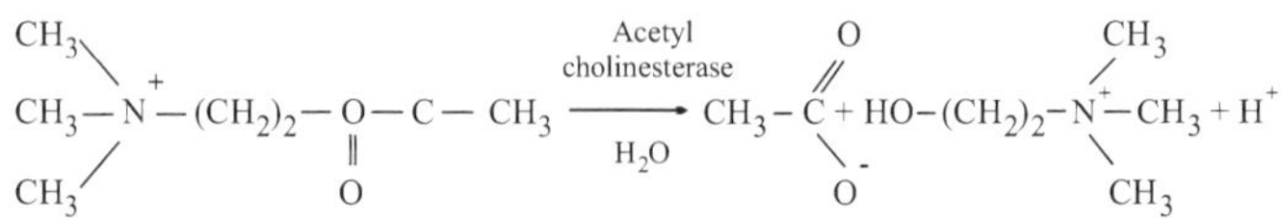

(9.20)

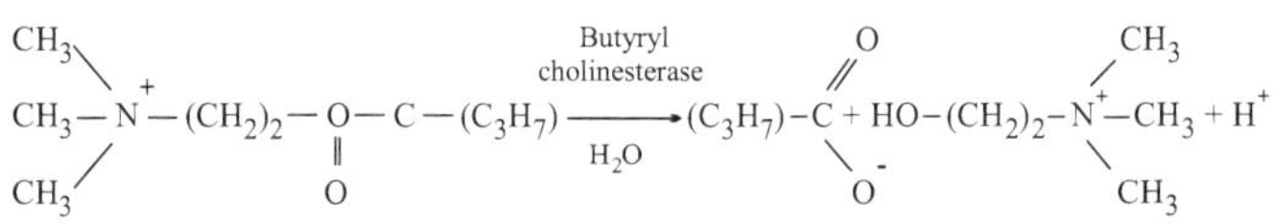

(9.21)

The determination of organophosphorus and carbamate compounds is based on their ability to inhibit cholinesterases by interaction with the serine –OH group in the enzyme-active site according to the following reaction (example of the organophosphorus compound trichlorfon):

$$\text{Enz-Ser-OH} + \underset{\text{trichlorfon}}{\text{CCl}_3\text{CH}_2\text{O-}\overset{\overset{\text{O}}{||}}{\text{P}}\text{-(OCH}_3)_2} \xrightarrow{\text{inhibition}} \text{Enz-Ser-O-}\overset{\overset{\text{O}}{||}}{\text{P}}\text{-(OCH}_3)_2 + \text{CCl}_3\text{CH}_2\text{OH} \quad (9.22)$$

The decrease in cholinesterase activity after its interaction with pesticides can be effectively monitored by the conductometric biosensors, facilitating the toxicity assessment of organophosphorus and carbamate pesticides.

The sensor sensitivity to different pesticides was investigated [73]. The calibration curves of a conductometric acetyl cholinesterase (AcChE) biosensor for different pesticides are shown in Fig. 9.14.

They are linear in a semilogarithmic plot for all pesticides tested. Detection limits determined as the inhibitor concentration giving a decrease of the substrate signal equal to three times the blank value were 5.0×10^{-11} M for diisopropyl fluorophosphate, 1.0×10^{-8} M

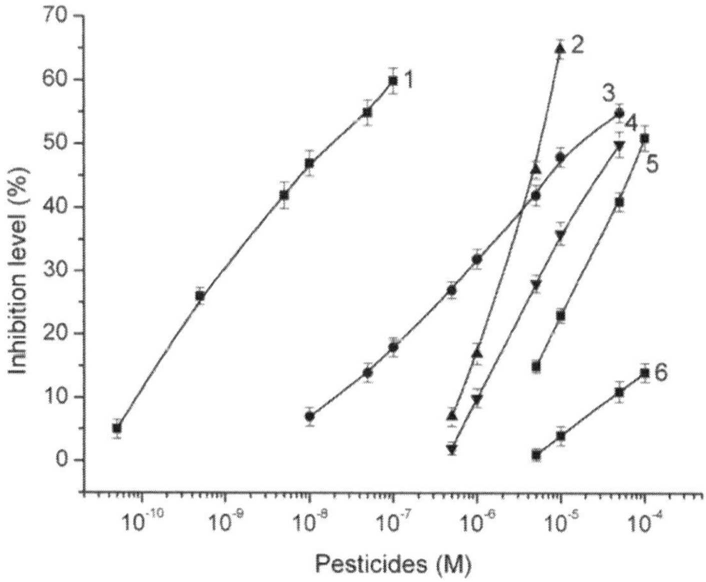

Figure 9.14 Calibration curves of a conductometric acetyl cholinesterase (AcChE) biosensor for diisopropyl fluorophosphate (1), trichlorphon (2), paraoxon-ethyl (3), paraoxon-methyl (4), carbofuran (5), and parathion-methyl (6). Measurements were conducted in triplicate in 5.0 mM phosphate buffer, pH 7.5, using 2.0 mM AcChCl and 20 min inhibition time.

for paraoxon-ethyl, 5.0×10^{-7} M for paraoxon-methyl, 3.0×10^{-7} M for trichlorfon, 5.0×10^{-6} M for parathion-methyl, and 2.0×10^{-6} M for carbofuran.

The conductometric biosensors applied to the analysis of total solution toxicity at parathion-methyl photodegradation were presented in Refs. [74, 75]. The results obtained were compared to the data from the traditional high-sensitivity method of HPLC and from the Lumistox device (LUMISTOX 300, Hach Lange, Germany) for toxicity determination. The solution toxicity was shown to increase dramatically as pesticide photodegradation began; the toxicity remained once the parathion-methyl dissociation had been completed.

This sensor could be used as a rapid technique ("early warning system") for measuring cholinesterase inhibitors present in the environment.

9.2.2.2 Biosensor for heavy metal ion determination

Heavy metals and their compounds in the environment are characterized by relatively high stability, solubility in atmosphere precipitations, and ability to be absorbed by soil and plants. They are accumulated in organisms, being toxic in any state for humans and animals, with a wide spectrum and variety of harmful effects.

The potential of a conductometric urease biosensor for the determination of heavy metal ions was demonstrated in Ref. [76]. This reaction is similar to the generation of metal sulphides; therefore, the metals capable of forming insoluble sulphides are strong urease inhibitors. The assay protocol included measurement of the biosensor response to a fixed concentration of urea before and after the incubation of the biosensor for a definite time in a solution containing heavy metal ions.

The calibration curves obtained with immobilized urease are presented in Fig. 9.15. The inhibition activities of heavy metals toward urease varied as follows: $Hg^{2+} > Cu^{2+} > Cd^{2+} > Co^{2+} > Pb^{2+} > Sr^{2+}$; reactivation of the inhibited enzyme with EDTA was shown to be probable. The immobilized urease, previously inhibited in 1 mM

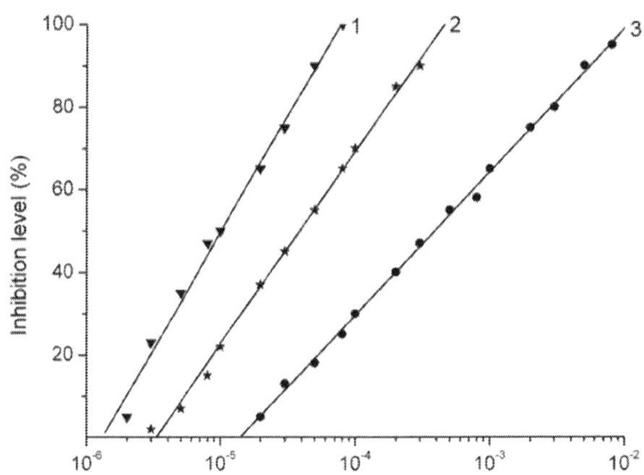

Figure 9.15 Calibration curves for determination of some heavy metal ions by a conductometric urease biosensor: (1) Hg^{2+}, (2) Cd^{2+}, and (3) Pb^{2+}. Measurements were performed in 5 mM potassium–phosphate buffer, pH 7.4. The time of preincubation in the presence of heavy metal ions was 10 min.

Cu^{2+} solution for 15 min, had its activity fully restored by soaking in 10 mM EDTA solution for 2 h.

In Ref. [77], a three-enzyme system, that is, invertase, mutoratase, and glucose oxidase, immobilized on the transducer surface was used as a bioselective element of a conductometric biosensor. An enzyme ratio in the membrane was found experimentally, considering the highest biosensor sensitivity both to the substrate (sucrose) and to heavy metals. The optimal concentration of sucrose for inhibitory analysis was 1.25 mM, and the incubation time in investigated solution amounted to 10–20 minutes. The developed biosensor demonstrated the best sensitivity toward ions Hg^{2+} and Ag^{+}. The principal possibility of biosensor reactivation by EDTA solution after inhibition with silver ions or by cysteine solution after inhibition with mercury ions was shown. The results of biosensor analysis of toxicants in real water samples were in good correlation with the results obtained by traditional methods.

9.2.2.3 Biosensor for surfactant determination

Surfactants are one of the widespread pollutants of the surroundings. Physical and chemical properties of surfactants lead to wide applications of these substances in industry, medicine, agriculture, and everyday household use. They are an integral part of the means of personal hygiene, numerous washing and cleansing agents, etc. After being used, surfactants are discharged into the environment in huge volumes, which results in contamination of water ecosystems. Furthermore, due to their ability to increase solubility of other pollutants, the latter appear in water in higher concentration.

A conductometric biosensor for determination of surfactants in aqueous solutions, based on inhibition of acetylcholinesterase, was studied [78].

The procedure of surfactant determination by conductometric biosensors is shown in Fig. 9.16. First, the response to the saturating concentration of substrate was obtained. Next, the surfactant was added and the signal decrease was recorded. The rate of bioselective element inhibition (μS/min) changed depending on the surfactant concentration, which allowed obtaining of calibration curves for surfactant determination in the tested sample.

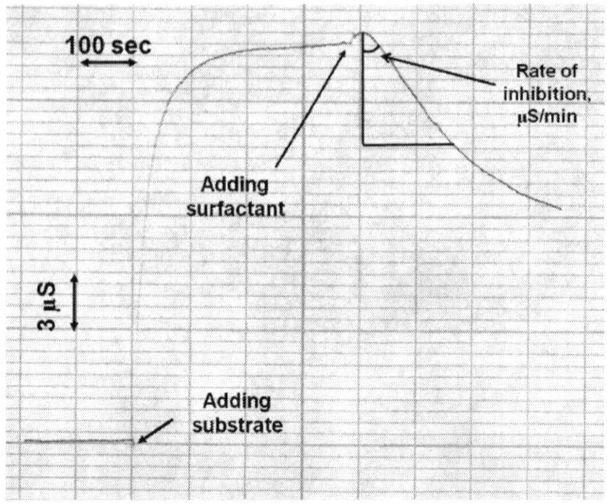

Figure 9.16 Procedure of surfactant determination by conductometric enzyme biosensors. Measurements were done in 5 mM phosphate buffer, pH 6.5, substrate concentration in the cell 3 mM, inhibitor concentration 8.75 mg/L.

Some variants of inhibitory determination of surfactants were examined; the biosensor sensitivity to benzalkonium chloride (0.061 g/L), sodium dodecyl sulfate (2.5 μM), and hexadecylpyridinium bromide (2.5 μM) was tested by the optimal procedure. The optimum value of pH was determined to be 6.0. The effect of ionic strength and buffer capacity on biosensor sensitivity to the substrate and inhibitor was analyzed, and the composition and concentration of the working buffer were optimized. The developed biosensor was proved to have high signal reproducibility at both direct measurement of the key substrate, acetylcholine, and inhibitory determination of benzalkonium chloride.

9.2.2.4 Biosensor for detection of diuron and atrazine

A conductometric biosensor for the detection of some toxic compounds, including diuron, atrazine, its main metabolites (desisopropylatrazine and deethylatrazine), and copper ions, was developed in Refs. [79, 80] on the basis of tyrosinase inhibition.

Under the experimental conditions employed using 4-chlorophenol as the enzyme substrate, and 30 minutes of contact with a

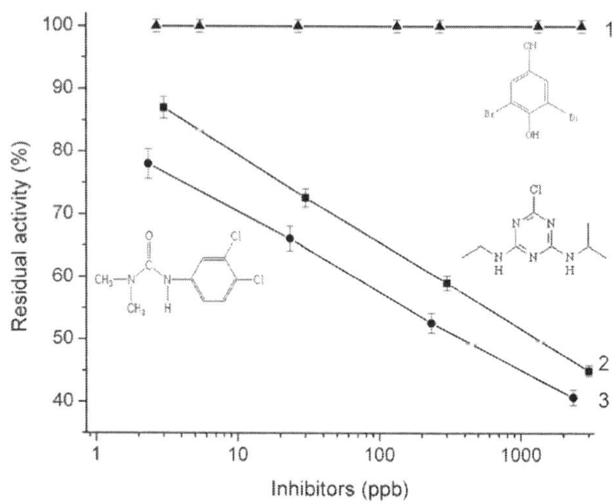

Figure 9.17 Dependence of tyrosinase residual activity on concentrations of bromoxynile (1), atrazine (2), and diuron (3). Response to 6 mM 4-chlorophenol was measured in 5 mM phosphate buffer, pH 6.0, inhibition for 30 min in solutions of toxins.

tyrosinase inhibitor, detection limits for diuron and atrazine were around 1 ppb and a dynamic range of 2.3–2330 ppb and 2.15–2150 ppb was obtained for diuron and atrazine, respectively (Fig. 9.17).

Generally, in the environment, pesticides are transformed into different metabolites by various degradation processes, including photo- and biodegradation. For example, atrazine commonly coexists with its two main metabolites, that is degradation products, desisopropylatrazine and diethylatrazine. It was shown that inhibition of tyrosinase by desisopropylatrazine and diethylatrazine was less than atrazine.

The different samples tested were solutions containing diuron, atrazine, copper, lead, and zinc ions, mixtures of copper/atrazine or copper/diuron, and real water samples coming from a Vietnamese river. In the last case, classical techniques such as gas chromatography–mass spectrometry (GC-MS) or atomic absorption spectrometry were used to estimate exact concentration of these species in real water samples. Results have shown that such a biosensor could be used as an early warning system for the detection of these pollutants, as no matrix effect coming from the real sample was observed and no synergetic or antagonist effects were found for the mixture of toxic compounds. In addition, results were coherent with the content of the tyrosinase inhibitors.

9.2.2.5 Biosensor for nitrite determination

Another work [81] described conductometric biosensors using a combination of peroxidase/catalase. A peroxidase (horseradish peroxidase [HRP]) was located in the inner layer, while the outer layer contained catalase. Catalase catalyzed the breakdown of H_2O_2 into H_2O and O_2, thus consuming totally H_2O_2, the substrate of HRP. The latter, in the presence of H_2O_2, generates a conductometric signal due to the reduction of H_2O_2. Nitrite was selected as an inhibitor of catalase. In the presence of H_2O_2, the nitrite addition blocked the H_2O_2 consumption by catalase. Since nitrite had no effect on HRP activity, its inhibitive effect on catalase leads to an increase in the conductometric signal. The bienzyme sensor exhibits an increase in conductometric response for nitrite concentration, leading to high values of conductivity. In both cases, the detection limit of nitrite is 0.3 μM, and for the bienzyme sensor the dynamic range is from 0.3 μM to 446 μM.

9.2.2.6 Biosensor for cyanide determination

Cyanide is continuously released in small doses in the environment and is considered as a violent poison that constitutes a real hazard for aquatic ecosystems. It is naturally produced by certain bacteria, fungi, and algae and is found in many foods and plants. Cyanide is also widely used in industry for the manufacture of synthetic fibers and plastics, as well as in electroplating baths and metal mining operations or pesticide agents and an intermediate in agricultural chemical production.

In Ref. [82] a conductometric biosensor for cyanide determination was prepared through immobilization of bovine liver catalase in a photoreticulated poly(vinyl alcohol) membrane. This biosensor was used to study the kinetics of the catalase–H_2O_2 reaction and its inhibition by cyanide.

Inhibition by cyanide was found to be noncompetitive. The decrease of the biosensor response by increasing cyanide concentration was linear up to 50 μM with a cyanide detection limit of 6 μM.

The apparent inhibition binding constant was determined from the Dixon plots for different H_2O_2 concentrations and was 13.9 μM. the inhibition coefficient I50 measured by conductometric biosensors was 24.9 μM. It showed great potential of conductometric biosensors for quantitative analysis of cyanides.

9.3 Conclusion

Application of conductometric measuring method to continuous recording in the course of enzyme processes is thoroughly examined and analyzed. Almost all electrochemical analytical methods are based on electrochemical reactions (potentiometry, voltamperometry, amperometry, coulometry). Conductometry is a method where there is either no electrochemical reaction on the electrodes at all or they are secondary ones and they can be neglected. Therefore in the conductometric method the most important property of the electrolytic solution, in the boundary layer, is its conductivity, which varies in accordance with quite a wide range of biological reactions.

The conductometric biosensors also have some advantages over other types of transducers. First, they can be produced through inexpensive thin-film standard technology. This, along with using an optimized method of immobilization of biological material, results in considerable decrease in both the primary cost of devices and the total price of analyses. For integral microbiosensors, it is easy to perform a differential measurement mode, thus compensating for external effects and considerably increasing measurement accuracy.

The data is convincing evidence of great potential of conductometric biosensors. However, it is still rather a novel trend in the field of biosensors, which is why the development of commercial devices has a promising future.

References

1. Coulet, P. R. (1991) Biosensor principles and application, in *What Is a Biosensor*, eds. L. J. Blum, P. R. Coulet (Marcel Dekker, New York), 1–6.
2. Hall, E. A. H. (1988) Recent progress in biosensor development, *Int. J. Biochem.*, **20**(4), 357–362.
3. Thevenot, D. R., Toth, K., Durst, R. A., Wilson, G. S. (1999) Electrochemical biosensors: recommended definitions and classification (technical report), *Pure Appl. Chem.*, **71**, 2333–2348.
4. Bartlett, P. N., Whitaker, R. G. (1987/88) Strategies for the development of amperometric enzyme electrodes, *Biosensors*, **3**, 359–379.
5. Mizutani, F., Yamanaka, T., Tanabe, Y., Tsuda, K. (1985) An enzyme electrode for L-lactate with chemically amplified electrode, *Anal. Chim. Acta*, **177**, 153–166.
6. Morrison, L. E. (1988) Time resolved detection of energy transfer: theory and application to immunoassays, *Anal. Biochem.*, **174**, 101–120.
7. Lee, H. A., Morgan, M. R. A. (1993) Food immunoassay: application of polyclonal, monoclonal and recombinant antibodies, *Trends Food Sci. Technol.*, **3**, 129–134.
8. Buch, R. M., Rechnitz, G. A. (1989) Neuronal biosensors, *Anal. Chem.* **61**(8), 533A–542A.
9. Lewis, R. (1989) Biological recognition repackaged, *BioScience*, **39**, 288–291.
10. Wiseman, A. (1992) Comparison of use of immobilized cells and immobilized enzymes for bioanalysis: considerations in determination of ethanol, *Trends Anal. Chem.*, **11**, 303–306.
11. Karube, I., Sode, K. (1991) Bioinstrumentation and biosensors, in *Microbial Sensors for Process and Environmental Control*, ed. D. L. Wise (Marcel Dekker, New York), 149–160.
12. Dumschat, C., Muller, H., Stein, K., Schwedt, G. (1991) Pesticide-sensitive ISFET based on enzyme inhibition, *Anal. Chim. Acta*, **252**, 7–9.
13. Bergveld, P. (2003) Thirty years of ISFETOLOGY. What happened in the past 30 years and what may happen in the next 30 years, *Sens. Actuators, B*, **88**, 1–20.
14. Jossinet, J., McAdams, E. T. (1991) The skin-electrode interface impedance, *Innov. Tech. Biol. Med.*, **12**, 21–31.
15. Macdonald, J. R. (1992) Impedance/admittance response of a binary electrolyte, *Electrochem. Acta*, **37**, 1007–1014.

16. Lorenzo, E., Pariente, F., Hernandez, L., et al. (1998) Analytical strategies for amperometric biosensors based on chemically modified electrodes, *Biosens. Bioelectronics*, **13**, 319–332.
17. Kell, D. B., Dave, C. L. (1990) Biosensors. A practical approach, in *Conductimetric and Impediometric Devices*, ed. A. F. Gass (IRL Press, Oxford), 125–153.
18. Cullen, D. C., Sethi, R. S., Lowe, C. R. (1990) Multi-analyte miniature conductance biosensor, *Anal. Chim. Acta*, **231**, 33–40.
19. Fawcett, N., Evans, J., Chien, L.-C., Flowers, N. (1988) Nucleic acid hybridization detected by piezoelectric resonance, *Anal. Lett.*, **21**, 1099–1114.
20. Roef, P. (1987) Attention focuses on optical fibre biosensors, *Sens. Rev.*, **7**(3), 127–132.
21. Smardzewski, R. R. (1988) Multi-element optical waveguide sensor: general concept and design, *Talanta*, **35**, 95–101.
22. Luong, J. H. T., Mulchandani, A., Guibault, G. G. (1988) Developments and application of biosensors, *Tibtech*, **6**, 310–316.
23. Dittmar, A., Pauchard, T., Delhomme, G., Vernet-Maury, E. (1992) A thermal conductivity sensor for measurement of skin blood-flow, *Sens. Actuators, B*, **7**, 327–331.
24. Dzyadevych, S. V. (2002) Amperometric biosensors. Modern technologies and commercial analyzers, *Biopol. Cell*, **18**, 363–376.
25. Dzyadevych, S. V. (2004) Biosensors based on ion-selective field-effect transistors: theory, technology and practice, *Biopol. Cell*, **20**, 7–16.
26. Dzyadevych, S. V. (2005) Conductometric enzyme biosensors: theory, technology and application, *Biopol. Cell*, **21**, 91–106.
27. Hall, E. (1991) *Biosensors* (Open University Press, Cambridge).
28. Tran Minh, C. (1993) *Biosensors* (Chapman & Hall, London).
29. Kress-Rogers, E. (1997) *Handbook of Biosensors and Electronic Noses: Medicine, Food, and Environment* (CRC Press, New York).
30. Fluitman, J. (1996) Microsystems technology: objectives, *Sens. Actuators, A*, **56**, 151–166.
31. Zinner, H. (1995) Microsystems: the European approach, *Sens. Actuators, A*, **46/47**, 1–7.
32. Abraham, M., Ehrfeld, W., Hessel, V., Kamper, K. P., Lacher, M., Picard, A. (1998) Microsystem technology: between research and industrial application, *Microelectron. Eng.*, **41/42**, 47–52.

33. Dorokhova, E. N., Prokhorova, G. V. (1991) *Analytical Chemistry. Physico-chemical Methods of Analysis* (Vysshaya shkola, Moscow).
34. Gopel, W., Jones, T. A., Kleitz, M., Lundstrom, J., Seiyama, T. (1991) Sensors. A comprehensive survey, in *Conductometry*, Vol. 2, Part I, eds. W. Gopel, J. Hesse, J. N. Zemel (VCH Verlagsgesellschaft, Weinheim), 314–337.
35. Kell, D. B. (1987) Biosensors: fundamentals and applications, in *The Principles and Potential of Electrical Admittance Spectroscopy: An Introduction*, eds. A. P. F. Turner, I. Karube, G. S. Wilson (Oxford University Press, Oxford), 427–468.
36. Chin, W. T., Kroontje, W. (1961) Conductivity method for determination of urea, *Anal. Chem.*, **33**, 1757–1760.
37. Pänke, O., Balkenhohl, T., Kafka, J., Schäfer, D., Lisdat, F. (2008) Impedance spectroscopy and biosensing, *Adv. Biochem. Eng. Biotechnol.*, **109**, 195–237.
38. Sheppard, N. F., Jr., Mears, D. J., Guiseppi-Elie, A. (1996) Model of an immobilized enzyme conductimetric urea biosensor, *Biosens. Bioelectron.*, **11**, 967–979.
39. Schmid, R. D., Karube, I. (1988) Biotechnology, in *Biosensors and "Bio-electronics,"* Vol. 6b, eds. H. J. Rehm, G. Reed (VCH Verlagsgesellschaft, Weinheim), 317–365.
40. Clark, L. C., Lyons, C. (1962) Electrode systems for continuous monitoring in cardiovascular surgery, *Ann. NY Acad. Sci.*, **102**, 29–45.
41. Shul'ga, A. A., Dzyadevich, S. V., Soldatkin, A. P., Patskovsky, S. V., Strikha, V. I. (1993) Conductometric biosensors for glucose and urea based on microfabricated thin-film interdigitated array-electrodes, *Biol. Ital.*, **23**(6), 40–45.
42. Shul'ga, A. A., Dzyadevich, S. V., Soldatkin, A. P., Patskovsky, S. V., Strikha, V. I., El'skaya, A. V. (1994) Thin-film conductometric biosensor for glucose and urea determination, *Biosens. Bioelectron.*, **9**, 217–223.
43. Dzyadevich, S. V., Arkhipova, V. N., Soldatkin, A. P., El'skaya, A. V., Shul'ga, A. A. (1998) Glucose conductometric biosensor with potassium hexacyanoferrate (III) as an oxidizing agent, *Anal. Chim. Acta*, **374**, 11–18.
44. Dzyadevich, S. V., Soldatkin, A. P., Arkhipova, V. N., Shul'ga, A. A., El'skaya, A. V. (1995) Conductometric enzyme glucosensor. search for ways to improve analytical characteristics, *Ukr. Biochem. J.*, **67**(6), 53–59.
45. Kerscher, L., Ziegnhorn, J. (1983) Methods of enzymatic analysis, in *Urea*, ed. H. U. Bergmeyer (Verlag Chemie, Weinheim), 444–453.

46. Watson, L. D., Maynard, P., Cullen, D. C., Sethi, R. S., Brettle, J., Lowe, C. R. (1987/88) A microelectronic conductometric biosensor, *Biosensors*, **3**, 101–115.

47. Hin, B. F. Y., Sethi, R. S., Lowe, C. R. (1990) Multi-analyte microelectronic biosensor, *Sens. Actuators, B*, **1**, 550–554.

48. Mikkelsen, S. K., Rechnitz, G. A. (1989) Conductometric transducers for enzyme-based biosensors, *Anal. Chem.*, **61**, 1737–1742.

49. Bilitewski, U., Drewes, W., Schmid, R. D. (1992) Thick film biosensors for urea, *Sens. Actuators, B*, **7**, 321–326.

50. Dzyadevich, S. V., Zhylyak, G. A., Soldatkin, A. P., El'skaya, A. V. (1996) Conductometric urease microbiosensor based on thin-films interdigitated electrodes for urea determination, *Biopol. Cell*, **12**, 53–57.

51. Won-Yong, L., Seung-Ryeol, K., Tae-Han, K., Kang Shin, L., Min-Chol, S., Je-Kyun, P. (2000) Sol–gel-derived thick-film conductometric biosensor for urea determination in serum, *Anal. Chim. Acta*, **404**, 195–203.

52. Nouira, W., Maaref, A., Vocanson, F., et al. (2012) Enhancement of enzymatic IDE biosensor response using gold nanoparticles. Example of the detection of urea, *Electroanalysis*, doi:10.1002/elan.201100681.

53. Saiapina, O. Y., Dzyadevych, S. V., Jaffrezic-Renault, N., Soldatkin, O. P., (2012) Development and optimization of a novel conductometric bi-enzyme biosensor for L-arginine determination, *Talanta*, **92**, 58–64.

54. Hnaien, M., Hassen, W. M., Abdelghani, A., et al. (2009) A conductometric biosensor for the estimation of the number of cleaving sites in peptides and proteins, *Electrochem. Commun.*, **11**, 165–168.

55. Hnaien, M., Ruffin, E., Bordes, C., et al. (2011) Integrity characterization of myoglobin released from poly(ε-caprolactone) microspheres using two analytical methods: UV/Vis spectrometry and conductometric bi-enzymatic biosensor., *Eur. J. Pharm. Biopharm.*, **78**, 298–305.

56. Gerberich, H. R., Seaman, G. C. (1994) Formaldehyde, in *Encyclopedia Chemical Technology* (John Wiley and Sons, New York), 929–951.

57. Schechter, D. S., Singer, P. C. (1995) Formation of aldehydes during ozonation, *Ozone Sci. Eng.*, **17**, 53–55.

58. Feron, V. J., Til, H. P., de Vrijer, F., et al. (1991) Aldehydes: occurrence, carcinogenic potential, mechanism of action and risk assessment, *Mutat. Res.*, **259**, 363–385.

59. Squire, R. A., Cameron, L. L. (1984) An analysis of potential carcinogenic risk from formaldehyde, *Regul. Toxicol. Pharmacol.*, **4**, 107–129.

60. Dzyadevych, S. V., Arkhypova, V. N., Korpan, Y. I., et al. (2001) Conductometric formaldehyde sensitive biosensor with specifically adapted analytical characteristics, *Anal. Chim. Acta*, **445**, 47–55.
61. EH40/94 (1994) *Occupational Exposure Limits 1994*, HSE Guidance Booklet (HMSO, London).
62. Marrakchi, M., Dzyadevych, S. V., Namour, Ph., Martelet, C., Jaffrezic-Renault, N. (2005) A novel proteinase K biosensor based on interdigitated conductometric electrodes for proteins determination in rivers and sewers water, *Sens. Actuators, B*, **111–112**, 390–395.
63. Marrakchi, M., Dzyadevych, S. V., Namour, Ph., Martelet, C., Jaffrezic-Renault, N. (2007) An enzyme biosensor based on gold interdigitated thin-film electrodes for water quality control, *Anal. Lett.*, **40**, 1307–1313.
64. Khadro, B., Namour, P., Bessueille, F., Leonard, D., Jaffrezic-Renault, N. (2009) Validation of a conductometric bienzyme biosensor for the detection of proteins as marker of organic matter in river samples, *J. Environ. Sci.*, **21**, 545–551.
65. Zhang, Z., Jaffrezic-Renault, N., Bessueille, F., et al. (2008) Development of a conductometric phosphate biosensor based on tri-layer maltose phosphorylase composite films, *Anal. Chim. Acta*, **615**, 73–79.
66. Zhang, Z., Xia, S., Leonard, D., et al. (2009) A novel nitrite biosensor based on conductometric electrode modified with cytochrome c nitrite reductase composite membrane, *Biosens. Bioelectronics*, **24**, 1574–1579.
67. Wang, X., Dzyadevych, S. V., Chovelon, J.-M., Jaffrezic-Renault, N., Ling, C., Siqing, X. (2006) Development of conductometric nitrate biosensor based on methyl viologen/Nafion® composite film, *Electrochem. Commun.*, **8**, 201–205.
68. Wang, X., Dzyadevych, S. V., Chovelon, J.-M., Jaffrezic-Renault, N., Ling, C., Siqing, X. (2006) Conductometric nitrate biosensor based on Methyl viologen/Nafion®/nitrate reductase interdigitated electrodes, *Talanta*, **69**, 450–455.
69. Soldatkin, O. O., Peshkova, V. M., Dzyadevych, S. V., Soldatkin, A. P., Jaffrezic-Renault, N., El'skaya, A. V. (2008) Novel sucrose three-enzyme conductometric biosensor, *Mater. Sci. Eng. C*, **28**, 959–964.
70. Pyeshkova, V. M., Saiapina, O. Y., Soldatkin, O. O., Kukla, O. L., Dzyadevych, S. V. (2008) Enzyme conductometric biosensor for lactose content determination, *Biotechnol. Kiev*, **1**(4), 76–84.

71. Pyeshkova, V. M., Saiapina, O. Y., Soldatkin, O. O., Dzyadevych, S. V. (2009) Enzyme conductometric biosensor for maltose determination, *Biopol. Cell*, **25**, 272–278.

72. Dzydevich, S. V., Shul'ga, A. A., Soldatkin, A. P., Nyamsi Hendji, A. M., Jaffrezic-Renault, N., Martelet, C. (1994) Application of conductometric biosensor based on the cholinesterases for sensitive detection of pesticides, *Electroanalysis*, **6**, 752–758.

73. Dzyadevych, S. V., Soldatkin, A. P., Arkhypova, V. N., et al. (2005) Early-warning electrochemical biosensor system for the environmental monitoring based on enzyme inhibition effect, *Sens. Actuators, B*, **55**, 81–87.

74. Dzyadevych, S. V., Chovelon, J.-M. (2002) A comparative photodegradation studies of methyl parathion by using lumistox test and conductometric biosensor technique, *Mater. Sci. Eng. C*, **21**, 55–60.

75. Dzyadevych, S. V., Soldatkin, A. P., Chovelon, J.-M. (2002) Assessment of the toxicity of parathion and its photodegradation products in water samples using conductometric enzyme biosensors, *Anal. Chim. Acta*, **459**, 33–41.

76. Zhylyak, G. A., Dzyadevich, S. V., Korpan, Y. I., Soldatkin, A. P., El'skaya, A. V. (1995) Application of urease conductometric biosensor for heavy-metal ion determination, *Sens. Actuators, B*, **24–25**, 145–148.

77. Soldatkin, O. O., Kucherenko, I. S., Pyeshkova, V. M., et al. (2012) Novel conductometric biosensor based on three-enzyme system for selective determination of heavy metal ions, *Bioelectrochemistry*, **83**, 25–30.

78. Kucherenko, I. S., Soldatkin, O. O., Arkhypova, V. M., Dzyadevych, S. V., Soldatkin, A. P. (2011) Conductometric biosensor based on acetylcholinesterase for cationic surfactants detection in water solution, *Biotechnol. Kiev*, **4**(5), 83–89.

79. Mai Anh, T., Dzyadevych, S. V., Chau Van, M., Jaffrezic-Renault, N., Duc Chien, N., Chovelon, J.-M. (2004) Conductometric tyrosinase biosensor for the detection of diuron, atrazine and its main metabolites, *Talanta*, **63**, 365–370.

80. Mai Anh, T., Dzyadevych, S. V., Prieur, N., et al. (2006) Detection of toxic compounds in real water samples using a conductometric tyrosinase biosensor. *Mater. Sci. Eng. C*, **26**, 453–456.

81. Zazoua, A., Hnaien, M., Cosnier, S., Jaffrezic-Renault, N., Kherrat, R. (2009) A new HRP/catalase biosensor based on microconductometric transduction for nitrite determination, *Mater. Sci. Eng. C*, **29**, 1919–1922.

82. Bouyahia, N., Hamlaoui, M. L., Hnaien, M., Lagarde, F., Jaffrezic-Renault, N. (2011) Impedance spectroscopy and conductometric biosensing for probing catalase reaction with cyanide as ligand and inhibitor, *Bioelectrochemistry*, **80**, 155–161.

Chapter 10

Impedance Immunosensors

Frank Davis and Séamus P. J. Higson
Cranfield Health, Cranfield University, Bedford, MK43 0AL, UK
s.p.j.higson@cranfield.ac.uk

This article is intended to firstly give a brief description of biosensing and immunosensing, followed by an overview of the technique of AC impedance, and then a detailed discussion of how AC impedance has been applied in the transduction step in the development of electrochemical immunosensors. Details of recent advances such as the incorporation of nanotechnology into these devices are also described. There will then follow a brief section on immunosensing electrode arrays and then a conclusion discussing future trends.

10.1 Introduction

A biosensor is a device that utilizes a biological moiety, such as an enzyme, an oligonucleotide, or an antibody as a selective recognition element to detect and quantify the presence of an analyte. There are many biochemical assays such as the classical

Electrochemical Biosensors
Edited by Serge Cosnier

ISBN 978-981-4411-46-2 (Hardcover), 978-981-4411-47-9 (eBook)
www.panstanford.com

enzyme-linked immunosorbent assay (ELISA) method, which can do this in solution; however, within a biosensor the recognition element is immobilized in some way onto a solid substrate. This means, there is a necessity for a transduction step to convert the biochemical interaction into a measurable signal. This can be a change in optical density, a change in mass, or most commonly a change in the electrochemical nature of the sensor, as demonstrated by the commonest device of this type in the world, the glucose biosensor. Glucose oxidase is the biorecognition element and is immobilized on a screen-printed carbon electrode, where it catalyzes the reaction of glucose and oxygen to give hydrogen peroxide and gluconolactone.

Many biosensors of this nature have been synthesized; however, one overriding challenge to these types of sensors is that there must be a suitable enzyme available to recognize the analyte of interest. In many cases no such enzyme exists and other recognition moieties must be used. One of the most popular is to utilize the strong selective binding properties of antibodies to their antigen. Antibodies to almost any species can be developed and then immobilized onto a wide variety of substrates such as metal or carbon electrodes. The resultant device is usually referred to as an immunosensor.

The simplest way of detecting a recognition event is whether, as in the case of the glucose electrode, electrochemically active species are produced or consumed. However, since antibody–antigen-binding events are a combination of such forces as hydrogen bonding and other forces such as van der Waals interactions, there is no production of electrons or active species. This challenge can be addressed by labeling one of the reaction components with an easily detectable species, such as a radioactive isotope, a fluorescent molecule, or an enzyme, which can then be used to catalyze a further chemical reaction, as utilized in ELISAs. Although this method has been highly successful, it does have the disadvantages that it introduces extra steps in the assay, requiring the use of further reagents and also increasing the time and expense of the protocol. This has led to the investigation and development of a range of label-free assays and sensors. One of the most prominent of these is the use of alternating current (AC) impedance methods.

10.2 The AC Impedance Method

Electrochemical sensors can use direct current (DC) or alternating current (AC). For AC impedance spectroscopy, an electrode is subjected to a sinusoidal potential stimulus, usually of an amplitude of 5–10 mV. The application of this varying potential both causes the flow of current and perturbs the electrode and its immediate environment. The impedance of the system is the ratio of the voltage to the current in the same way as the resistance of a DC circuit is calculated from $R = V/I$. The impedance of any electrode is affected by the presence of any film on the electrode surface, and any modification of this film, for example, by binding of a species from solution, will be reflected by a change in the impedance characteristics. A detailed description of the theory of AC impedance is beyond the scope of this article, but this subject has been extensively reviewed elsewhere [1, 2].

The measurement of the current in response to the potential waveform is what enables determination of the impedance. For a measurement of impedance at least two electrodes are required; however, more commonly three are used. The working electrode is the electrode onto which the sensing film is immobilized. A second electrode, the reference electrode, is used to maintain a fixed electrode potential with respect to the solution into which the electrodes are immersed. Common reference electrodes include calomel and silver/silver chloride electrodes. Finally a counterelectrode is utilized, which exchanges current with the solution and thereby enables the reference electrode to maintain its fixed potential. A number of materials have been utilized for counter- and working electrodes, such as metals and carbon. One vital factor is that during measurement the distance between the electrodes is constant, else the solution impedance will become another variable. One widely used solution to this is to screen-print the electrodes onto suitable ceramic or plastic substrates; this enables the rapid construction of a large number of identical electrode arrays at relatively low expense.

Usually impedance is measured using a potentiostat, which allows measurement of current flow, whilst maintaining the required voltage at the working electrode. One advantage of AC over DC measurements is that whereas in DC only the potential can be varied, in AC measurements the frequency of the potential waveform can also be varied: A typical range can be from less than 1 Hz up to megahertz, although single-frequency measurements can also be made. Since the voltages are commonly very small, a few millivolts, they usually do not have major effects on the sensing film themselves. Higher voltages such as those often used in DC measurements (amperometry or cyclic voltammetry) could potentially interfere with the interactions between antibodies and antigens, perhaps repelling charged species from the electrode surface or otherwise disrupting its structure. The use of DC can also be combined with AC in that a bias can be applied to the electrode with the AC waveform being superimposed on this potential. Care must be taken that this does not disrupt the binding interactions or damage the film.

One advantage of modern technology has been that potentiostats are now almost inclusively computer controlled; this not only enables a range of DC and AC interrogation regimes to be easily applied to electrodes but also modeling of the response. Using AC impedance methods allows the determination of several factors influencing the electrode response. Impedance of an electrode immersed in solution is commonly a combination of several factors: the solution impedance and that of any cables (which is normally constant and relatively small and so in these cases can be ignored), along with the resistance of the cell (known as the real or faradic component), as well as the capacitance of the cell (often called the imaginary component). Numerous mathematical models of these systems exist; one of the most commonly used is the Randles equivalent circuit [3], a schematic of which is shown in Fig. 10.1. This

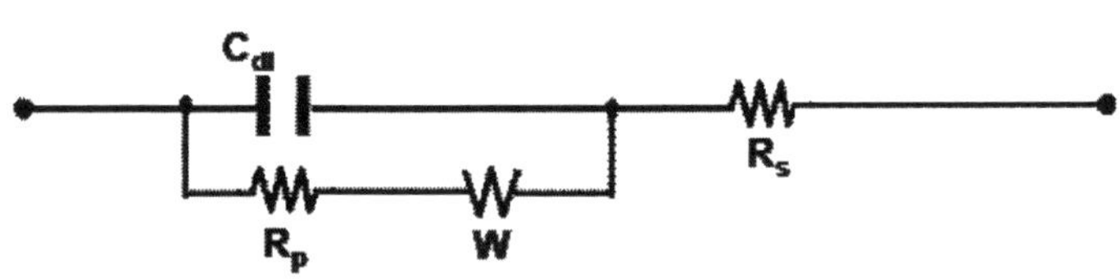

Figure 10.1 A Randles equivalent circuit representing each component at the interface and in the solution during an electrochemical reaction. C_{dl}, double-layer capacitor; R_p, polarization resistor; W, Warburg resistor; R_s, solution resistor.

model circuit consists of a double-layer capacitor in parallel with a polarization resistor and a Warburg impedance connected in series with a resistor that models the resistance of the electrolyte solution. Modern instrumentation allows the electrode to be scanned over a range of discrete frequencies and uses a frequency response analyzer to obtain detailed measurements of the electrochemical system [1, 2]. Most commercial instruments also come supplied with dedicated software capable of fitting the responses obtained to mathematical models such as the Randles circuit.

Utilizing AC impedance as a detection method confers several advantages. It is label free, meaning that there are no additional labeling steps that may require expensive reagents and lengthy reaction times. Although label-free methods may not be as sensitive as labeled methods, they are usually quicker and easier to perform [1, 2]. Ease of use and rapid response times are especially important if the resultant device is to be used in a point-of-care situation such as a doctor's surgery or in the field rather than in a specialized laboratory.

One technique for enhancing the sensitivity of the AC impedance technique is to utilize a redox couple in solution; typical couples include ferrocene derivatives, ferri-/ferrocyanide mixtures, or ruthenium salts. When combined with a suitable DC bias, this couple can undergo reversible redox reactions at the electrode surface, leading to electron transfer between the probe and the electrode. This process is known as faradic impedance spectroscopy; however, one issue is that using such a probe may complicate the use of this technique in point-of-care applications.

There are a number of methods used to display the data obtained from an AC impedance method. One of the simplest is the so-called Bode plot, where total impedance is plotted as a function of frequency (Fig. 10.2a). A second method is to plot the phase angle against frequency (Fig. 10.2b). The phase angle describes the relationship between the sinusoidal voltage waveform and the waveform of the resultant current and can vary from 0° (the two waveforms are completely in phase with each other) through 90° where they are completely out of phase. However, most commonly, impedance data is plotted as a Nyquist plot, where the imaginary component of the impedance is plotted on the *y* axis against the

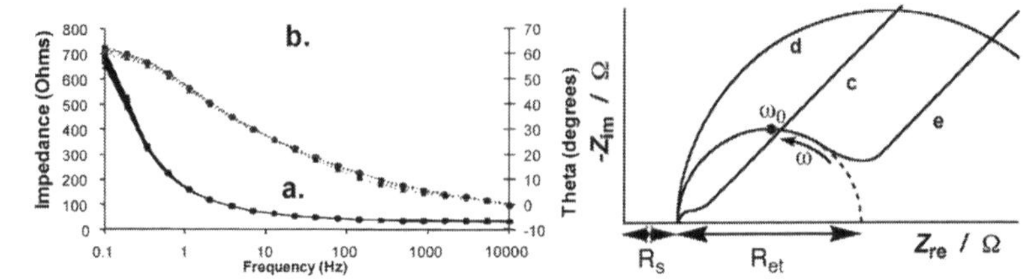

Figure 10.2 A Bode plot (a) and a phase angle plot (b) for a typical AC impedance experiment. Nyquist plots for (c) a system that is diffusion controlled, (d) a system that is controlled by electron transfer, and (e) an intermediate system that is controlled by electron transfer (high frequencies) and diffusion (low frequencies). Reproduced from Katz and Willner (2003) with permission from Wiley-VCH [1].

real component (x axis). Three examples of Nyquist plots are shown (Fig. 10.2c–e).

Nyquist plots allow details of the kinetics of the various processes occurring at the electrode surface to be determined, especially in faradic impedance spectroscopy. When electron transfer between the redox probe and the electrode is very fast, the diffusion of probe molecules to the surface is the rate-determining step and the resulting plot is practically linear in nature (Fig. 10.2c). Figure 10.2d shows the type of curve where the electron transfer is much slower; in this case the curve is semicircular in nature: the real part of the capacitance rises as frequency falls, whereas the imaginary component reaches a maximum and then falls again. Figure 10.2e shows a system controlled by both factors: At the high-frequency end of the spectrum (closest to the origin) electron transfer kinetics between the probe and the electrode dominate the response and the Nyquist plot is again semicircular; however, as the frequency falls, diffusion of the redox species begins to dominate the process and the curve becomes linear. By extrapolating the semicircle portion of the spectrum to where it crosses the x axis, certain parameters of the system can be obtained. The value at which the curve crosses the x axis at the high-frequency end is the solution resistance (R_S), whereas the diameter of the semicircle gives the electron transfer resistance (R_{et}) between the probe and the electrode. A much more detailed description of these processes has been published elsewhere [1].

The AC impedance of a system is affected by a number of factors such as the nature of the electrode itself, the presence of any film on the surface of the electrode, and the nature of the surface–solution interface. Should the nature of any film on the electrode change, this will have an effect on the impedance, therefore allowing detection of any recognition event such as antibody–antigen binding. This makes AC impedance of use in interrogating electrochemical immunosensors. For example, if a bulky analyte such as a protein is bound to a film, this could hinder the access of any redox probe to the electrode and have an insulating effect. Other binding events could have different effects. For instance, binding a cationic species to a surface could lead to increased attraction of an anionic probe, thereby reducing impedance. However, if the bound species and the probe molecule have the same type of charge, there will be a repulsion [1, 2]. Also many antibodies show changes in conformation upon binding of the complementary antigens, so an antibody-loaded film can show changes in impedance even if the target is small and uncharged.

10.3 Impedance Immunosensors

The commonest type of immunosensor appears to be of the type shown in Fig. 10.3, where antibodies are immobilized by a suitable chemical or physical method and exposed to a solution of antigen.

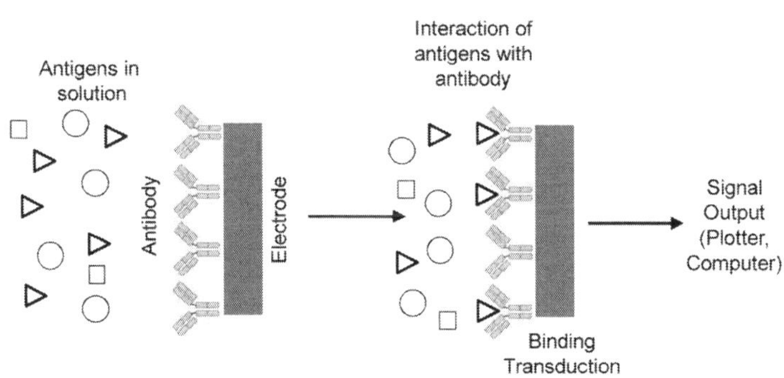

Figure 10.3 Schematic of an electrochemical immunosensor.

Binding of the antigen occurs, thereby changing the nature of the film and the AC characteristics of the cell. Early immunosensors were based on interdigitated electrodes such as in this work, where a composite of human IgG and crosslinked protein was deposited and interrogated by AC impedance before and after exposure to the antigen. Limits of detection as low as 50 nm mL^{-1} were obtained, and nonspecific binding could be accounted for by use of a control electrode containing no antibody [4]. A similar system using a polysiloxane matrix and anti-α-fetoprotein antibodies on silicon electrodes allowed detection of the antigen in the range of 10–150 ng mL^{-1} [5].

Gold electrodes can be modified with a number of thiol monolayers, which can then be used to immobilize antibodies. For example, impedimetric immunosensors to α-fetoprotein [6] with linear ranges of 0.01–10 mg mL^{-1} and a detection limit of 10 μg mL^{-1} could be constructed. Other workers compared chronoamperometry with AC impedance for immunosensors to the protein interferon-γ [7]. The AC method proved much more sensitive with extremely low limits of detection (0.02 fg mL^{-1}), and either 0.1 M KCl or electrochemical pulsing could regenerate the sensor. However, nonspecific binding was a challenge.

Early work did not utilize redox probes, that is, it was nonfaradic in nature. However, use of a ferri-/ferrocyanide redox couple was demonstrated to amplify the signal. Antibodies to IgG were immobilized on thiol-modified gold electrodes and shown to facilitate quantification of the antigen with detection limits as low as 5–10 pg mL^{-1} [8]. Similarly antihemoglobin could be immobilized using avidin–biotin interactions at thiol-modified gold electrodes and the resultant immunosensors interrogated by faradic impedance at low AC frequencies to detect the antigen with a detection limit of 10^{-10} mol L^{-1} [9].

A number of other groups have utilized AC impedance within immunosensing applications, and within this chapter we will describe some of the most recent works. Amongst the most recent investigations studies is where an analogue of the pesticide atrazine was immobilized onto a gold electrode, and this was used in a competitive assay, where the electrode was exposed to a mixed solution containing both the antibody and free atrazine. The lower the level of pesticide in solution, the more the antibody binds to the sensor, allowing detection of atrazine with a detection limit of 40 ng mL^{-1} in water or 190 ng mL^{-1} in red wine [10]. Similarly, antibodies to *Salmonella* could be cast onto gold electrodes and crosslinked with glutaraldehyde. This gave an immunosensor that when interrogated using AC impedance with a redox couple was capable of detecting *Salmonella typhimurium* in milk [11] without any need for extensive sample pretreatment. Other groups have also used these methods to detect bacteria; gold screen-printed electrodes could be used as substrates for immobilization of antibodies that had been thiolated or bound via a crosslinker to give a label-free impedimetric sensor for *E. coli* in river and tap water [12]. Thiol-modified gold electrodes could be used in the detection of much smaller molecules. For example, antibodies to the antibiotic enrofloxacin were covalently immobilized onto these substrates to give an immunosensor with picogram per milliliter limits of detection [13].

Conducting polymers have been widely used as a method of immobilizing biological species to construct biosensors and immunosensors. For example, luteinizing hormone could be entrapped within a matrix of conducting polypyrrole and binding of the resultant antigen detected using impedance spectroscopy [14]. A similar codeposition protocol could be used to entrap antibodies into polypyrrole films to give AC impedance immunosensors for bovine serum albumin (BSA) and digoxin [15]. Biotinylated polypyrrole films could be used to immobilize anti–human IgG by making use of biotin–avidin-binding protocols to give AC impedance immunosensors that had a detection limit of 10 pg mL^{-1} and a linear range of 10–80 ng mL^{-1} for IgG [16].

A composite of polyaniline/poly(styrene sulphonate) could be electrodeposited and used as a platform to immobilize aflatoxin B1 antibodies, which gave rise to a label-free immunosensor with a detection limit of 0.1 μg mL^{-1} [17]. Gold electrodes could be used as a substrate to electrodeposit polytyramine and then glutaraldehyde used to couple a suitable antibody to give a label-free capacitative sensor for HSA with a usable range of 1.84–368.6 ng mL^{-1} and a limit of detection of 1.60 ng mL^{-1} [18]. This array is stable to acidic washing, allowing the antibody–antigen complex to be disrupted and the sensor to be regenerated. Other workers used poly(1,2-diaminobenzene) to immobilize antibodies to give an immunosensor to transferrin [19].

One challenge with sensors formulated by electrodeposition of conductive polymers and biomolecules is that the recognition species can be buried on the film, thereby preventing its availability for binding. The amino groups present in polyaniline can be chemically modified after deposition to give a conductive polymer film substituted with biotin groups. Avidin can be adsorbed onto this surface, and then biotinylated antibodies can be immobilized on top of this via the strong avidin–biotin interaction. Antibodies to digoxin were immobilized onto polyaniline using this method as well as simple entrapment into polymeric films [20]. It was shown that the sensors produced using the affinity protocol had limits of detection 2–3 orders of magnitude lower than those from entrapment. A range of other antibodies were also utilized, allowing the construction of immunosensors for myelin basic protein [21], ciprofloxacin [22, 23], and internalin B, a marker for *Listeria* [24]. These sensors all had limits of detection of 1 ng mL^{-1} of the antigen or lower, and in the case of ciprofloxacin, quantification could also be obtained in a complex organic media, namely, milk [23].

Most of the immunosensors described in this paper operate in a label-free protocol; however, sensitivity of impedimetric sensors can be improved by labeling. For example, the enzyme transglutaminase can be immobilized onto polystyrene sulphonate–coated gold electrodes and then used to bind antitransglutaminase from human serum samples. The assay was then developed by using a peroxidase-labeled secondary antibody, which was then used to oxidize amino-9-ethyl carbazole, and the product of this reaction formed an insoluble precipitate on the surface, which could be determined by AC impedance [25].

10.4 Immunosensors Containing Nanomaterials

Nanotechnology has become one of the buzz words in many fields of science, and the world of biosensors is no exception. There has been a wide range of research carried out into incorporating nanosized materials into these sensing systems. There are many potential benefits for the use of such materials as metal nanoparticles or carbon nanotubes (CNTs) in sensors, one of the simplest being that they increase the surface area. Other benefits include the fact that due to their similar sizes, nanomaterials and biological species can interact intimately, allowing for such processes as electron transfer to occur more efficiently, as well as increases in electrical conductivity and also catalytic effects due to the novel chemistry of these nanomaterials. Much of the work on the use of nanomaterials in electrochemical immunosensors has been recently reviewed [26].

Gold nanoparticles have proved popular in biosensing applications, and some have been incorporated into electrochemical immunosensors. Gold nanoparticles could be electrochemically deposited onto electrodes coated with a sol–gel and used as a substrate for the immobilization of antibodies to transferrin. The resultant immunosensor had a linear range of 1–75 ng mL^{-1} and a detection limit of 0.05 ng mL^{-1}, much lower than the analogue immunosensor (1 ng mL^{-1}) that did not contain gold nanoparticles. Similarly a glassy carbon electrode was coated with gold nanoparticles and then modified with an allergen probe. This was then exposed to solutions of the corresponding antibody interrogated by faradic impedance to determine antibody concentrations [27]. Other workers [28] electropolymerized 2-aminobenzene thiol and used the thiol moiety to immobilize gold nanoparticles onto the electrode and used this as a substrate for immobilizing antibodies. The resultant device was highly sensitive, being capable of determining transferrin [29] with a limit of detection of 80 pg mL^{-1}, along with a relatively large dynamic range of 0.125–100 ng mL^{-1} and a high signal-to-noise ratio. Gold nanoparticles were also assembled onto glassy carbon electrodes and used to immobilize antibodies to *Salmonella* [30]. The resultant sensor was interrogated by AC impedance using a ferricyanide probe and the capacitance shown to be proportional to the logarithmic concentration of the bacterium between 1.0×10^2 and 1.0×10^5 CFU mL^{-1}. The sensor could be regenerated by washing with glycine/HCl buffer and was successfully used to detect *Salmonella* spp. in lab-processed commercial pork samples.

CNTs have also been widely studied within biosensors and in many cases display enhanced detection sensitivity compared to sensors that have been fabricated using other carbon species. Possible reasons for this are enhanced surface area, their ability to catalyze electrochemical reactions, and their high electrical conductivity [31]. Immunosensors containing CNTs have been described. For example, composite polyanisidine/CNT electrodes were shown to have improved characteristics within impedimetric biosensors compared to electrodes containing carbon black [32].

10.5 Immunosensor Arrays

Many of the immunosensors described above use electrodes that are a few square millimeters in size. However, there has been much interest in the miniaturization of electrodes to give microelectrodes and even nanoelectrodes. Such small electrodes display several advantages over their larger brethren. For example, whereas for large planar electrodes, diffusion to and from the surface can be the rate-limiting step (as demonstrated by AC impedance measurements amongst others), the smaller electrodes demonstrate hemispherical diffusion profiles, leading to enhanced diffusion and, amongst other things, rendering them relatively stir independent. One challenge, though, is that small electrodes lead to small signals. A solution to this is to construct an array of microelectrodes, each of which displays typical microelectrode behavior, and then combine their responses.

Within our group we have carried out extensive investigations into the use of microelectrode arrays as a basis for impedimetric immunosensors. A schematic of our process is shown in Fig. 10.4. The initial platform is a gold or screen-printed carbon electrode onto which we electrodeposit an insulating film of poly(1,2-diaminobenzene) [20]. This film is then sonochemically ablated in water; the ultrasonic irradiation leads to the formation of superheated bubbles within the water, which then collapse, sending out jets that ablate holes in the insulating layer.

Figure 10.4d shows the layer of poly(1,2-diaminobenzene) after ablation. As can be seen there are a number of randomly distributed pores ranging from 0.1 to 3 μm in diameter "punched" through the film, exposing the underlying conductive surface.

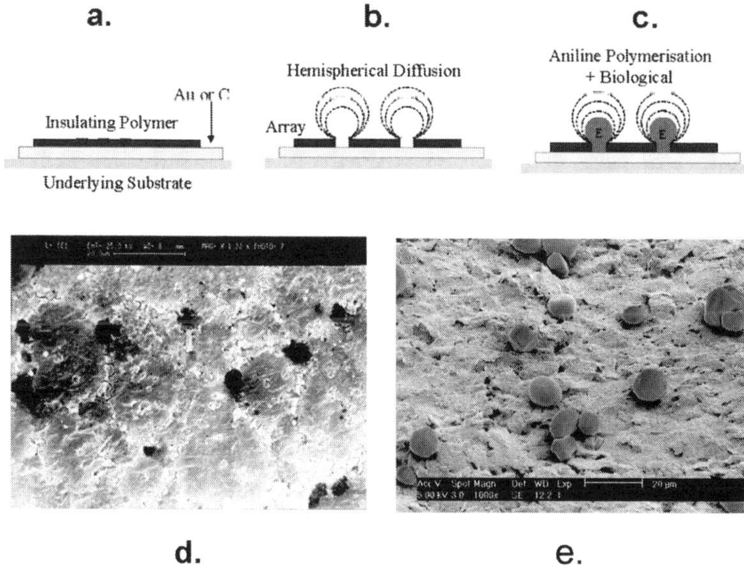

Figure 10.4 Polyaniline microarrays for immunosensing (a) electrodeposition of a poly(1,2-diaminobenzene) insulating layer, (b) sonochemical ablated pores, (c) electropolymerization to form polyaniline electrodes, and scanning electron micrographs of (d) sonochemically formed pores and (e) polyaniline protrusions.

After formation of the array, it is then possible to carry out a further electropolymerization, this time of aniline, from the exposed areas to form an array of conductive polymer protrusions (Fig. 10.4e). Biological molecules, such as enzymes or antibodies, can be entrapped within the protrusions during this growing process, or alternatively simple polyaniline protrusions can be formed and chemically modified with a biotinylating agent. This can then be used as a platform for immobilizing biotinylated antibodies via a common avidin-to-biotin affinity protocol [20].

Within initial works, polyaniline containing entrapped antibodies to BSA was deposited, both in the form of a simple planer film and as a sonochemically fabricated array of polyaniline microelectrodes. Interrogation of both systems by AC impedance before and after exposure to solutions of antigen allowed the determination of calibration profiles. It was demonstrated that whereas planar films had working ranges of 1–75 μg mL^{-1}, the microelectrode arrays had working ranges of 1–200 ng mL^{-1} and also gave much lower detection limits than simple planar polyaniline electrodes. Since other work described above demonstrated that affinity-based protocols led to much higher sensitivities than entrapment protocols for anti-digoxin-based immunosensors, the next stage was to deposit polyaniline microarrays and then use them as a support for antibodies by avidin–biotin affinity. This combining of affinity binding with a microarray support led to further increases in sensitivity and allowed the construction of an immunosensor for prostate-specific antigen, a marker for prostate cancer, with detection limits as low as 1 pg mL^{-1} [33]. Other advantages of the affinity protocol included much longer storage lifetimes (at least 12 weeks dry at 4°C) and also the potential to regenerate the immunosensors several times by washing with dilute HCl. Entrapped antibody sensors were much less temporally stable, significant drops in activity being obtained after six weeks, and activity was destroyed by a single washing step. Other immunosensors with similar behavior, picogram per milliliter limits of detection and dynamic ranges of 1–100 pg mL^{-1}, even in the presence of large excesses of potentially interfering proteins, were successfully constructed for the stroke marker proteins neuron-specific enolase [34] and S100[β] [35].

A variety of other methods have been used to fabricate micro-/nanoarray sensors suitable for immunological sensors. Interdigitated gold microelectrode arrays with a finger thickness of 5 μm could be constructed using photolithographic techniques, substituted with thiol, and then used as a platform to bind monoclonal antibodies to cortisol. AC impedance could then be used to give label-free detection of the antigen in the range of 1 pM–100 nM [36]. Other workers also used interdigitated electrodes to construct immunosensors for H5N1 avian influenza virus [37] with a linear range between 10^3 and 10^7 EID_{50} mL^{-1}.

Chemical vapor deposition could be used to deposit vertically aligned CNTs up to 2 mm long onto a silicon surface [38] in 100 × 100 μm blocks to give the microelectrode array shown in Fig. 10.5. These arrays could be protected by encapsulating in epoxy resin

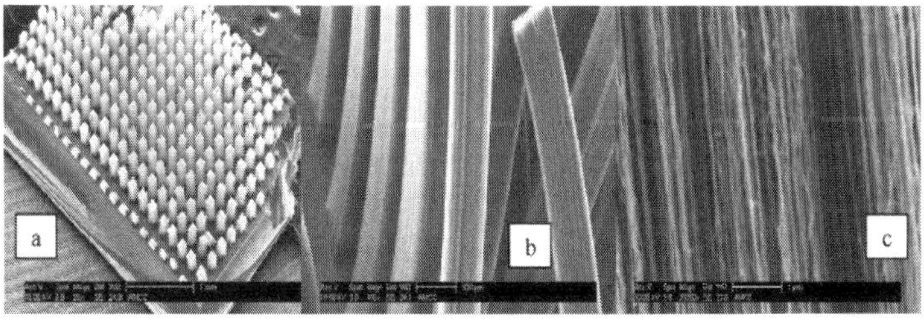

Figure 10.5 ESEM images of (a) aligned multiwalled carbon nanotube–patterned arrays on a Si substrate, (b) side view of the nanotube array, and (c) high-resolution side view of the nanotube array. Reproduced from Yun et al. (2007) with permission from Elsevier.

and then polishing to expose the ends. This could then serve as a substrate to immobilize antibodies to mouse IgG and AC impedance used to determine the electron transfer resistance in order to give an immunosensor with a detection limit of 200 ng mL^{-1} and a dynamic range up to 100 μg mL^{-1}.

The immunosensors described so far have usually been analyzing for a single target. Even the arrays shown so far generally measure the sum of responses of the individual microelectrodes. One large advantage of an array of sensors is that it could well prove possible to both modify and address each sensor individually. This would allow each sensor to be modified with a different antibody, enabling simultaneous determination of a number of analytes such as a range of cancer markers. This would greatly reduce the time and cost required to carry out an analysis and make the device more suitable for point-of-care applications. One example of this is in this work [39], where the authors developed an immunosensor with several working electrodes individually modified with antibodies to IgG (human, rat, and goat). AC impedance could then be utilized to selectively determine which antigens were present in a sample. Commercial devices such as the Combimatrix device, which contains large numbers of individually addressable microelectrodes, have now made their way into the market and could be used as platforms for these types of assays. A number of labeled and amperometric assays for multiple analytes have also been developed [2].

10.6 Conclusion and Outlook

We have within this chapter attempted to review the principles of impedance immunosensing and given an overview of the most recent work in this field. Immunosensing is based on the exquisite specificity of antibodies for their antigens. This specificity combined with the high sensitivity of AC impedance methods has allowed selective determination of many antigens at picograms per milliliter levels and lower. Also AC measurements offer the advantages of relative rapidity since they are label free, meaning that sample pretreatment is often not required, thereby reducing time and expense. The use of inexpensive screen-printed single-shot sensors will remove the need for washing or regeneration steps.

One highly desired device would be a handheld point-of-care sensor capable of taking a sample and simultaneously interrogating it for a wide variety of analytes such as markers for stroke and myocardial infarction, etc., which would be of great use to paramedics. Multiple simultaneous assays for a range of cancer markers is another potential application or a device that simultaneously analyzes for a range of bacterial or viral pathogens such as those found in sepsis, a major killer in the U.K. and the rest of the world.

Challenges still exist, for example, the reproducible and accurate detection of targets at low concentrations in samples such as blood or saliva that contain high levels of potential interfering compounds. Stability and batch-to-batch reproducibility of these types of systems can also be an issue.

We predict more research into this field, and these types of devices will be driven by the point-of-care market. Attempts will be made to enhance sensitivity and selectivity, reduce interference and nonspecific binding effects, and minimize size, cost, and power requirements. Impedance immunosensors fit these criteria, especially because of their label-free nature and potential for single-use electrodes. Sensing may be enhanced and size and power reduced by the use of nanotechnology and nanosized materials. Integration with other technologies such as microfluidics will allow simultaneous testing for a range of analytes and take these assays from the laboratory into the clinical, food, and environmental assay fields.

References

1. Daniels, J. S., Pourmand, N. (2007) Label-free impedance biosensors: opportunities and challenges, *Electroanalysis*, **19,** 1239–1257.
2. Katz, E., Willner, I. (2003) Probing biomolecular interactions at conductive and semiconductive surfaces by impedance spectroscopy: routes to impedimetric immunosensors, DNA-sensors, and enzyme biosensors, *Electroanalysis*, **15,** 913–947.
3. Randles, J. E. B. (1947) Kinetics of rapid electrode reactions, *Discuss. Faraday Soc.*, **1**, 11–19.
4. Taylor, R. F., Marenchic, I. G., Spencer, R. H. (1991) Antibody- and receptor-based biosensors for detection and process control. *Anal. Chim. Acta,* **249**, 67–70.
5. Maupas, H., Saby, C., Martelet, C., et al. (1996) Impedance analysis of Si/SiO2 heterostructures grafted with antibodies: an approach for immunosensor development, *Electroanal. Chem.*, **406**, 53–58.
6. Limbut, W., Kanatharana, P., Mattiasson, B., Asawatreratanakul, P., Thavarungkul, P. (2006) A comparative study of capacitive immunosensors based on self-assembled monolayers formed from thiourea, thioctic acid, and 3-mercaptopropionic acid, *Biosens. Bioelectron.*, **22**, 233–240.
7. Dijksma, M., Kamp, B., Hoogvliet, J. C., van Bennekom, W. P. (2001) Development of an electrochemical immunosensor for direct detection of interferon-γ at the attomolar level, *Anal. Chem.*, **73**, 901–907.
8. Ameur, S., Martelet, C., Jaffrezic-Renault, N., Chovelon, J. M. (2000) Sensitive immunodetection through impedance measurements onto gold functionalized electrodes, *Appl. Biochem. Biotechnol.*, **89**, 161–170.
9. Hays, H. C. W., Millner, P. A., Prodromidis, M. I. (2006) Development of capacitance based immunosensors on mixed self-assembled monolayers, *Sens. Actuators, B*, **114**, 1064–1070.
10. Ramón-Azcón, J., Valera, E., Rodríguez, Á., et al. (2007) An impedimetric immunosensor based on interdigitated microelectrodes (IDμE) for the determination of atrazine residues in food samples, *Biosens. Bioelectron.*, **23**, 1367–1373.
11. Pournaras, A. V., Koraki, T., Prodromidis, M. I. (2008) Development of an impedimetric immunosensor based on electropolymerized polytyramine films for the direct detection of salmonella typhimurium in pure cultures of type strains and inoculated real samples, *Anal. Chim. Acta*, **624**, 301–307.

12. Escamilla-Gómez, V., Campuzano, S., Pedrero, M., Pingarrón, J. M. (2009) Gold screen-printed-based impedimetric immunobiosensors for direct and sensitive Escherichia coli quantisation, *Biosens. Bioelectron.*, **24**, 3365–3371.
13. Wu, C. C., Lin, C. H., Wang, W. S. (2009) Development of an enrofloxacin immunosensor based on label-free electrochemical impedance spectroscopy, *Talanta*, **79**, 62–67.
14. Farace, G., Lillie, G., Hianik, T., Payne, P., Vadgama, P. (2002) Reagentless biosensing using electrochemical impedance spectroscopy, *Bioelectrochemistry*, **55**, 1–3.
15. Grant, S., Davis, F., Law, K. A., et al. (2005) Reagentless immunosensor for the detection of bsa at platinum electrodes by an AC impedance protocol, *Anal. Chim. Acta*, **537**, 163–168.
16. Ouerghi, C., Touhami, A., Jaffrezic-Renault, N., Martelet, C., Ben Ouada, H., Cosnier S. (2004) Electrodeposited biotinylated polypyrrole as an immobilization method for impedimetric immunosensors, *IEEE Sens. J*, **4**, 559–567.
17. Owino, J. H. O., Ignaszak, A., Al-Ahmed, A., et al. (2007) Modelling of the impedimetric responses of an aflatoxin B1 immunosensor prepared on an electrosynthetic polyaniline platform, *Anal. Bioanal. Chem.*, **388**, 1069–1074.
18. Wu, Z.-S., Li, J.-S., Deng, T., Luo, M.-L., Shen, G.-L., Yu, R.-Q. (2005) A sensitive immunoassay based on electropolymerized films by capacitance measurements for direct detection of immunospecies, *Anal. Biochem.*, **337**, 308–315.
19. Lu, Z., Huang, S., Jiang, D., Liu, B., Kong, J. (2004) A novel capacitive immunosensor using electropolymerized insulating poly(o-phenylenediamine) film on a glass carbon electrode for probing transferrin, *Anal. Lett.*, **37**, 2283–2301.
20. Barton, A. C., Collyer, S. D., Davis, F., et al. (2009) Labeless AC impedimetric antibody-based sensors with pg mL^{-1} sensitivities for point-of-care biomedical applications, *Biosens. Bioelectron.*, **24**, 1090–1095.
21. Tsekenis, G., Garifallou, G. Z., Davis, F., Millner, P. A., Gibson, T. D., Higson, S. P. J. (2008) Label-less immunosensor assay for myelin basic protein based upon an AC impedance protocol, *Anal. Chem.*, **20**, 2058–2062.
22. Garifallou, G.-Z., Tsekenis, G., Davis, F., et al. (2007) Labeless immunosensor assay for fluoroquinolone antibiotics based upon an AC impedance protocol, *Anal. Lett.*, **40**, 1412–1422.

23. Tsekenis, G., Garifallou, G.-Z., Davis, F. et al. (2008) Detection of fluoroquinolone antibiotics in milk via a labeless immunoassay based upon an alternating current impedance protocol, *Anal. Chem.*, **80**, 9233–9239.

24. Tully, E., Higson, S. P. J., O'Kennedy, R. (2008) The development of a "labeless" immunosensor for the detection of listeria monocytogenes cell surface protein, internalin B, *Biosens. Bioelectron.*, **23**, 906–912.

25. Balkenhohl, T., Lisdat, F. (2007) Screen-printed electrodes as impedimetric immunosensors for the detection of anti-transglutaminase antibodies in human sera, *Anal. Chim. Acta*, **597**, 50–57.

26. Holford, T. R. J., Davis, F., Higson, S. P. J. (2012) Recent trends in antibody based sensors, *Biosens. Bioelectron.*, **34**, 12–24.

27. Huang, H. Z., Liu, Z. G., Yang, X. R. (2006) Application of electrochemical impedance spectroscopy for monitoring allergen–antibody reactions using gold nanoparticle-based biomolecular immobilization method, *Anal. Biochem.*, **356**, 208–214.

28. Hu, S. Q., Xie, Z. M., Lei, C. X., Shen, G. L., Yu, R. Q. (2005) The integration of gold nanoparticles with semi-conductive oligomer layer for development of capacitive immunosensor, *Sens. Actuators, B*, **106**, 641–647.

29. Yin, T., Wei, W., Yang, L., Gao, X., Gao, Y. (2006) A novel capacitive immunosensor for transferrin detection based on ultrathin alumina sol-gel-derived films and gold nanoparticles, *Sens. Actuators, B*, **117**, 286–294.

30. Yang, G. J., Huang, J. L., Meng, W. J., Shen, M., Jiao, X. A. (2009) A reusable capacitive immunosensor for detection of Salmonella sbased on grafted ethylene diamine and self-assembled gold nanoparticle monolayers, *Anal. Chim. Acta*, **647**, 159–166.

31. Dumitrescu, I., Unwin, P. R., Macpherson, J. V. (2009) Electrochemistry at carbon nanotubes: perspective and issues, *Chem. Commun.*, 6886–6901.

32. Carrara, S., Bavastrello, V., Ricci, D., Stura, E., Nicolini, C. (2005) Improved nanocomposite materials for biosensor applications investigated by electrochemical impedance spectroscopy, *Sens. Actuators, B*, **109**, 221–226.

33. Barton, A. C., Davis, F., Higson, S. P. J. (2008) Labeless immunosensor assay for prostate specific antigen with pg ml^{-1} limits of detection based upon an AC impedance protocol, *Anal. Chem.*, **80**, 6198–6205.

34. Barton, A. C., Davis, F., Higson, S. P. J. (2008) Labeless immunosensor assay for the stroke marker protein neuron specific enolase based upon an AC impedance protocol, *Anal. Chem.*, **80**, 9411–9416.

35. Barton, A. C., Davis, F., Higson, S. P. J. (2010) Labeless immunosensor assay for the stroke marker protein s-100[β] based upon an AC impedance protocol, *Anal. Lett.*, **43,** 2160–2170.

36. Arya, S. K., Chornokura, G., Venugopalb, M., Bhansali, S. (2010) Dithiobis(succinimidyl propionate) modified gold microarray electrode based electrochemical immunosensor for ultrasensitive detection of cortisol, *Biosens. Bioelectron.*, **20**, 2296–2301.

37. Wang, R., Wang, Y., Lassiter, K., et al. (2009) Interdigitated array microelectrode based impedance immunosensor for detection of avian influenza virus H5N1, *Talanta*, **79**, 159–164.

38. Yun, Y. H., Bange, A., Heineman, W. R., et al. (2007) A nanotube array immunosensor for direct electrochemical detection of antigen–antibody binding, *Sens. Actuators, B*, **123**, 177–182.

39. Yu, X. B., Xu, D. W., Xu, D. K., Lv, R., Liu, Z. H. (2006)An impedance biosensor array for label-free detection of multiple antigen-antibody reactions, *Front. Biosci.*, **11**, 983–990.

Chapter 11

Transduction of Biochemical Reactions by Use of Quantum Dots and Photocurrent Detection

Fred Lisdat and Daniel Schäfer

Biosystems Technology, Technical University of Applied Sciences Wildau, 15745 Wildau, Germany

flisdat@th-wildau.de

This chapter is devoted to a particular kind of nanostructures—quantum dots, which are semiconducting nanoparticles showing not only interesting optical properties but also electrical features modulated by light. These particles can be fixed on electrode surfaces and thus act as a light-switchable layer. To use quantum dot electrodes and photocurrent measurements for signal transduction in bioanalysis, the surface properties of the nanoparticles have to be adapted and defined reaction schemes have to be established. On this basis it can be shown that analytical signal chains can be constructed that allow light-controlled detection of an analyte molecule. Provided that the biochemical reaction can be confined to a certain surface region it gives access to a spatially resolved readout of a sensor electrode.

Electrochemical Biosensors

Edited by Serge Cosnier

ISBN 978-981-4411-46-2 (Hardcover), 978-981-4411-47-9 (eBook)

www.panstanford.com

11.1 Introduction to Quantum Dots

Quantum dots (QDs) are nanoparticles (NPs) of a semiconductor material. Often the term "nanoparticle" is applied to a system with a size of <100 nm. For QDs this definition can be given more precisely on the basis of quantum confinement. As these particles consist of a semiconductor there is a gap between the energetic state of electrons in the valance band and in the conduction band. By interaction with light electrons can acquire a higher-energy state. Subsequently a free electron and a hole (or defect electron) are created. The hole and the electron are called excitons, which possess a certain lifetime and radius. The latter is termed as "exciton Bohr radius." Quantum confinement occurs when at least one dimension of the NP is smaller than the exciton diameter. One can observe that in these cases the energetic structure and thus the bandgap become a function of the size of the NP.

This electronic situation has some consequences for the properties of semiconducting NPs. Most obvious are consequences for optical properties, which appear to be also dependent on the particle size. Namely, these are absorption and luminescence properties [1, 2]. QDs show a rather broad range of wavelengths where absorption of light is possible but with a clear absorption maximum. This shifts to higher wavelengths with increasing size of the QDs (Fig. 11.1I).

Since after photoexcitation the NPs return to the original state the wavelength of emitted photons is again size dependent. In contrast to absorption the emission spectra of QDs show a rather

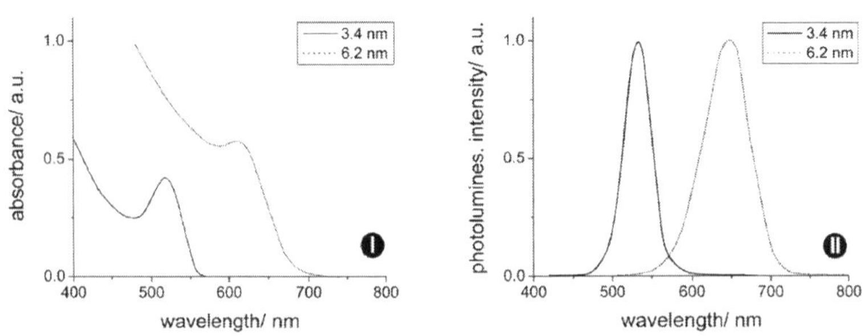

Figure 11.1 Absorption (I) and emission (II) spectra of CdTe quantum dots as a function of particle size, according to Li and Murase [3].

sharp peak, provided the size distribution of the prepared NPs is not very broad. The fluorescence spectra of QDs with different sizes show a well-defined blue shift of emission with decreasing size (Fig. 11.1II). Thus, QDs have become very popular labels in biology and bioanalysis since molecules coupled to QDs of different sizes can be easily distinguished. Therefore a multiplexed analysis is possible. In addition, it is advantageous that the same coupling strategy for biomolecule immobilization can be used for all these fluorescence labels of different wavelengths. Compared to organic fluorophores the emission spectra of QDs are more symmetric and photobleaching is much less pronounced. Besides the size of the QDs the composition is also a factor that modifies optical properties [4, 5].

The confinement model does not consider surface states. However, a large fraction of atoms in such nanoclusters are surface atoms. This results in a rather high number of surface states that can act as a trap for charge carriers. This will also lead to decreased fluorescence. Besides the use of improved synthesis protocols and organic capping agents, QDs can also be encapsulated with a thin layer of another semiconductor of higher bandgap. CdSe particles with a ZnS shell are typical examples of this kind of particles. By such core–shell particles the fluorescence properties can be significantly improved [2, 4].

Since no free charge carriers are present in QDs they are electrically nonconductive. However, by thermal- or photoactivation charge carriers can be generated, as explained above. This allows the generation of a photocurrent when QDs are fixed on electrodes. Depending on the potential applied on the electrode it is possible to generate a cathodic as well as an anodic photocurrent. This means that electron transfer from the electrode to holes in the valence band occurs in one case and transfer of excited electrons from the conduction band to the electrode in the other case [6–8]. This is schematically shown in Fig. 11.2. When this model is correct then the photocurrent should follow the absorption spectra of the NPs. This can be shown and is exemplified in Fig. 11.3 for CdSe/ZnS QDs.

Different directions of research can be seen with respect to the photoexcitation of charge-carriers in QDs:

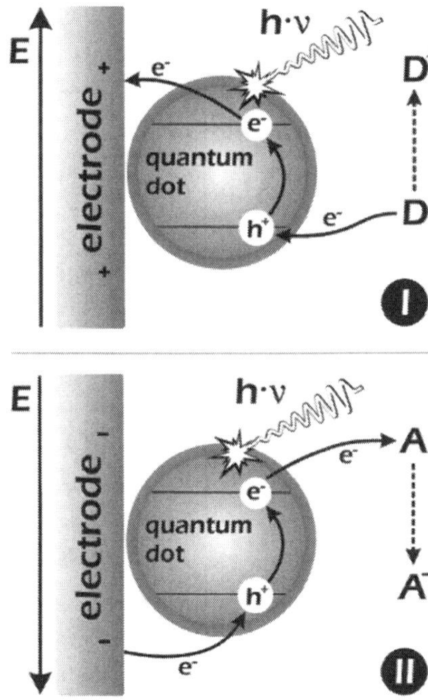

Figure 11.2 Electron transfer steps at a quantum dot–modified electrode after illumination. (I) Oxidation cascade and (II) reduction cascade. "D" represents a donor molecule, which will be oxidized, and "A" represents an acceptor, which is reduced at the QD surface.

- Investigation of charge transport through molecules that are used to fix the QDs to the electrode [10, 11]
- Conversion of light energy into electrical energy (solar cells)
- Electroluminescence (stimulation of light emission by electrical power)
- Photocatalysis (induction of defined reactions on the excited QDs in contact with a solution)
- Light triggered (bio)analysis

In all these systems the generation of charge carriers is connected to electron transfer reactions with the electrode and/or molecules that are in solution or immobilized on the electrode system. This means that not only can the electrode act as a reaction

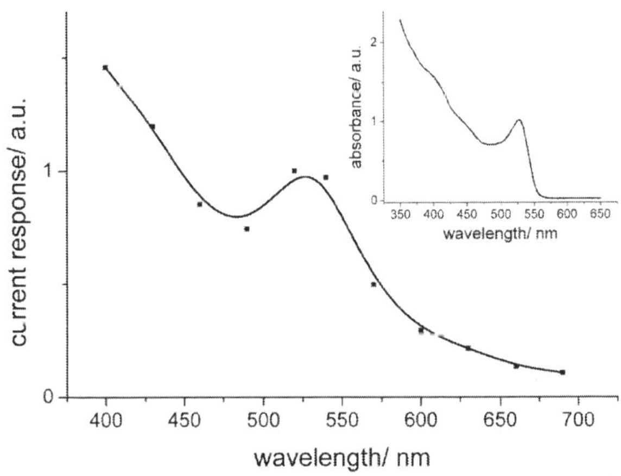

Figure 11.3 Wavelength-dependent photocurrent of a CdSe/ZnS electrode and absorption of a CdSe/ZnS quantum dot solution, according to Stoll et al. [9].

partner but depending on the redox properties of the respective substance electron transfer from and to the QDs is also feasible. This can be used to facilitate certain reactions that, for example, may result in the decomposition of organic molecules as in photocatalysis [12, 13] or simply the oxidation of a redox mediator in a solar cell [14, 15]. For bioanalysis a defined reaction cascade has to be established that can be read out by a photocurrent measurement. This is a rather new area compared to the use of QDs as fluorescence labels or as photocatalytic labels, which can induce easily detectable reactions after illumination.

11.2 Concepts of Using Quantum Dots in Electrochemical Bioanalysis

When QDs are immobilized on an electrode surface a rather insulating layer is prepared, hindering electrochemical conversions to occur. By illuminating such a QD electrode charge carriers are generated, and thus current flow is possible, which is normally termed as "photocurrent." This current is influenced not only by the polarization and magnitude of the voltage applied on the electrode

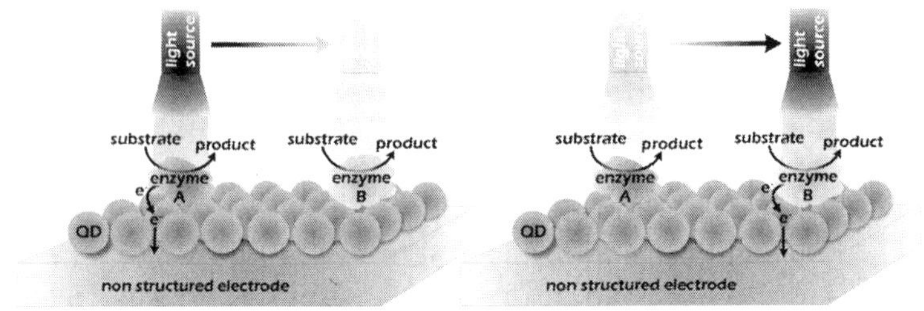

Figure 11.4 Spatially resolved readout of a sensor surface with different biochemical systems immobilized by illumination of the respective electrode area.

but also by the presence of substances that can act as electron donors or acceptors. This means, under fixed potential conditions concentration analysis is possible by photocurrent measurements. Such a reaction scheme is illustrated in Fig. 11.2.

In comparison to an unmodified electrode the QD electrode provides an additional tool to regulate a certain kind of reaction—light. This also means that only the area of the electrode is analyzed, which is illuminated. When one immobilizes different biochemical systems at different areas of the electrode surface then parallel analysis becomes feasible. One analyte molecule can only be detected when the respective area is photoexcited. After this light has to be focused on another area, and the next reaction will be evaluated. Figure 11.4 illustrates the concept of this kind of measurement.

Resolution is dependent on the immobilization of the respective biocomponent and the light system used. These measurements reflect the concept of light-addressable sensors, which have been developed in the area of chemically sensitive semiconductor devices and which are called light-addressable potentiometric sensors (LAPSs) [16, 17]. In contrast to LAPSs, QD electrodes are using only a thin, semiconducting particle layer and allowing electron transfer reactions to occur. As for all kinds of amperometric sensing systems, this provides a better resolution for the analyte detection and gives access to detection schemes for which no potential forming step is available as required for a LAPS. This largely increases the applicability of such systems.

The generation of a photocurrent can also be used as a transduction principle for the detection of a biorecognition event. Here the function of the QD is that of a label. Provided that the binding event brings the QDs in close contact with the electrode a photocurrent can be generated [18]. Further amplification by the addition of a donor or an acceptor compound to the solution is feasible. It is also possible to create a multilayered network on the electrode, provided several binding molecules are fixed on the QD surface. When the distance between the QDs is small enough then interparticle electron transfer seems to be possible, greatly enhancing the photocurrent to be detected. Figure 11.5 is illustrating the principle. For rather small binding molecules such as DNA this system is nicely working [19]; limitations, however, can be seen when multiple protein–QD layers are formed on an electrode. The proteins can hinder the interparticle electron transfer and thus will result in rather small photocurrents [20].

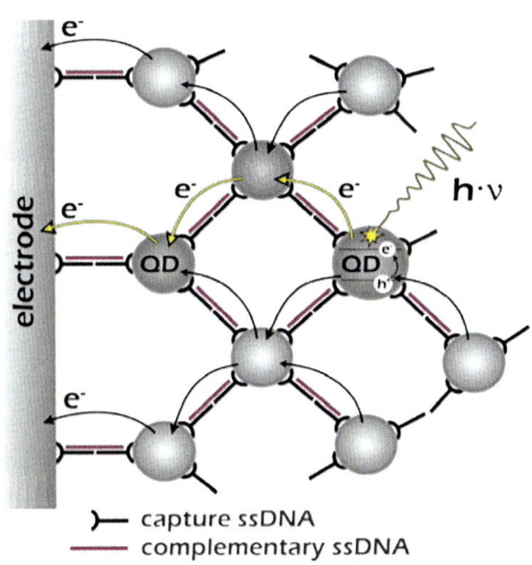

Figure 11.5 QD/DNA superstructure formation by DNA-modified QDs and analyte DNA (complementary ssDNA). The analyte DNA (red) can bridge the QDs on the electrode surface by hybridization and results in significant photocurrent generation upon excitation by multiple electron transfer steps. Modified according to Willner et al. [6].

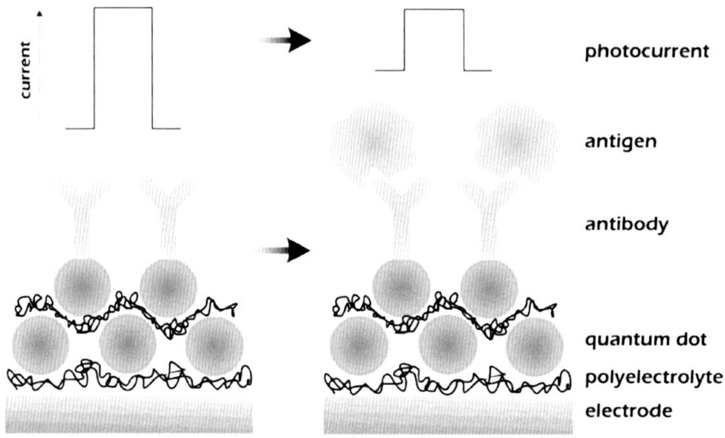

Figure 11.6 Photocurrent change of a QD electrode with immobilized capture molecules by analyte (antigen) binding. Modified according to Zhang [22].

For the detection of binding events also another approach has been developed. Here the QDs are used as the transduction layer as for enzymatic systems. The QD electrode is modified by a specific recognition element (DNA or antibody), and then the change in photocurrent upon binding is evaluated [21]. The principle is illustrated in Fig. 11.6. Although this is a rather unspecific approach, the response can be enhanced by a second binding reaction bringing Au NPs close to the QD electrode and thus allowing energy transfer to occur [23]. It has also been demonstrated that the specific attachment of cells to surfaces can be detected by such an approach [22, 24].

An intensively investigated area of analytical research is the application of QDs as labels on the basis of the very sensitive electrochemical detection of the metal ions the particles consist of. This is done by dissolution of the surface-attached QDs and voltammetric stripping analysis [25, 26]. Another direction of label detection relies on the possibility of electrochemiluminescence generation by electrode-fixed QDs. Electrochemical reactions are applied to result in QD excitation, and thus the emitted light appears to be the analytical signal that corresponds to the presence or absence of a certain analyte molecule [27, 28]. Since for both approaches light is not used as a tool for controlling the QD properties and signal generation these systems will not be the focus of this chapter.

11.3 Immobilization of Quantum Dots on Electrodes

To provide defined conditions for electron transfer of photoexcited QDs with the electrode a reproducible immobilization is essential. For this purpose mainly dithiol or disulfide compounds have been used. In principle two different procedures can be seen here:

(i) The first relates to a modification of the electrode surface in such a way that the bifunctional thiols react with one end of the molecule via chemisorptions to the electrode and expose the other end to the solution. The QDs are then immobilized via ligand exchange directly on the surface, replacing the hydrophobic capping agent from the synthesis by the thiol. This procedure results only in a partial replacement of the organic capping agent, and a subsequent second ligand exchange can follow.

(ii) The second approach performs first a complete ligand exchange at the NPs in solution and then transfers the thiol-modified particles to a gold surface.

Different thiol compounds have been used for anchoring of the QDs: 1,4-dithiane, 1,4-dithiobenzene, bisphenyldithiol, *trans*-stilbenedithiol, etc. The amount of deposited QDs can be measured by quartz crystal microbalance, indicating that the overall amount can be adjusted to be in the range of a monolayer [29, 30]. This, however, does not mean that full monolayer coverage exists at all places on the surface. To verify the presence of NPs on the surface photocurrent measurements in a buffer solution under varying potential can be performed. In dependence on the electrode bias anodic and cathodic photocurrents can be detected (see Fig. 11.7). This allows also the evaluation of a potential with no photocurrent generation, which can serve as a formal indicator of the redox properties of the particles. However, buffer composition and pH will have an influence on this value.

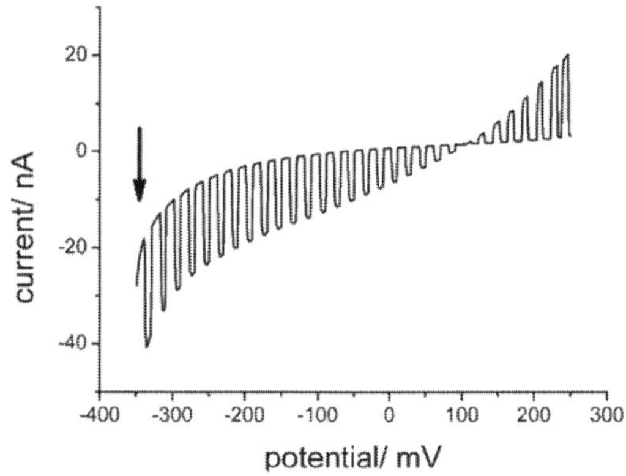

Figure 11.7 Dependence of the photocurrent of a CdSe/ZnS electrode on the applied electrode potential (100 mM HEPES, linear sweep voltammetry 1 mV/s with overlayed light pulses).

To show that the photocurrent generated is really connected to the excitation of NPs by light a wavelength-dependent measurement can be performed. It is shown, for example, that the magnitude of the photocurrent follows the absorption properties of the QDs investigated in solution (Fig. 11.3). This ensures that observed current effects, which are detected, can be correlated to the activation of the QD interlayer and not of molecules in solution.

The idea of a photoswitchable layer can be visualized by impedance measurements of the electrode with and without illumination. Here a clearly decreased surface impedance can be shown for the photoexcited QD electrode [26].

The quality of the thiol interlayer between the electrode and the QD will influence the electron transfer reactions between the nanocrystals and the electrode. Recently the influence of the dithiol layer on the photocurrent has been studied in more detail by X-ray photoelectron spectroscopy (XPS) and near-edge X-ray absorption fine structure (NEXAFS) measurements [31]. *Trans*-stilbenedithiol layers prepared at higher temperature show high ordering and no oxidation products of the thiol group on the gold surface and subsequently also high photocurrent values after QD immobilization.

11.4 QD Electrodes for Detection of Small Molecules in Solution

When the scheme in Fig. 11.2 is correct, then it should be possible to detect different concentrations of redox-active substances in solution by choosing a proper potential and illumination of the electrode surface. This has been shown with the simple ferri-/ferrocyanide system. In these cases an enhanced photocurrent is detected. For example, a light-induced reduction current is found for a CdSe/ZnS electrode at −250 mV versus Ag/AgCl when hexacyanoferrate (III) ions are present in solution. Thus, an electron transfer from the electrode via the QDs to the ions in solution—acting here as an electron acceptor—is achieved (see Fig. 11.8).

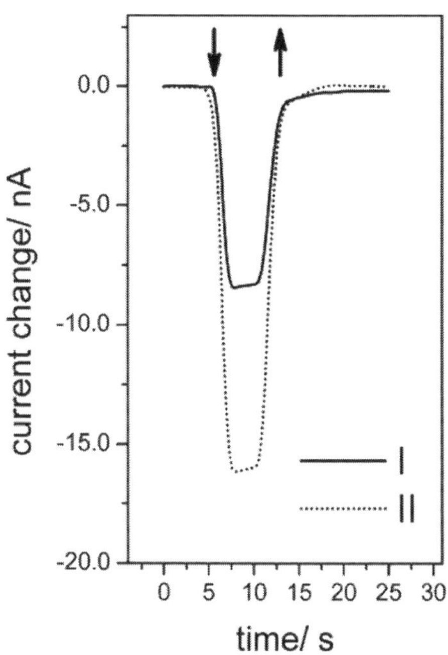

Figure 11.8 Photocurrent behavior of a CdSe/ZnS QD layer fixed on a gold electrode by dithiane in the absence (I) and presence (II) of an electron acceptor—50 μM hexacyanoferrat (III)—in solution, according to Stoll et al. [7]. ($E = -250$ mV vs. Ag/AgCl.)

As expected, no enhanced photocurrent is measurable at +250 mV versus Ag/AgCl, since all the ions are already in the oxidized state and thus not able to transfer electrons toward the CdSe/ZnS NPs. Hence, by applying a negative potential in combination with a light pulse it is possible to control the reduction of hexacyanoferrate (III) ions [7].

Hydrogen peroxide is a biologically active molecule, which is also the byproduct of several enzyme reactions. Thus, it is tested whether this compound can be oxidized at CdS and CdSe/ZnS QDs. However, no photocurrent change can be detected in the presence of the molecule. To introduce a new functionality into the light-switchable QD layer two different approaches have been tested. For these experiments one can make use of the catalytic activity of another kind of NPs: FePt toward hydrogen peroxide conversion.

(i) The combination of CdS and FePt NPs by a simple co-immobilization on the Au electrode surface leads to a photocurrent, the amplitude of which depends on the H_2O_2 concentration in the buffer above the electrode. The reduction of H_2O_2 involves H^+ and e^-, which account for a dependence on pH and bias potential, respectively. Thus, it is found that reduction is preferred in the acidic pH region (pH 5) and can be facilitated by decreasing the electrode potential from −0.2 to −0.6 V versus Ag/AgCl [31].

(ii) Modern synthesis protocols also allow for the production of hybrid NPs, such as CdS NPs grown on top of FePt NPs [32, 33]. In this way, a combination of two materials with different functionalities in one particle is possible. In comparison to the mixed assembly of CdS and FePt NPs, see (i), a two to three times' higher response to H_2O_2 could be observed (see Fig. 11.9). Upon reduction of H_2O_2, electrons need to be transferred from the FePt NPs to H_2O_2. Sources of the electrons are the CdS QDs, which in turn receive electrons from the Au electrode via the conductive dithiol self-assembled monolayer (SAM). Effectively, electrons are injected from the Au electrode for reduction of H_2O_2, which results in a photocurrent with a negative sign. In the case of CdS/FePt NPs, electrons can flow directly from the CdS to the FePt domain without having to cross

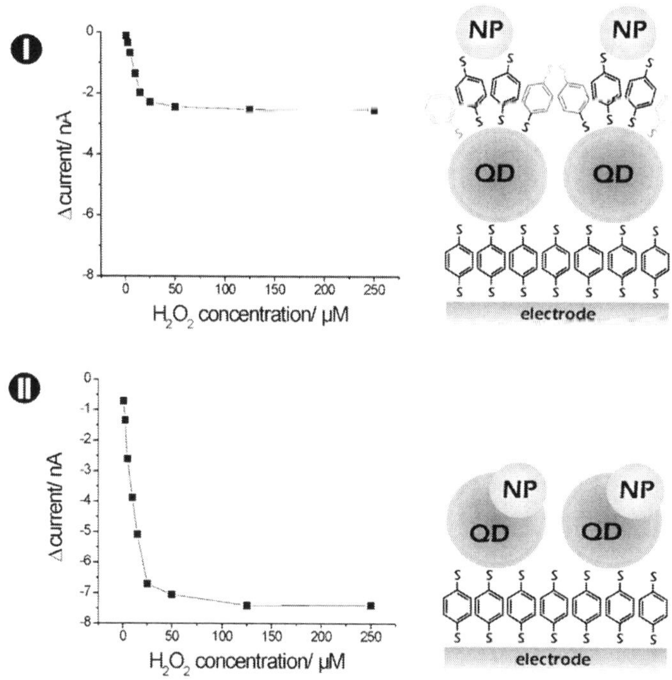

Figure 11.9 Detection of H_2O_2 with different electrode assemblies. (I) FePt nanoparticles (NPs) are fixed via a dithiol on top of a CdS quantum dot layer on an electrode. (II) Hybrid CdS/FePt NPs are fixed via a dithiol on an electrode (photocurrent behavior in dependence on hydrogen peroxide concentration in solution, $E = -200$ mV vs Ag/AgCl, phosphate buffer pH 7.5).

an external interface. In case of coassembly of CdS and FePt NPs, electrons need to be transferred from the CdS NPs to the FePt NPs via BDT molecules, which reduces the intensity of the photocurrent. Sensitivity can be provided in a range of 1 μM to about 30 μM [31].

The example illustrates that one is not limited to the chemical/electrochemical properties of QDs that appear from synthesis protocols, but one can design these NPs in such a way that a desired conversion is much facilitated. Of course here preparation conditions have to be carefully controlled.

With respect to small molecules it has also been demonstrated that the detection of copper and silver ions is feasible on an indium tin oxide (ITO) electrode modified with CdS NPs. Here the photocurrent is found to decrease with increasing metal ion concentration because of the formation of new compounds on the QD surface (e.g., Cu_xS) with energy states within the bandgap of CdS [34]. However, for such an approach long-term stability will be an issue.

11.5 QD Electrodes and Enzymatic Signal Chains Based on First Sensor Generation

Classical biosensors of the first generation exploit the specificity of an enzyme reaction and detect this conversion by following the formation of an electroactive product or the consumption of a cosubstrate [35]. Following this idea one can couple an enzymatic reaction to the QD electrode, provided that a product or cosubstrate can be effectively converted at the illuminated QDs. Several systems have been established, meanwhile, demonstrating the power of this approach.

Substrates of the enzyme acetylcholine esterase can be detected in such a way [36]. The enzyme catalyzes the hydrolysis of acetylthiocholine, liberating thiocholine and acetate. The former can be oxidized at the illuminated CdS QD electrode. The photocurrent reflects the substrate concentration in solution and is modified by the presence of enzyme inhibitors (at constant substrate concentration). Since acetylthiocholine is an artificial substrate this approach is mainly directed to the analysis of potential inhibitors of the enzyme acetylcholine esterase. The scheme in Fig. 11.10I illustrates the reaction cascade.

Another example is the electrocatalytic oxidation of NADH [30]. Here CdSe/ZnS QDs have been studied. NADH detection is possible in a rather wide potential range around 0 V versus Ag/AgCl. At an electrode potential where no photocurrent can be detected under illumination an anodic photocurrent occurs after addition of NADH. At more positive potentials the anodic photocurrent increases significantly in the presence of NADH. Thus, electrons are obviously

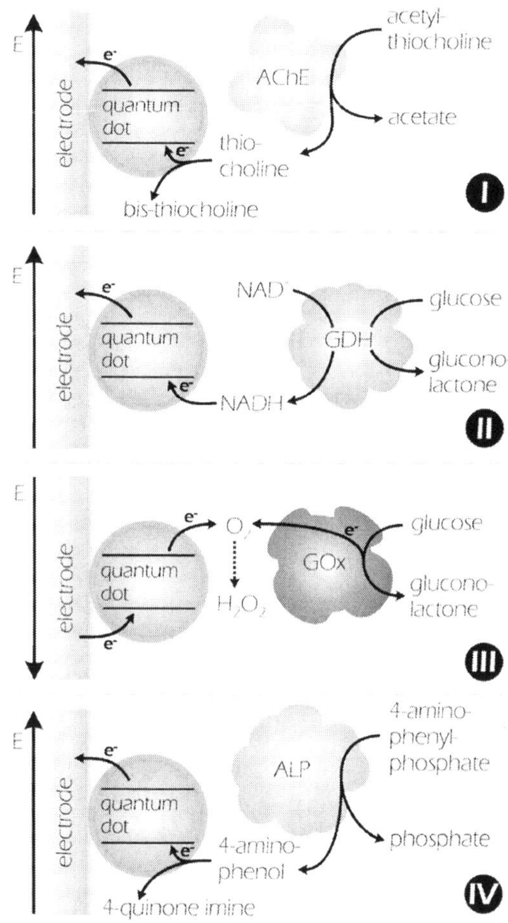

Figure 11.10 Different enzymatic signal chains on quantum dot electrodes. These systems rely on the enzymatic generation of a product that can be converted at the illuminated QDs (I, II, IV) or on competition of reactions at the QDs and the enzyme (III).

transferred from NADH to the excited QD layer. At potentials where a cathodic photocurrent is generated (i.e., below +100 mV vs. Ag/AgCl) this current decreases in the presence of NADH, indicating a depressed charge transfer between the gold electrode and the QD layer. Thus, electrons transferred from the electrode to the nanocrystals seem to compete with electrons from NADH. This also means that the direction of the photocurrent generated can be reversed by a chemical reaction. For NADH eventually a cathodic photocurrent can be transformed into an anodic current at an appropriate potential when the NADH concentration is high enough.

Since NADH is involved in many dehydrogenase reactions the QD electrode system can be used to follow such catalytic reactions. The amount of NADH produced in an enzymatic catalysis depends directly on the concentration of the substrate. The feasibility of the concept is shown with glucose dehydrogenase (*Pseudomonas* sp.) as a biocatalyst. This dehydrogenase catalyzes the reaction of β-D-glucose to D-glucono-1,5-lactone while reducing its cofactor NAD^+ to NADH. This principle is illustrated in Fig. 11.10II.

It has to be mentioned here that glucose cannot be oxidized directly on the QD gold electrode at low potential. But in the presence of the enzyme glucose dehydrogenase and its cofactor NAD^+ a concentration-dependent change of the photocurrent can be detected with the NP-based system. The glucose signal is converted to NADH by electron transfer via the enzyme, and subsequently NADH is detected by electron transfer to the illuminated QDs. The concentration range for glucose detection is rather similar to that for NADH. Sensitivity for glucose is observed up to 1 mM.

As it is found that the photoluminescence of QDs is influenced by the oxygen concentration in solution [37–40] it is studied whether oxygen can act as an electron acceptor for excited QDs. At CdSe/ZnS QD electrodes a cathodic photocurrent is measured in an air-saturated and argon-purged buffer when the electrode is negatively polarized. However, the photocurrent is clearly smaller in an oxygen-free solution. This points to the fact that oxygen can be reduced by the generated electrons during illumination of the nanocrystals. The hole generated in the valence band of the QDs can be filled through electron transfer from the electrode. It can be observed that not only the overall photocurrent but also the oxygen-dependent part increases with decreasing potential. Figure 11.2II illustrates schematically the electron transfer steps during the illumination of the QD electrode [41].

Measurements in basic and neutral pH solutions result in the highest values for the oxygen-dependent photocurrent. In acidic pH, however, the results show a smaller current—this is valid not only for the overall photocurrent but also for the oxygen-dependent part. Obviously the recombination of the charge carriers is enhanced under these conditions, probably on the particle surface. The reaction product of the reduction process is not evaluated yet. It could be superoxide since the presence of protons does not enhance the oxygen conversion.

The observed oxygen sensitivity can be coupled with an oxidase in order to analyze the substrate of the respective oxidase reaction (see Fig. 11.10III). This has been first exemplified with glucose oxidase [41]. For the coupling different approaches can be used, for example, covalent crosslinking of GOD or a layer-by-layer deposition of the enzyme by means of an oppositely charged polyelectrolyte, poly(allylamine hydrochloride) (PAH). In both cases a rather high enzyme density has to be provided in order to reach an oxygen depletion in front of the QD surface and subsequently a decreased photocurrent signal in the presence of substrate. The layer-by-layer deposition provides the additional advantage that the surface concentration can be tuned by the number of deposition steps. Optimal conditions have been found for layer structures with four enzyme layers (see Fig. 11.11). Sensitivity can be provided in the range of 100 μM to 5 mM glucose. Since no substances have to be added during the analysis this is a real reagent-less sensor device. Furthermore the detection is rather fast, and stable photocurrents can be detected after a few tens of seconds after sample addition.

These experiments provide the basis for coupling other oxidases in an analogous way to the QD electrode such that they become an integral part of the device. This has already been demonstrated for sarcosin oxidase and sarcosin detection [42].

Another system has been developed on the basis of the oxidation of phenolic compounds on the surface of illuminated CdS NPs (see Fig. 11.10IV). Here *p*-aminophenol (pAP) has been selected since it is the reaction product of the enzymatic action of alkaline phosphatase (ALP) to its substrate *p*-aminophenylphosphate (pAPP). For this purpose the electrode potential is varied and the current is measured under pulsed illumination. A clear response of the photocurrent to the presence of pAP is found, indicating that the QD electrode provides a suitable surface for the oxidation of the

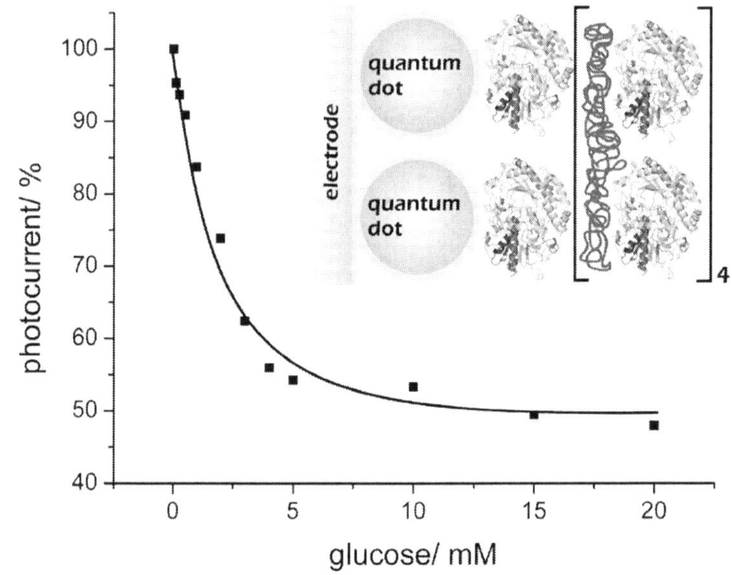

Figure 11.11 Schematic illustration of a GOD multilayer formed by means of a polyelectrolyte on a quantum dot electrode and photocurrent behavior of such an electrode in solutions of different glucose concentrations ($E = -0.35$V vs. Ag/AgCl). The photocurrent is depressed because of a decreased oxygen concentration near the QDs, according to Tanne et al. [41].

phenolic enzyme product under positive polarization. A maximum photocurrent is detected for an applied bias potential of +200 mV versus Ag/AgCl. Thus, the QDs can be used as a light-triggered interlayer to transfer electrons from the redox couple in solution to the electrode.

To create a photobioelectrochemical sensor for pAPP the enzyme needs to be immobilized on the photosensitive electrode. The layer-by-layer approach in depositing protein molecules is also used here as a favorable technique, allowing control of the deposited amount in one layer but also on the whole assembly by the number of deposition steps. So much the deposit ALP again the positively charged polyelectrolyte PAH is used. By comparing assemblies of ALP and PAH with one and two layers on the QD electrode it is found that the bilayer system results in rather small current values, which illustrates that the current signal change cannot always be enhanced by increasing the layer numbers. Particularly the diffusion of the substrate into the layered structure and the diffusion of the product within the system have to be considered here. However, with an ALP monolayer fixed on the QDs a very well-defined response to the enzyme substrate is obtained. Sensitivity for pAPP detection can be provided in the range of 25 μM to 1.5 mM [43].

11.6 QD Electrodes and Enzymatic Signal Chains Based on Mediated Electron Transfer

Enzyme electrodes of the so-called second generation apply a mediator between the biocatalyst and the electrode and thus allow new sensing schemes with rather low potential. With QD electrodes not much research has been devoted to this topic. One example uses glucose oxidase and the tris-1,10-phenantroline complex of cobalt $[Co(Phen)_3]Cl_2$ [44]. A layered assembly on CdSe/CdS core–shell NPs on a TiO_2-modified electrode has been created. A rather high potential has been used for the analysis (+0.4 V vs. Ag/AgCl). Thus, the generated photocurrent is only partly attributed to the mediated electron transfer but influenced by the direct sugar oxidation.

11.7 QD Electrodes and Enzymatic Signal Chains Based on Direct Protein Electron Transfer

As shown in the previous paragraphs it is possible to enforce the reaction of several molecules on the surface of QDs when they are light-excited. When this is used for biosensor construction one relies on the detection of small shuttle molecules—mostly reaction products of an enzymatic conversion. The progress in direct protein electrochemistry, particularly in the last decade, has shown the benefits of a direct signal transfer from the converting biomolecule to the electrode by establishing a heterogeneous direct electron transfer (DET) between the electrode and the redox center of the protein [45–47].

Particularly nanostructures have been shown to be beneficial as modifiers of electrode surfaces since their size and properties are often suited for an oriented interaction with the biomolecule and thus may facilitate DET reactions. Mainly gold NPs and carbon nanotubes, graphene, etc., have been used [48–51].

But QDs as semiconducting NPs have also been studied in this respect. It has been found that the surface properties are very essential for the observation of a heterogeneous electron transfer with a protein. After synthesis the QDs are capped with an organic ligand. For example, CdSe/ZnS NPs are coated with trioctylphosphine oxide (TOPO). When these particles are fixed on an electrode and investigated with the redox protein cytochrome c no change in the photocurrent can be found regardless of which potential is applied on the electrode. However, when the original layer of the hydrophobic surfactant is converted to a hydrophilic one by ligand exchange using, for example, mercaptopropionic acid or mercaptosuccinic acid a completely different behavior is detected. When oxidized cytochrome c is added and a negative potential is applied an enhanced photocurrent is recorded. The current can follow the protein concentration in a nonlinear fashion for μ molar concentrations and is a result of protein reduction by the excited nanocrystals (Fig. 11.12). Addition of reduced cytochrome c will not change the photocurrent (with negative polarization), and the same is valid for the use of a positive polarization with addition of oxidized

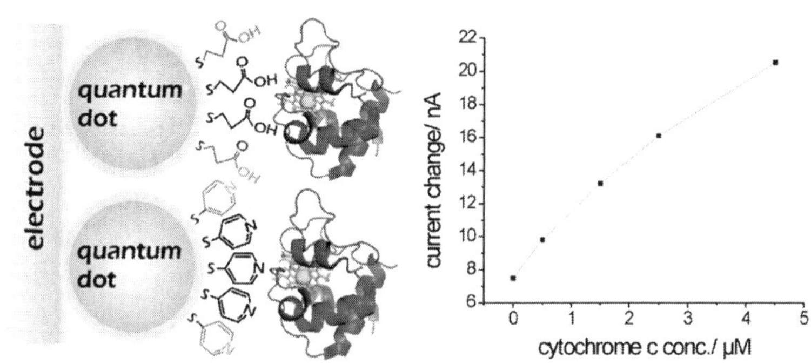

Figure 11.12 QD electrodes such as CdSe/ZnS can be decorated with different thiols, providing an interface for direct cytochrome c conversion. The photocurrent of a CdSe/ZnS electrode modified with mercaptosuccinic acid shows a defined dependence on the concentration of oxidized cytochrome c ($E = -0.25$V vs. Ag/AgCl), according to Stoll et al. [7].

protein. This shows the defined basis of the electron transfer steps observed [8].

Cytochrome c is the first redox protein for which DET to an activated semiconducting NP has been observed. It provides the basis for the construction of analytical signal chains that allow the quantification of analyte molecules in solution.

As a first example the simple but well-defined reaction of superoxide radicals with the redox protein has been used [9]. A superoxide results from one electron transfer to molecular oxygen and thus belongs to the family of reactive oxygen species that are produced under several physiological but also pathophysiological conditions. Subsequently detection of this short-lived species is an intensively studied area [52, 53].

When the superoxide radical is produced in solution it can quickly interact with the redox protein, probably guided by electrostatic interaction to the heme center where reduction occurs [54]. Then the reduced protein can be oxidized at the QD electrode, provided the QDs are photoexcited and the electrode potential is high enough to subtract electrons from the valance band (see Fig. 11.13I). This results in an enhanced photocurrent when the radical is produced. It has been shown that with mercaptopyridine-modified CdSe/ZnS QDs on gold and cytochrome c different radical concentrations in the nanomolar range can be followed in solution [9].

Another direction for the construction of signal chains is based on the reaction of cytochrome c with enzyme molecules. Such biocatalytic cascades can amplify the photoelectrochemical process. One example is the coupling of a QD electrode with cytochrome c and the enzymatic system of lactate dehydrogenase (LDH) and lactate [29]. The anodic photocurrent is enhanced as the lactate concentration increases, and it levels off to saturation at a lactate concentration of about 70 mM. The saturated photocurrent value in the system is about ninefold higher than the value observed with reduced cytochrome c only, verifying that the photocurrent is indeed amplified. Control experiments show that no enhanced photocurrent is observed in the system in the absence of LDH.

A photoswitchable bioelectrocatalytic system of the opposite direction of current flow can be created by combining a QD electrode

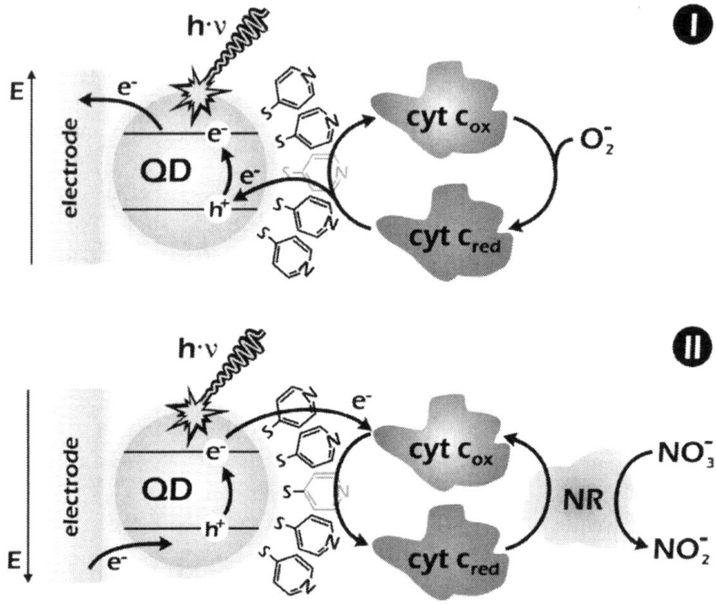

Figure 11.13 Light-triggered signal chains on quantum dot electrodes that are based on direct protein electrochemistry. (I) A superoxide can be detected by an anodic photocurrent via cytochrome c, according to Stoll et al. [9]. (II) The conversion of cytochrome c can be amplified by the enzyme nitrate reductase (NR), which can oxidize cytochrome c in the presence of nitrate. This results in a nitrate-dependent cathodic photocurrent, according to Katz et al. [29].

with cytochrome c and nitrate reductase (NR) [29], as illustrated in Fig. 11.13II. Also here an increased photocurrent is found with increasing substrate (NO_3^-) concentrations. The saturated cathodic photocurrent is about eightfold higher than in the presence of oxidized cytochrome c only. The amplification of the cathodic photocurrents in the presence of NR/NO_3^- is attributed to the biocatalytic regeneration of oxidized cytochrome c at the CdS NP interface.

It has been also suggested that the embedding of glutamate dehydrogenase in a composite of multiwalled carbon nanotubes, poly-diallyldimethylammonium chloride, and CdS NPs can result in a photoelectrochemical sensor for glutamate, which can avoid the addition of the enzyme cofactor NAD^+ [55]. This is based on an earlier work using formaldehyde dehydrogenase as a biocatalyst [8]. However, in such a complex system the clarification of the electron pathway needs further investigation since substances can also interact directly with the excited QDs, as has been shown, for example, with glucose [44].

11.8 Summary

This chapter introduced into basic properties of semiconductor NPs, so-called QDs. Focus was on the application of such particles in combination with electrodes for bioanalysis. Illumination of QDs with an appropriate wavelength can induce charge carrier generation, namely, electrons and holes. This allows electron transfer reactions with an electrode. Depending on the applied potential anodic and cathodic photocurrent generation is feasible. Furthermore reactions with substances in solution can occur. Also here electron transfer from the excited NPs to acceptor molecules in solution and from donor compounds to excited particles is possible. This allows the construction of analytical signal chains or even photobioelectrochemical sensors. The main advantage is that besides the electrode potential an additional control tool is available—light. Spatially resolved, multiplexed analysis becomes feasible without the need of preparing a miniaturized electrode array.

Different principles have been followed for enzymatic analysis. Most research has concentrated on sensing schemes relying on the detection of enzymatic products or cosubstrates. However, first examples of DET from a protein to QDs have been shown.

Another direction of actual research uses the photocurrent generation to monitor binding reactions. This can make use of the accumulation of QDs on the electrode surface and thus an enhanced photocurrent or the modulation of the photocurrent of a QD electrode by a binding event, provided a specific recognition layer has been constructed.

The chapter illustrated that besides the unique luminescence properties of QDs, which have been well recognized and are intensively used nowadays in bioanalysis, the photoelectrochemical properties still represent an area that is worth to study and exploit for practical analytical demands.

References

1. Timp, G. L. (1998) Chemical approaches to semiconductor nanocrystals and nanocrystal materials, in *Nanotechnology*, ed. Louis Brus (Springer, New York).
2. Vollath, D. (2008) An introduction to synthesis, properties and application, in *Nanomaterials* (Wiley-VCH, Weinheim).
3. Li, C., Murase, N. (2004) Synthesis of highly luminescent glasses incorporating CdTe nanocrystals through sol–gel processing, *Langmuir*, **20**(1), 1–4.
4. Alivisatos, A. P. (1996) Perspectives on the physical chemistry of semiconductor nanocrystals, *J. Phys. Chem.*, **100**(31), 13226–13239.
5. Medintz, I. L., Uyeda, H. T., Goldman, E. R., Mattoussi, H. (2005) Quantum dot bioconjugates for imaging, labelling and sensing, *Nat. Mater.*, **4**(6), 435–446.
6. Willner, I., Katz, E. (2005) From theory to applications. 11. Molecular optobioelectronics, in *Bioelectronics*, eds. E. Katz, A. N. Shipway (Wiley-VCH, Weinheim).
7. Stoll, C., Kudera, S., Parak, W. J., Lisdat, F. (2006) Quantum dots on gold: electrodes for photoswitchable cytochrome c electrochemistry, *Small*, **2**(6), 741–743.
8. Curri, M. L., Agostiano, A., Leo, G., et al. (2002) Development of a novel enzyme/semiconductor nanoparticles system for biosensor application, *Mater. Sci. Eng. C*, **22**(2), 449–452.
9. Stoll, C., Gehring, C., Schubert, K., Zanella, M., Parak, W. J., Lisdat, F. (2008) Pho signal chain based on quantum dots on gold–sensitive to superoxide radicals in solution, *Biosens. Bioelectron.*, **24**(2), 260–265.
10. Bakkers, E. P. A. M., Roest, A. L., Marsman, A. W., et al. (2000) Characterization of photoinduced electron tunneling in Gold/SAM/Q-CdSe systems by time-resolved photoelectrochemistry, *J. Phys. Chem. B*, **104**(31), 7266–7272.
11. Miyake, M., Torimoto, T., Nishizawa, M., Sakata, T., Mori, H., Yoneyama, H. (1999) Effects of surface charges and surface states of chemically modified cadmium sulfide nanoparticles immobilized to gold electrode

substrate on photoinduced charge transfers, *Langmuir*, **15**(8), 2714–2718.

12. Rajeshwar, K., Tacconi, N. R. de, Chenthamarakshan, C. R. (2001) Semiconductor-based composite materials: preparation, properties, and performance, *Chem. Mater.*, **13**(9), 2765–2782.
13. Tristão, J. C., Magalhães, F., Corio, P., Sansiviero, M. T. C. (2006) Electronic characterization and photocatalytic properties of CdS/TiO2 semiconductor composite, *J. Photochem. Photobiol. A*, **181**(2–3), 152–157.
14. Günes, S., Sariciftci, N. S. (2008) Hybrid solar cells, *Inorg. Chim. Acta*, **361**(3), 581–588.
15. Tada, H., Fujishima, M., Kobayashi, H. (2011) Photodeposition of metal sulfide quantum dots on titanium(iv) dioxide and the applications to solar energy conversion, *Chem. Soc. Rev.*, **40**(7), 4232–4243.
16. Hafeman, D., Parce, J., McConnell, H. (1988) Light-addressable potentiometric sensor for biochemical systems, *Science*, **240**(4856), 1182–1185.
17. Bratov, A., Abramova, N., Ipatov, A. (2010) Recent trends in potentiometric sensor arrays: a review, *Anal. Chim. Acta*, **678**(2), 149–159.
18. Golub, E., Pelossof, G., Freeman, R., Zhang, H., Willner, I. (2009) Electrochemical, photoelectrochemical, and surface plasmon resonance detection of cocaine using supramolecular aptamer complexes and metallic or semiconductor nanoparticles, *Anal. Chem.*, **81**(22), 9291–9298.
19. Willner, I., Patolsky, F., Wasserman, J. (2001) Photoelectrochemistry with controlled DNA-cross-linked CdS nanoparticle arrays, *Angew. Chem., Int. Ed. Engl.*, **40**(10), 1861–1864.
20. Göbel, G., Schubert, K., Schubart, I. W., Khalid, W., Parak, W. J., Lisdat, F. (2011) Enhanced photocurrent generation with quantum dots containing multilayers on gold, *Electrochim. Acta*, **56**(18), 6397–6400.
21. Kang, Q., Yang, L., Chen, Y., et al. (2010) Photoelectrochemical detection of pentachlorophenol with a multiple hybrid $CdSe_xTe_{1-x}$/TiO2 nanotube structure-based label-ree immunosensor, *Anal. Chem.*, **82**(23), 9749–9754.
22. Zhang, X., Li, S., Jin, X., Li, X. (2011) Aptamer based photoelectrochemical cytosensor with layer-by-layer assembly of semiconductor nanoparticles as photoelectrochemically active species, *Biosens. Bioelectron.*, **26**(8), 3674–3678.

23. Zhao, W.-W., Wang, J., Xu, J.-J., Chen, H.-Y. (2011) Energy transfer between CdS quantum dots and Au nanoparticles in photoelectrochemical detection, *Chem. Commun.*, **47**(39), 10990.
24. Qian, Z., Bai, H.-J., Wang, G.-L., Xu, J.-J., Chen, H.-Y. (2010) A photoelectrochemical sensor based on CdS-polyamidoamine nano-composite film for cell capture and detection, *Biosens. Bioelectron.*, **25**(9), 2045–2050.
25. Wang, J., Liu, G., Merkoçi, A. (2003) Electrochemical coding technology for simultaneous detection of multiple DNA targets, *J. Am. Chem. Soc.*, **125**(11), 3214–3215.
26. Kuçur, E., Boldt, F. M., Cavaliere-Jaricot, S., Ziegler, J., Nann, T. (2007) Quantitative analysis of cadmium selenide nanocrystal concentration by comparative techniques, *Anal. Chem.*, **79**(23), 8987–8993.
27. Huang, H., Li, J., Zhu, J.-J. (2011) Electrochemiluminescence based on quantum dots and their analytical application, *Anal. Methods*, **3**(1), 33.
28. Bertoncello, P., Forster, R. J. (2009) Nanostructured materials for electrochemiluminescence (ECL)-based detection methods: recent advances and future perspectives, *Biosens. Bioelectron.*, **24**(11), 3191–3200.
29. Katz, E., Zayats, M., Willner, I., Lisdat, F. (2006) Controlling the direction of photocurrents by means of CdS nanoparticles and cytochrome c-mediated biocatalytic cascades, *Chem. Commun.*, (13), 1395–1397.
30. Schubert, K., Khalid, W., Yue, Z., Parak, W. J., Lisdat, F. (2010) Quantum-dot-modified electrode in combination with nadh-dependent dehydrogenase reactions for substrate analysis, *Langmuir*, **26**(2), 1395–1400.
31. Khalid, W., El Helou, M., Murböck, T., Yue, Z., Montenegro, J.-M., Schubert, K. (2011) Immobilization of quantum dots via conjugated self-assembled monolayers and their application as a light-controlled sensor for the detection of hydrogen peroxide, *ACS Nano*, **5**(12), 9870–9876.
32. Gu, H., Zheng, R., Zhang, X. X., Xu, B. (2004) Facile one-pot synthesis of bifunctional heterodimers of nanoparticles: a conjaugate of quantum dot and magnetic nanoparticles, *J. Am. Chem. Soc.*, **126**, 5664–5665.
33. Zanella, M., Falqui, A., Kudera, S., Manna, L., Casula, M. F., Parak, W. J. (2008) Growth of colloidal nanoparticles of group II–VI and IV–VI semiconductors on top of magnetic iron-platinum nanocrystals, *J. Mater. Chem.*, **18**, 4311–4317.
34. Wang, G.-L., Xu, J.-J., Chen, H.-Y. (2010) Selective detection of trace amount of Cu2+ using semiconductor nanoparticles in photoelectrochemical analysis, *Nanoscale*, **2**(7), 1112.

35. Scheller, F. W., Schubert, F., Wollenberger, U. (1989) *Biosensoren* (Birkhäuser, Basel).

36. Pardo-Yissar, V., Katz, E., Wasserman, J., Willner, I. (2003) Acetylcholine esterase-labeled CdS nanoparticles on electrodes: photoelectrochemical sensing of the enzyme inhibitors, *J. Am. Chem. Soc.*, **125**(3), 622–623.

37. van Sark, W. G. J. H. M., Frederix, P. L. T. M., van den Heuvel, D. J., Gerritsen, H. C., Bol, A. A., van Lingen, J. N. J. (2001) Photooxidation and photobleaching of single CdSe/ZnS quantum dots probed by room-temperature time-resolved spectroscopy, *J. Phys. Chem. B*, **105**(35), 8281–8284.

38. Frederix, P. L. T. M., Bol, A. A., Gerritsen, H. C., Meijerink, A. (2002) Blueing, bleaching, and blinking of single CdSe/ZnS quantum dots, *ChemPhysChem*, **3**(10), 871–879.

39. Koberling, F., Mews, A., Basche, T. (2001) Oxygen-induced blinking of single CdSe nanocrystals, *Adv. Mater.*, **13**(9), 672–676.

40. Kloepfer, J. A., Bradforth, S. E., Nadeau, J. L. (2005) Photophysical properties of biologically compatible CdSe quantum dot structures, *J. Phys. Chem. B*, **109**(20), 9996–10003.

41. Tanne, J., Schäfer, D., Khalid, W., Parak, W. J., Lisdat, F. (2011) Light-controlled bioelectrochemical sensor based on CdSe/ZnS quantum dots, *Anal. Chem.*, **83**(20), 7778–7785.

42. Riedel, M., Göbel, G., Parak, W. J., Lisdat, F., in preparation.

43. Khalid, W., Göbel, G., Hühn, D., et al. (2011) Light triggered detection of aminophenyl phosphate with a quantum dot based enzyme electrode, *J. Nanobiotechnol.*, **9**(1), 46.

44. Zheng, M., Cui, Y., Li, X., Liu, S., Tang, Z. (2011) Photoelectrochemical sensing of glucose based on quantum dot and enzyme nanocomposites, *J. Electroanal. Chem.*, **656**(1–2), 167–173.

45. Armstrong, F. (2000) Recent developments in faradaic bioelectrochemistry, *Electrochim. Acta*, **45**(15–16), 2623–2645.

46. Murphy, L. (2006) Biosensors and bioelectrochemistry, *Curr. Opin. Chem. Biol.*, **10**(2), 177–184.

47. Lisdat, F., Dronov, R., Möhwald, H., Scheller, F. W., Kurth, D. G. (2009) Self-assembly of electro-active protein architectures on electrodes for the construction of biomimetic signal chains, *Chem. Commun.*, (3), 274–283.

48. Pumera, M., Sánchez, S., Ichinose, I., Tang, J. (2007) Electrochemical nanobiosensors, *Sens. Actuators, B*, **123**(2), 1195–1205.

49. Rivas, G., Rubianes, M., Rodriguez, M., Ferryra, N., Lugue, G., Pedano, M. (2007) Carbon nanotubes for electrochemical biosensing, *Talanta*, **74**(3), 291–307.
50. Shao, Y., Wang, J., Wu, H., Liu, J., Aksay, I. A., Lin, Y. (2010) Graphene based electrochemical sensors and biosensors: a review, *Electroanalysis*, **22**(10), 1027–1036.
51. Willner, I., Baron, R., Willner, B. (2007) Integrated nanoparticle–biomolecule systems for biosensing and bioelectronics, *Biosens. Bioelectron.*, **22**(9–10), 1841–1852.
52. Valko, M., Leibfritz, D., Moncol, J., Cronin, M. T. D., Mazur, M., Telser, J. (2007) Free radicals and antioxidants in normal physiological functions and human disease, *Int. J. Biochem. Cell Biol.*, **39**(1), 44–84.
53. Ge, B., Lisdat, F. Superoxide sensor based on cytochrome c immobilized on mixed-thiol SAM with a new calibration method, *Anal. Chim. Acta*, **454**(1), 53–64.
54. Wegerich, F., Turano, P., Allegrozzi, M., Möhwald, H., Lisdat, F. (2009) Cytochrome C mutants for superoxide biosensors, *Anal. Chem.*, **81**(8), 2976–2984.
55. Tang, L., Zhu, Y., Yang, X., Sun, J., Li, C. (2008) Self-assembled CNTs/CdS/dehydrogenase hybrid-based amperometric biosensor triggered by photovoltaic effect, *Biosens. Bioelectron.*, **24**(2), 319–323.

Chapter 12

Biosensors Based on Electrochemiluminescence

Huangxian Ju
State Key Laboratory of Analytical Chemistry for Life Science, Nanjing University, Nanjing 210093, China
hxju@nju.edu.cn

Electrochemiluminescence (ECL) is an electrochemical technique combining the electrochemical reaction with luminescent detection. The electrochemical reaction can lead to the formation of an exited species, which then emits an optical signal for the detection of the analyte and study of the ECL mechanism. Since the discovery of light emission of Grigard compounds in 1927 and luminol in 1929 at different applied potentials, many ECL emitters have been presented, which promotes the quick development of ECL investigation. These emitters can be used as labels to couple with different biorecognition events, leading to a very powerful biosensing technique. This technique integrates the advantages of high sensitivity of luminescent detection, high specificity of biological recognition, and controllability of electrochemistry and thus has extensively been applied in different fields such as clinic diagnostics, food and environmental monitoring, and even

Electrochemical Biosensors
Edited by Serge Cosnier

ISBN 978-981-4411-46-2 (Hardcover), 978-981-4411-47-9 (eBook)
www.panstanford.com

biowarfare agent detection. This chapter covers the development of ECL biosensing techniques, including the principle of ECL from different labels, the strategy of ECL biosensor design, and the application of ECL detection.

12.1 Introduction

Electrochemiluminescence (ECL), also known as electrogenerated chemiluminescence, is an electrochemical technique coupled with optical detection. The electrochemical process generates oxidized or reduced species at electrodes to initiate high-energy electron transfer reactions for the formation of exited states, which then emit optical signals for the detection of the analyte and study of the ECL mechanism. Although the light emission during electrolysis of Grigard compounds and luminol at different applied potentials was reported in 1927 and 1929, respectively [1, 2], the first detailed ECL studies were described in the mid-1960s [3, 4]. Up to now, many ECL emitters such as polyaromatic hydrocarbons [5], metal complexes [6], Si nanoparticles (NPs) [7], and quantum dots (QDs) [8] have been presented, which makes ECL detection a very powerful analytical technique and can be widely used in different fields such as clinic diagnostics, food and environmental monitoring, and even biowarfare agent detection.

As a method of producing light at an electrode, ECL integrates the advantages of luminescent detection and electrochemistry. This attributes ECL with many distinct advantages over other spectroscopy-based detection methods. For example, as ECL does not involve a light source, as fluorescence methods, the problems of scattered light and luminescent impurities are absent, which greatly lowers the background signal and improves sensitivity. The electrochemical reaction allows the time and position of the light-emitting reaction to be controlled, which can further improve the sensitivity by increasing the signal-to-noise ratio and extend the ECL technique for multiplexed analysis by combining with biosensor arrays. By employing ECL-active species as labels on biological molecules, ECL has been endowed with the high specificity of biological recognition events, such as antibody–antigen recognition for immunoassays, DNA–DNA interaction for DNA analyses, and aptamer–target recognition for aptasensing. In addition, ECL emitters can be regenerated after the ECL emission, which leads to good repeatability of ECL biosensors. Thus ECL detection has been quickly developed as an important electrochemical biosensing technique with outstanding applications. Focusing on these applications, over thousands of papers, patents, and book chapters have been published, which have increased exponentially over the past 20 years. This chapter covers the development of the ECL biosensing technique, including the principle of ECL from different labels, the design of ECL biosensors and signal amplification strategies, and the applications of ECL detection.

12.2 Principles of ECL Biosensing

ECL is a means of converting electrical energy into radiative energy. The electrochemical reaction firstly produces reactive intermediates from a stable reactant at the electrode surface. These intermediates then directly emit light or react with different species, including the analytes, to form excited states that emit a detectable optical signal. ECL biosensing is a technique integrating the ECL process with specific biological recognition by using ECL-active species to label biomolecules. Thus different ECL mechanisms have been employed in ECL biosensing, which mainly include two types, ion annihilation and coreactant ECL. The early ECL mechanism is an ion annihilation process [9]. The ECL from luminol is a special process [10], in which luminol itself is directly oxidized at the electrode to produce light, and once oxidized, luminal cannot be regenerated upon light emission. Since the ECL from $Ru(bpy)_3{}^{2+}$ was first reported in 1972 [6], coreactant ECL has attracted considerable interest. At present, all commercially available ECL analytical instruments and biosensing systems are based on coreactant ECL technology. With the discovery of ECL emission from Si NPs [6] and QDs [8] and the presentation of the first QD-based ECL biosensor [11, 12], some new NP-based ECL emitters have been proposed for ECL biosensing. These biosensing systems generate ECL emission from both annihilation and coreactant processes.

12.2.1 *Ion Annihilation Mechanism*

Ion annihilation ECL involves the formation of an excited state as a result of an exergonic electron transfer between electrochemically generated species, often radical cations and radical anions, at the surface of an electrode. As shown in Scheme 12.1, after the emitter (R) is electrochemically oxidized (Eq. 12.1) and reduced (Eq. 12.2) by alternate pulsing of the electrode potential, the newly formed radical cation ($R^{\bullet+}$) and anion ($R^{\bullet-}$) are annihilated to form the excited-state species (R^*) (Eq. 12.3) that emits light (Eq. 12.4) [9, 13].

Scheme 12.1:

$$R - e \rightarrow R^{\bullet+} \quad \text{(oxidation at electrode)} \tag{12.1}$$

$$R + e \rightarrow R^{\bullet-} \quad \text{(reduction at electrode)} \tag{12.2}$$

$$R^{\bullet-} + R^{\bullet+} \rightarrow R^* + R \quad \text{(excited-state formation)} \tag{12.3}$$

$$R^* \rightarrow R + h\gamma \tag{12.4}$$

Annihilation reactions (Eq. 12.3) also can occur in "mixed systems" via "cross reactions," where the radical cation and radical anion are produced from different molecules (Scheme 12.2) [14]. Equations 12.7 and 12.8 are a chemiluminescence process that produces the excited singlet state with the excess energy.

Scheme 12.2:

$$A - e \rightarrow A^{\bullet+} \quad \text{(oxidation at electrode)} \tag{12.5}$$

$$D + e \rightarrow D^{\bullet-} \quad \text{(reduction at electrode)} \tag{12.6}$$

$$D^{\bullet-} + A^{\bullet+} \rightarrow A^* + D \; (\text{or } A + D^*) \quad \text{(excited-state formation)} \tag{12.7}$$

$$A^* \; (\text{or } D^*) \rightarrow R + h\gamma \tag{12.8}$$

ECL from NPs can occur in the annihilation routine via a redox chemistry to inject holes or electrons into NPs [15]. The resulting charged states must be sufficiently stable to survive until colliding with oppositely charged species in an annihilation reaction. From this point, it is clear that not only sufficient applied electrode potentials to generate both negatively and positively charged species but also sufficient energy in the electron transfer reaction to produce the excited state are required. Thus, no ECL is observed through annihilation-type mechanism 1 for thioglycerol-capped CdS QDs due to the instability of the electrogenerated reactants [16].

12.2.2 *Coreactant ECL*

ECL can be generated in a single potential step using a coreactant. The coreactant is a species that produces an intermediate upon electrochemical oxidation or reduction to react with an ECL luminophore to form excited states. Thus the coreactant routine only involves electron transfer between electrochemically generated species and an ECL luminophore. The ECL coreactant should be easily oxidized or reduced by the luminophore species at the electrode to form an intermediate that has sufficient reducing or oxidizing energy to react with the oxidized or reduced luminophore to form the excited state.

Up to now, a wide variety of molecules have been found to exhibit ECL. However, the overwhelming majority of publications concerned with coreactant ECL and its analytical applications use $Ru(bpy)_3{}^{2+}$, or its derivatives, as the emitting species due to their excellent chemical, electrochemical, and photochemical properties. Thus here $Ru(bpy)_3{}^{2+}$/coreactant ECL systems are first discussed:

- *Oxalate* ($C_2O_4{}^{2-}$) *system*: Oxalate is the first account of a coreactant ECL system reported by Bard's group in 1977 [17]. This system can form a strong and stable reducing agent ($CO_2{}^{\bullet-}$) upon electrochemical oxidation (Scheme 12.3). The electrochemical products react to produce an excited state capable of emitting light at about 620 nm.

Scheme 12.3:

$$Ru(bpy)_3{}^{2+} - e \rightarrow Ru(bpy)_3{}^{3+} \tag{12.9}$$

$$C_2O_4{}^{2-} - e \rightarrow C_2O_4{}^{\bullet-} \rightarrow CO_2{}^{\bullet-} + CO_2 \tag{12.10}$$

$$Ru(bpy)_3{}^{3+} + CO_2{}^{\bullet-} \rightarrow Ru(bpy)_3{}^{2+*} + CO_2 \tag{12.11}$$

$$Ru(bpy)_3{}^{2+*} \rightarrow Ru(bpy)_3{}^{2+} + h\gamma \tag{12.12}$$

The direct oxidation of oxalate at the electrode depends on the surface property of the electrode and the solvent. In an aqueous solution, oxalate is oxidized by the formed $Ru(bpy)_3{}^{3+}$, leading to another ECL process (Scheme 12.4) [18]:

Scheme 12.4:

$$Ru(bpy)_3{}^{3+} + C_2O_4{}^{2-} \rightarrow Ru(bpy)_3{}^{2+} + C_2O_4{}^{\bullet-} \quad (12.13)$$

$$C_2O_4{}^{\bullet-} \rightarrow CO_2{}^{\bullet-} + CO_2 \quad (12.14)$$

$$Ru(bpy)_3{}^{3+} + CO_2{}^{\bullet-} \rightarrow Ru(bpy)_3{}^{2+*} + CO_2 \quad (12.15)$$

$$Ru(bpy)_3{}^{2+} + CO_2{}^{\bullet-} \rightarrow Ru(bpy)_3{}^{+} + CO_2 \quad (12.16)$$

$$Ru(bpy)_3{}^{3+} + Ru(bpy)_3{}^{+} \rightarrow Ru(bpy)_3{}^{2+*} + Ru(bpy)_3{}^{2+} \quad (12.17)$$

$$Ru(bpy)_3{}^{2+*} \rightarrow Ru(bpy)_3{}^{2+} + h\gamma \quad (12.18)$$

- *Tri-n-propylamine (TPrA) system*: TPrA is another important popular "oxidative reductive" coreactant for ECL systems, especially $Ru(bpy)_3{}^{2+}$. The electrochemical oxidation of TPrA can produce the short-lived TPrA radical cation ($TPrA^{\bullet+}$), which loses a proton from an α-carbon to form the strongly reducing intermediate $TPrA^{\bullet}$. ECL emission results from the reaction between $Ru(bpy)_3{}^{3+}$ and $TPrA^{\bullet}$ to produce the excited $Ru(bpy)_3{}^{2+*}$ [19].

The ECL intensity of the $Ru(bpy)_3{}^{2+}$/TPrA system strongly depends on the solution pH with dramatic increases at pH $>\sim 5.5$ and a maximum value at pH 7.5 [20]. This may be associated with the deprotonation reaction as well as the stability of the intermediates formed. Solubility decrease of TPrA at high pH can be another reason to produce the highest ECL intensity at pH 7.5. The pH condition makes this system widely applicable in biosensing.

- *Peroxydisulfate (persulfate,* $S_2O_8{}^{2-}$*) system*: Peroxydisulfate is the first example of a so-called reductive oxidation coreactant reported previously [21]. The reduction of peroxydisulfate produces the strong oxidant $SO_4{}^{\bullet-}$, which reacts with an ECL luminophore, such as the reduction product of $Ru(bpy)_3{}^{2+}$, to generate light. The general mechanism for the peroxydisulfate system can be drawn as Scheme 12.5.

Scheme 12.5:

$$Ru(bpy)_3{}^{2+} + e \rightarrow Ru(bpy)_3{}^{+} \tag{12.19}$$

$$S_2O_8{}^{2-} + e \rightarrow S_2O_8{}^{3\bullet-} \rightarrow SO_4{}^{2-} + SO_4{}^{\bullet-} \tag{12.20}$$

$$Ru(bpy)_3{}^{+} + SO_4{}^{\bullet-} \rightarrow Ru(bpy)_3{}^{2+*} + SO_4{}^{2-} \tag{12.21}$$

12.2.3 *ECL from NPs*

The annihilation routine via a redox chemistry to inject holes or electrons into NPs is an important mechanism of ECL from NPs. The electron and hole injections for NPs ECL favorably occur at the surface with a large surface area. As the electrochemical potentials can provide data for the bandgap of the NPs, stepwise addition (or removal) of charge from NPs by the electrochemical method can yield information about the energy required for electron transfer and ECL emission. With different energies required for the ECL emissions, there are two generation types for the ECL reactions of NPs One type is ECL originated from surface states of NPs. The spectra of this type of NP ECL show red-shifted ECL maxima with respect to their PL spectra because less energy is required for the electron (or hole) injections. This phenomenon occurs in the ECL emission from Si [7], CdSe [8], and Ge [22] NPs. Due to the unpassivated surface of NPs it is easier to emit light from the surface. Another type of NP ECL is bandgap ECL. It is mainly corresponding to the bulk in NPs with the size-dependent and tunable ECL spectrum matching the photoluminescence (PL) spectrum. [23–26]. To achieve bandgap ECL, a common method is to passivate the surface state of NPs by capping the NPs with different stabilizers such as thioglycolic acid [24] and mercaptopropionic acid [25, 26].

Recently, coreactant NP ECL systems have also been extensively applied. By introducing the coreactants for $Ru(bpy)_3{}^{2+}$ ECL systems into NP ECL or immobilizing $Ru(bpy)_3{}^{2+}$ or its coreactants on NPs, these $Ru(bpy)_3{}^{2+}$/coreactant systems can conveniently be used for biosensing. For example, Bard and coworkers added excess $C_2O_4{}^{2-}$ to an NP solution to obtain coreactant anodic ECL for Si NPs by the electron transfer between $CO_2{}^{\bullet-}$ and oxidized Si NPs [8]. Wang and coworkers [27] demonstrated anodic ECL emission from 3-mercaptopropionic acid-capped CdTe/CdS QDs with TPrA as the coreactant in aqueous solution. $SO_4^{\bullet -}$ can react with the negatively charged NPs by injecting a hole into the highest occupied molecular orbital (HOMO) and producing an excited state of the NPs; thus peroxydisulfate has been used as an efficient coreactant for ECL emission from NPs [28–33].

Many biosensing applications of ECL have used hydrogen peroxide as a coreactant. Zou and Ju firstly demonstrated that the electron transfer reaction between electrochemically reduced nanocrystal species and H_2O_2 or reduced dissolved oxygen can result in ECL emission from NPs [11]. The general mechanism for the H_2O_2 (or dissolved oxygen) system can be drawn as Scheme 12.6.

Scheme 12.6:

$$\mathrm{NP} + \mathrm{e} \rightarrow \mathrm{NP}^{\bullet -} \tag{12.22}$$

$$\mathrm{O_2} + \mathrm{H_2O} + 2\mathrm{e} \rightarrow \mathrm{OOH^-} + \mathrm{OH^-} \tag{12.23}$$

$$2\mathrm{NP}^{\bullet -} + \mathrm{OOH^-} + \mathrm{H_2O} \rightarrow 3\mathrm{OH^-} + 2\mathrm{NP}^* \tag{12.24}$$

or

$$2\mathrm{NP}^{\bullet -} + \mathrm{H_2O_2} \rightarrow 2\mathrm{OH^-} + 2\mathrm{NP}^* \tag{12.25}$$

This mechanism has been used for different NPs [11, 33–39], and many NP ECL biosensors for different analytes have also been developed on the basis of the H_2O_2 system [40, 41]. In a later study Jiang and Ju [24] observed the presence of $OH^{\bullet}$ during the ECL process. Thus they presented the $1S_e$–$1S_h$ transition emission following Scheme 12.7, which was demonstrated with $OH^{\bullet}$ radical scavengers using γ-L-glutamyl-L-cysteine-glycine or L-cysteine as model molecules

Scheme 12.7:

$$\mathrm{NP} + \mathrm{e} \rightarrow \mathrm{NP}^{1\mathrm{Se}\bullet -} \tag{12.26}$$

$$\mathrm{NP}^{1\mathrm{Se}\bullet -} + \mathrm{H_2O_2} \rightarrow \mathrm{NP} + \mathrm{OH}^{\bullet} + \mathrm{OH^-} \tag{12.27}$$

$$\mathrm{OH}^{\bullet} + \mathrm{NP} \rightarrow \mathrm{NP}^{1\mathrm{Sh}\bullet +} + \mathrm{OH^-} \tag{12.28}$$

$$\mathrm{NP}^{1\mathrm{Se}\bullet -} + \mathrm{NP}^{1\mathrm{Sh}\bullet +} \rightarrow \mathrm{NP}^* \tag{12.29}$$

and

$$NP^{1Se\bullet-} + OH^{\bullet} \rightarrow NP^{*} + OH^{-} \quad (12.30)$$

Sulfite ($SO_3{}^{2-}$) has been demonstrated to be a new corcactant for enhancing the anodic ECL of MPA-cappled CdTe QDs [25]. This system shows high sensitivity for ECL biosensing. The proposed mechanism may be drawn as Scheme 12.8.

Scheme 12.8:

$$NP - e \rightarrow NP^{1Sh\bullet+} \quad (12.31)$$

$$SO_3{}^{2-} - e \rightarrow SO_3{}^{\bullet-} \quad (12.32)$$

$$2OH^{-} + SO_3{}^{\bullet-} + 2O_2 \rightarrow O_2{}^{\bullet-} + SO_4{}^{2-} + H_2O \quad (12.33)$$

$$NP + O_2{}^{\bullet} \rightarrow NP^{1Se\bullet-} + O_2 \quad (12.34)$$

$$NP^{1Se\bullet-} + NP^{1Sh\bullet+} \rightarrow NP^{*} \quad (12.35)$$

12.3 ECL Biosensor and Its Applications

ECL biosensors are generally prepared by integrating the ECL process with specific biological recognition by using ECL-active species to label biomolecules. Thus the labeled biomolecules are usually immobilized or captured on the sensor surface by a recognition process for targets. Theoretically, both ECL emitters and coreactants can act as labels for the design of ECL biosensors because ECL emission intensity is usually proportional to the concentration of the emitter or coreactant. Earlier ECL biosensors are based on the immobilization of ECL emitters such as $Ru(bpy)_3{}^{2+}$. This not only cuts the consumption of $Ru(bpy)_3{}^{2+}$ but also allows simpler instrumentation. This way has been used for monitoring of coreactants such as small amine biomolecules and for enzymatic reactions by the generation or consumption of an ECL coreactant. After the ECL emitters are covalently attached to one of the species involved in an affinity binding reaction, this technique has been used for immunoassays and DNA analysis. By coupling with nanotechnology, enzymatic cycle, and molecular biological techniques ECL emission can be greatly enhanced to achieve signal amplification.

12.3.1 *Immobilization of Ru(bpy)$_3$$^{2+}$ and Its Derivatives for Biosensing*

Ru(bpy)$_3$$^{2+}$ and its derivatives are firstly immobilized on the electrode surface via electrostatic attachment in either the cation exchange polymer Nafion® [42] or an electropolymer. This can be conveniently performed by coating Nafion® or electrochemically copolymerizing a layer of film on electrodes. Nafion® film has drawbacks of slow mass transfer through the film and the partition of Ru(bpy)$_3$$^{2+}$ into the more hydrophobic regions of Nafion®. These properties can be improved by adding silica nanocomposites [43] or titania [44] gel to Nafion®, which can accelerate the diffusion of cations and stabilize ECL emitters in the membrane. Thus the modified films exhibit faster response and higher sensitivity for ECL detection. The gel membrane can directly immobilize Ru(bpy)$_3$$^{2+}$ on the electrode surface [45]. However a pure gel membrane cannot avoid the loss of Ru(bpy)$_3$$^{2+}$. Thus chitosan [46] and Triton X-100 [47] have been mixed with sol–gel to immobilize Ru(bpy)$_3$$^{2+}$, which can inhibit the cracking of the gel membrane. To further stabilize the Ru(bpy)$_3$$^{2+}$ on the sensor surface, Eastman-AQ55D and nanomaterials such as zeolite Y have been entrapped in silica [48] and titania [49] sol–gel membranes, respectively. On the basis of the ion exchange properties of zeolite Y, Ru(bpy)$_3$$^{2+}$ can be entrapped in the supercages of zeolite Y [50].

Ru(bpy)$_3$$^{2+}$ can be immobilized onto the electrode surface as a monolayer via Langmuir–Blodgett [51] or by forming a self-assembled monolayer [52] with the derivatives of Ru(bpy)$_3$$^{2+}$. However, the limited amount of ruthenium complex attached to the electrode surface limits the sensitivity of the resulting biosensors. Thus, NP-based composite films and the signal amplification of ECL emission have been applied in ECL biosensing. A simple method is to use zeolite Y–modified carbon paste to immobilize the Ru(bpy)$_3$$^{2+}$ on the electrode surface [53]. The resulting sensor shows good reproducibility and long-term stability for flow injection analysis. The advantages include:

(1) The sufficient conductivity allows fast electron transfers occurring among the electrode, $Ru(bpy)_3{}^{2+}$, and the ECL coreactant or analytes
(2) The surface of the sensor can be renewed by a simple polish step.

Introduction of carbon nanotubes (CNTs) [54] or Au or Pt NPs [55] into the composite films can significantly increase the film conductivity and the amount of $Ru(bpy)_3{}^{2+}$ and decrease the leaching of the emitting species from the films.

12.3.2 *Enzymatic ECL Biosensing with Immobilized Emitters*

Some enzymatic reactions involve the participation (generation or consumption) of coreactants such as amines, DNAH or NADPH, and H_2O_2. Thus they can be used for ECL biosensing of amines and the substrates of dehydrogenases and oxidases.

Amine groups are prevalent in numerous biologically and pharmacologically important compounds including alkylamines, antibiotics, antihistamines, opiates, nicotinamide, and the reduced form of NADH; thus reactions between $Ru(bpy)_3{}^{3+}$ and tertiary amines have led to the development of ECL-based detection devices for a variety of biologically important molecules [56]. As a general rule, the ECL signal from alkylamine coreactants follows the order tertiary > secondary > primary. Therefore, primary and secondary amines are often derivatized to tertiary amines prior to analysis [57]. The reagent and the derivative can be separated with a simple step. A variety of derivatization reagents such as dansyl chloride [58], 3-(diethylamino)propionic acid [59], and acetaldehyde [57] have been developed for $Ru(bpy)_3{}^{2+}$-based primary amine detection. Acetaldehyde does not interfere with the determination of amino acids since it does not emit detectable ECL signals [57]. The ECL efficiency can be increased by covalent attachment of a β lactamase substrate to a $Ru(bpy)_3{}^{2+}$ derivative [6].

NADH contains an amine moiety that can act as a coreactant for $Ru(bpy)_3{}^{2+}$, while NAD^+, the oxidized form of NADH, is not a coreactant. Thus numerous NADH-dependent dehyrogenases can be used for ECL biosensing of a variety of analytes including glucose, carbon dioxide, ethanol, and lactate [6], by immobilizing the dehyrogenases on the electrode surface. The linear ranges for glucose, ethanol, and lactate are 10 μM–3 mM, 25 μM–1.5 mM, and 50 μM–2.5 mM, respectively. When using $NADP^+$ to replace NAD^+, the linear range for glucose is 0.2 μM–3 mM.

ECL of luminol can be produced by both its electrochemical oxidation at the electrode surface and its oxidation by H_2O_2. Thus, by immobilizing oxidases on the electrode surface the substrates of oxidases can be detected with the ECL emission by adding luminal to the detection solution. For example, choline oxidase has been immobilized on a screen-printed carbon electrode for ECL biosensing of choline [61]. An ECL biosensor for glucose has been reported by immobilizing glucose oxidase in a carbon paste electrode [62]. The ECL reactions can achieve the maximum emission within 20 seconds.

12.3.3 *Labels with $Ru(bpy)_3{}^{2+}$ and Its Derivatives for Biosensing*

$Ru(bpy)_3{}^{2+}$ and its derivatives can be covalently linked to one of the binding partners. The labeled partner then recognizes the analyte to produce ECL emission on the biosensor surface in the presence of an added ECL coreactant, typically TPrA. The binding partners include antibody–antigen, enzyme–inhibitor, carbohydrate–lectin, and nucleic acid–complementary nucleic acid. Figure 12.1 illustrates eight examples of ECL biosensing in which panels a–c, d–f, and g–h are DNA, antibody–antigen, and peptide related, respectively [4]. The first case uses an ECL emitter–labeled target single-stranded DNA (ssDNA) to hybridize with an immobilized ssDNA probe (Fig. 12.1a) [63]. Obviously, the ECL produced from the complementary ssDNA hybridization is higher than that obtained from a noncomplementary one. Sandwich hybridization format can be performed using a labeled probe and an immobilized probe, each specific for a different sequence of a target ssDNA (Fig. 12.1b), which has been employed for the sensitive detection and quantification of DNA and RNA amplification products [64]. Figure 12.1c shows the detection process of the ligation activity of HIV integrase [65].

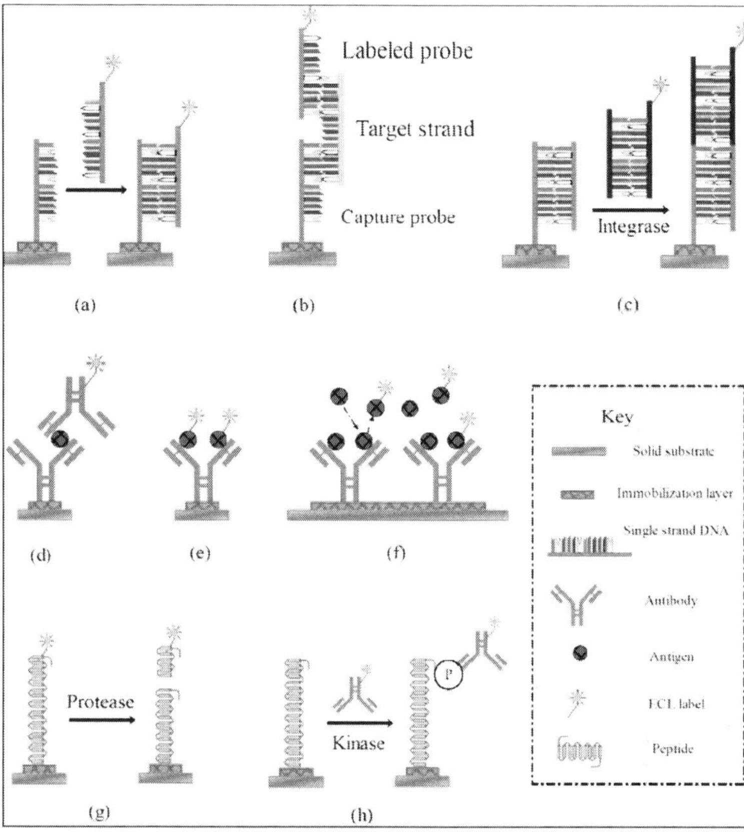

Figure 12.1 Examples of ECL biosensing, (a) DNA hybridization assay using immobilized ssDNA and labeled target ssDNA, (b) sandwich-type DNA biosensor, (c) integrase activity test with immobilized and free labeled dsDNA, (d) sandwich-type immunoassay, (e) direct immunoassay, (f) competitive immunoassay with immobilized antibody, (g) protease activity assay in which cleavage of the immobilized peptide results in the decrease in ECL emission due to the removal of the ECL label, and (h) kinase activity assay using a labeled antibody to recognize the phosphorylated product. Reprinted with permission from Miao [4].

The most commonly used immunoassay format of antigens is shown in Fig. 12.1d using an ECL emitter–labeled secondary antibody and an immobilized capture antibody, which can also be used in a competitive immunoassay (Fig. 12.1f) by the competition binding of the unlabeled analyte antigen and labeled antigen to the antibody. Figure 12.1g,h show the ECL biosensing of the activity of enzymes (e.g., protease, kinase) [4].

Instead of $Ru(bpy)_3^{2+}$, the coreactant 4-(dimethylamino)butyric acid has also been used as an ECL label for immunoassays by attaching it to immobilized bovine serum albumin (BSA) and anti-IgG. In this case the biosensor should be in contact with the $Ru(bpy)_3^{2+}$ solution after immunoreactions [66]. The ECL signal can be amplified with Au NPs.

In labeling ECL biosensing, signal amplification can be achieved by using polystyrene microspheres, microbeads, different nanomaterials and even liposomes as the carriers of a large number of ECL emitters such as $Ru(bpy)_3^{2+}$. This has been demonstrated for ultrasensitive DNA hybridization detection (Fig. 12.2) [64, 67]. The complementary target ssDNA-coated polystyrene bead (PSB) contains a large number of ECL labels. The loading capacity of the ECL labels per PSB can be as high as $\sim 10^9$ molecules without complicated treatment. After hybridization with the probe ssDNA, the aggregate is magnetically separated from the reaction mixture and transferred to a MeCN solution, in which the ECL label is released for ECL measurement in MeCN in the presence of TPrA,

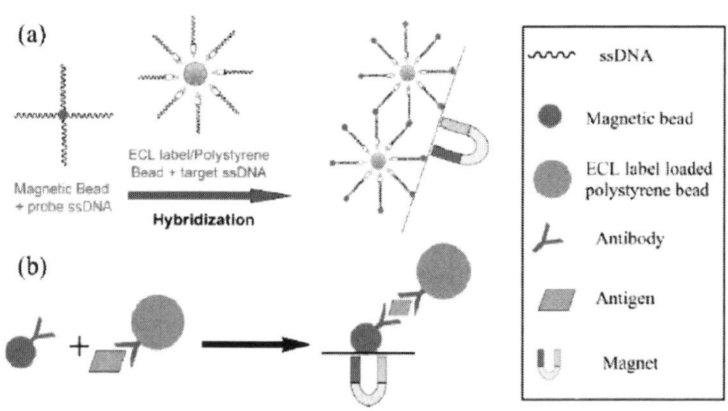

Figure 12.2 Schematic diagrams of (a) DNA hybridization and (b) sandwich-type immunoassay using a polystyrene bead as the ECL label carrier and a magnetic bead for separation of the analyte containing the ECL label/polystyrene bead. Reprinted with permission from Miao [4, 64, 67].

along with a ~100-fold improvement in sensitivity [64]. A similar approach can be used for a sandwich-type immunoassay. When using liposomes as carriers, the ECL label can be released in 0.1 M phosphate-buffered saline (PBS) (pH 7.6) containing 0.1 M NaCl and 1% (v/v) Triton X-100 [68]. Because the ECL detection is performed in solution, high stability and the possibility of multiple measurements of ECL without loss of signal can be offered.

12.3.4 *Biosensing Based on ECL Emission from QDs*

ECL emission from QDs is a relatively mature NP-based ECL biosensing system. The reported biosensing strategies based on the ECL emission from QDs include four ways: 1) generation or consumption of coreactants such as H_2O_2 on the biosensor surface upon biological recognition of the target and potential scanning, which changes the ECL emission of the QD-co-immobilized biosensor; 2) quenching effect of analytes on the ECL from QDs; 3) change of electron transfer kinetics between the electrode and QDs upon biological recognition of the target; and 4) use of QDs to label one of the binding partners such as the antibody, DNA and lectin.

12.3.4.1 Generation or consumption of coreactants

H_2O_2 is an important biological molecule and an NP ECL coreactant. The first example of a QD-based ECL sensor is fabricated for detection of H_2O_2 by depositing a CdSe QDs film on a graphite electrode [11]. In the presence of H_2O_2, the cathodic scanning of the QD film–modified electrode produces sensitive ECL emission linearly depending on the H_2O_2 concentration. The sensor displays acceptable fabrication reproducibility [11]. By co-immobilizing QDs and oxidase on the electrode surface, an ECL biosensor for the oxidase substrate, based on the generation of H_2O_2 and consumption of dissolved oxygen as a coreactant has been proposed [12]. As shown in Scheme 12.9, O_2 can capture more electrons from electrochemically reduced QDs than H_2O_2, and their reaction rate with the reduced QDs also differ. As a consequence, the dissolved oxygen is a more efficient coreactant than H_2O_2 to enhance the ECL emission from QDs. Thus when the GOD catalytic cycle consumes dissolved oxygen to produce equimolar hydrogen peroxide, the ECL intensity decreases. This strategy can be applied in more bioanalytical systems for oxidase substrates.

Scheme 12.9:

$$\text{CdSe/Cd(TGA) QDs} + \text{ne} \rightarrow \text{nR}^{\bullet -} \quad (12.36)$$

$$\text{Catalyzed by GOD: Glucose} + \text{O}_2 \rightarrow \text{Gluconate} + \text{H}_2\text{O}_2 \quad (12.37)$$

$$\text{For oxygen}: \text{O}_2 + 2\text{R}^{\bullet -} + 2\text{H}^+ \rightarrow 2\text{R}^* + \text{H}_2\text{O}_2 \quad (12.38)$$

$$\text{H}_2\text{O}_2 + 2\text{R}^{\bullet -} \rightarrow 2\text{R}^* + 2\text{OH}^- \quad (12.39)$$

$$\text{For peroxide}: \text{H}_2\text{O}_2 + 2\text{R}^{\bullet -} \rightarrow 2\text{R}^* + 2\text{OH}^- \quad (12.40)$$

$$\text{Then R}^* \rightarrow \text{R} + h\nu \quad (12.41)$$

With the same mechanism, a disposable ECL biosensor has been fabricated by co-immobilizing GOD and surface-unpassived CdTe QDs on a screen-printed carbon electrode (Fig. 12.3) [69]. The QDs show strong ECL emission at −0.89 V in pH 9.0 HCl-Tris buffer. The ECL intensity using dissolved oxygen as a coreactant is twice that with H_2O_2 as a coreactant at the same concentration. It shows

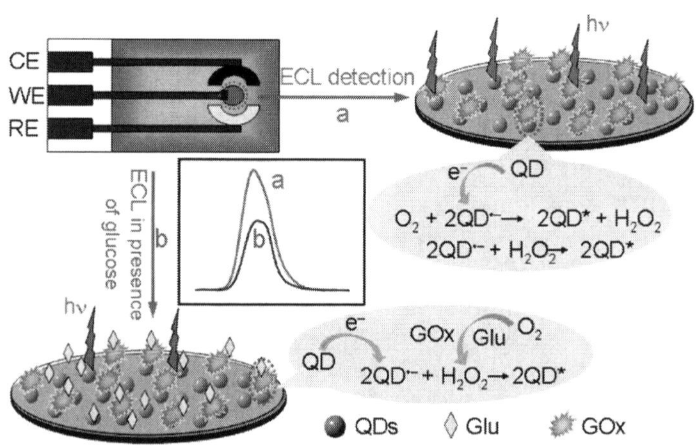

Figure 12.3 Schematic diagram of the biosensor structure and ECL mechanism for glucose detection. Reprinted with permission from Cheng and Ju [69].

rapid response to glucose with a linear range of 0.8–100 μM and a detection limit of 0.3 μM.

Another sensor for H_2O_2 based on CdS QDs and hemoglobin multilayer film has been proposed [70]. Due to the highly stable ECL of emission from $\{Hb/CdS\}_n$ and the excellent electrocatalytical ability of Hb to reduction of H_2O_2, this multilayer film sensor can be used for preparation of biosensors in life and environmental sciences. Using H_2O_2 as a coreactant of MPA-capped CdTe QDs, the ECL biosensor can detect H_2O_2 down to 0.06 μM [33]. ZnSe QDs have been used for ECL sensing of H_2O_2 in pig kidney cells, veto cells, and mineral water, respectively [71]. The biosensor shows good reproducibility.

During cathodic scanning, H_2O_2 can be produced from electrochemical reduction of dissolved oxygen. On the basis of the quenching of ECL emission from QDs by consumption of produced H_2O_2 in the hydroquinone (HQ)–horseradish peroxidase–H_2O_2 system, a reagentless phenolic ECL biosensor for HQ has been constructed (Fig. 12.4) [72].

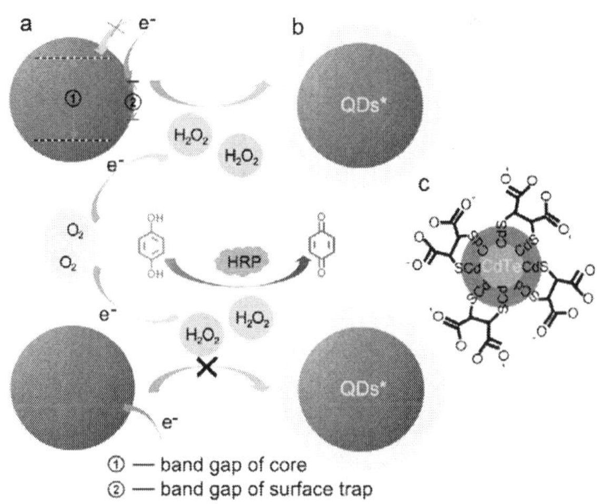

Figure 12.4 Cathodic ECL mechanism of GCE/QDs/chitosan and biosensing mechanism of the biosensor for HQ. (a) Electron injection to surface trap of QDs, (b) ECL emission from excited QDs*, and (c) chemical banding structure of QDs. Reprinted with permission from Cheng and Ju [72].

Some NPs such as multiwalled carbon nanotubes (MWCNTs), nitrogen-doped carbon nanotubes (NCNTs), and room-temperature ionic liquid can enhance the ECL emission from QDs. For example, CdSe QDs composited with NCNTs show a cathodic ECL emission that is five times stronger than that from pure CdSe QDs and three times stronger than that from CdSe QDs composited with CNTs [73]. This result leads to a sensitive ECL sensing of H_2O_2 with good stability and acceptable reproducibility. An ECL enzyme biosensor has been constructed using glutaraldehyde to crosslink choline oxidase and/or acetylcholine esterase to CdS NPs formed in situ on MWCNTs. With H_2O_2 as the coreactant the biosensor can be used for detection of choline and acetylcholine [35]. Compared to pure CdS NPs, the MWCNT–CdS QDs enhance the ECL intensity by 5.3-fold and move the onset ECL potential more positively for about 400 mV.

Besides the II–VI QDs, many nanostructures with low biotoxicity, such as Au_{25} [74] and Ag [75] nanoclusters, C-dots [29], and ZnO [76] and TiO_2 NPs [39], have shown excellent ECL emission. The ECL emission from TiO_2 nanotubes with H_2O_2 as the coreactant has also been observed [77]. On the basis of the "coreactant inhibition" mechanism, the ECL quenchers such as hemoglobin can be measured by inhibiting the transformation of O_2 to H_2O_2. The methodology for detection of a "coreactant inhibition"-related quencher shows acceptable sensitivity.

On the basis of the consumption of coreactant H_2O_2, the ECL emission from QDs has been used for immunosensing. For example, in the presence of electrochemically produced H_2O_2, bidentate chelate CdTe QDs exhibit strong ECL emission. Upon introduction of HQ and horseradish peroxidase to the solution, the ECL can be quenched [72]. Thus the QDs can be co-immobilized with an antigen on the electrode surface to construct an immunosensor [78]. Upon the immunorecognition of the immobilized antigen to its antibody labeled with horseradish peroxidase, the enzyme may be introduced to the immunosensor. In the presence of HQ, the enzymatic cycle consumes the self-produced coreactant H_2O_2, leading to a competitive immunoassay method. This facile immunosensing strategy opens a new avenue for the detection of proteins and application of QDs in ECL biosensing.

Dissolved O_2 can strongly adsorb to NCNTs; thus the presence of NCNTs can accelerate the electrochemical reduction of O_2 to produce $O_2^{\bullet-}$, which acts as a coreactant for promoting the ECL emission from QDs [79]. These properties have been used for construction of an immunosensing strategy by functionalizing NCNTs with polystyrene sulfonate for labeling signal antibodies and co-immobilizing QDs and capture antibodies on the electrode surface.

12.3.4.2 Quenching effect of analytes

This strategy is mainly based on the inhibition effect of the analyte on NP ECL. It includes four approaches: 1) interaction of the analyte with QDs, 2) consumption of the coreactant or intermediate of the ECL process by the reaction with the analyte, 3) energy transfer from exited QDs to the analyte or acceptor, and 4) electrochemical oxidation inhibition of the analyte. The ECL emission from QDs greatly depends on the surface properties of the QDs. When the QDs are capped with thiol compounds, other metal cations existing in the solution can competitively combine the stabilizer to change the surface properties of the QDs, which leads to a quenching effect on the ECL emission. Thus a sensor for detection of metal ions has been proposed using surface-unpassivated CdTe QDs with meso-2,3-dimercaptosuccinic acid as a stabilizer [38]. With a similar mechanism, a sensitive and selective method for Cu^{2+} determination has also been presented by the ECL of CdTe QDs [80].

The second approach has been used for design of a detection method for scavengers of hydroxyl radicals. For example, in a TGA-capped CdSe QDs film/peroxide ECL system, the intermediate $OH^{\bullet}$ radical is the key species for producing holeinjected QDs. Thus the ECL sensors for biologically important scavengers of hydroxyl radicals are developed [24].

The energy transfer from exited QDs to the analyte has been demonstrated with the anodic ECL emission from MPA-capped CdTe QDs at an indium tin oxide electrode. During anodic scanning the QDs can be oxidized at +1.15 V to produce cation radical $QD^{\bullet+}$ by a hole injection; meanwhile, the superoxide ion produced at the electrode surface can inject an electron into the 1Se quantum-

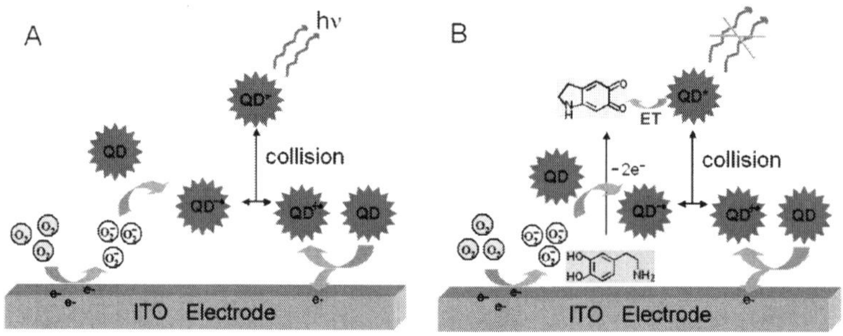

Figure 12.5 Anodic ECL mechanism of QDs (A) and its quenching procedure by the oxidation product of catechol derivatives (B). Reprinted with permission from Liu and Ju [26].

confined orbital of CdTe to form QDs anions (Fig. 12.5). The collision between $QD^{\bullet-}$ and $QD^{\bullet+}$ leads to the formation of exited QDs and ECL emission [26]. The ECL energy transfer from the exited QDs to the quencher produces a methodology for detection of catechol derivatives that are electro-oxidized to form the quencher.

In the presence of dissolved oxygen, tyrosinase can catalyze the oxidation of tyrosine to produce *o*-quinone. By combining the enzymatic cycle of trace tyrosinase to produce the oxidized product with the energy transfer mechanism, an extremely sensitive method for ECL detection of tyrosine has been achieved [25]. The anodic ECL sensing strategy based on a TGA-capped CdSe QD/sulfite system in a neutral medium has also been developed in a similar way for detecting ECL quenchers using dopamine as a model molecule [81].

Another energy transfer way is from the exited QDs to Au NPs as a label for biosensing. The amount of Au NPs captured on the electrode surface is related to the analytes. For example, ECL energy transfer from the CdS:Mn NP film to proximal Au NPs that are labeled at one end of the molecular beacon can quench ECL emission from the CdS:Mn NPs. Upon the recognition of the co-immobilized molecular beacon to target DNA (t-DNA), the distance between Au NPs and CdS:Mn NPs enlarges and hence enhances the ECL emission from the CdS:Mn NP film, leading to a biosensor for detection of t-DNA [82].

Some compounds such as nitrite have been proved to be electroactive ECL quenchers in the $Ru(bpy)_3{}^{2+}$ system via an electrochemical oxidation inhibition mechanism [83]. The presence of these compounds produces a larger *IR* drop and makes the practical potential E_{w} less than the applied potential E [84]:

$$E_{\mathrm{w}} = E - IR = E - (I_{\mathrm{q}} + I_{\mathrm{emitter}})R \qquad (12.42)$$

where I_{q} and I_{emitter} are the oxidation currents of the quencher and the light emitterm respectively. The lower E_{w} decreases the oxidation of the light emitter, leading to weaker ECL emission. Thus after addition of nitrite to the detection solution, the ECL of the CdSe QDs/sulfite system at +0.927 V was attenuated, which leads to a rapid analytical method for nitrite detection down to 0.1 μM [84].

12.3.4.3 Electron transfer kinetics

This strategy is mainly used for immunosensing. The capture antibody and QDs are co-immobilized on the electrode surface. Upon the formation of an immunocomplex, the electron transfer kinetics among the electrode, QDs, and coreactant decrease due to increased resistance, leading to inhibited ECL emission. With this strategy, several label-free QD ECL immunoassay methods have been proposed [31, 32, 85]. The ECL immunosensor prepared with CdSe QDs/MWNTs–chitosan/3-aminoproprpyl-triethoxysilane shows high sensitivity for the detection of human IgG [85]. Similarly, another label-free immunosensor based on ECL emission from CdSe QDs/Au NPs/anti-PAB has been reported for the detection of human prealbumin (PAB; antigen) [32]. With self-assembly and gold NP amplification techniques, an ECL biosensor has also been prepared for detection of low-density lipoprotein in the concentration range of 0.025–16 ng/mL with a detection limit of 0.006 ng/mL [31].

ECL biosensing based on the change of electron transfer kinetics upon recognition of immobilized capture biomolecules to the target has been used for detecting the cell number and monitoring the change of cell surface carbohydrate expression. Ju and coworkers [86] reported an ECL biosensor by immobilizing thioglycolic acid–capped CdSe QDs on a chitosan–Au NP composite–modified electrode and then covalently binding lectins to the CdSe QDs. The biofunctional films showed excellent ECL behavior. After the specific recognition of QD-bound lectins to cell surface carbohydrates, the

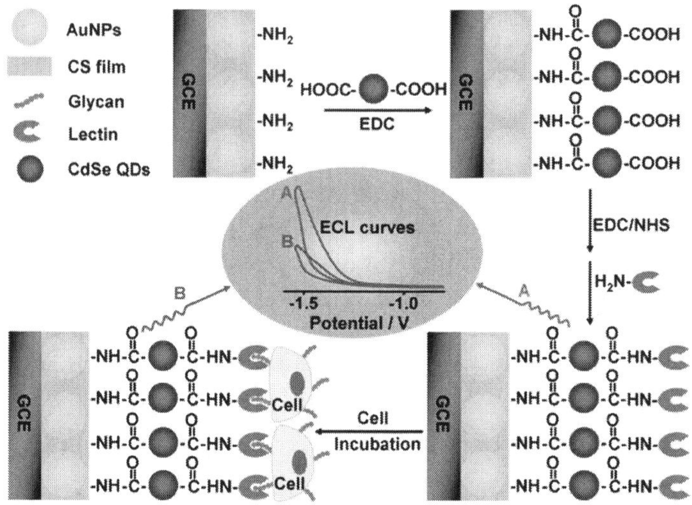

Figure 12.6 Schematic representation of ECL cytosensor for monitoring cell surface carbohydrate expression. Reprinted with permission from Ju and coworkers [86].

captured cells lowered the electron transfer rate, leading to a decrease of ECL intensity (Fig. 12.6), which provides a simple and highly sensitive way for ECL cytosensing and dynamic monitoring of cell surface carbohydrate expression corresponding to the lectins. By integrating the specific recognition of lectins to carbohydrates with carbohydrate-functionalized CdS QDs, a facile ECL strategy for in situ label-free monitoring of carbohydrate expression on living cells has also been designed [87]. The functionalized CdS QDs are immobilized on a CNT–modified electrode for competitive recognition to concanavalin A with the target cells in solution, which leads to a change in ECL intensity due to the resistance of concanavalin A, producing a method to in situ dynamically evaluate cell surface glycan.

12.3.4.4 ECL biosensing using QDs to label binding partners

QDs can be used as labels for construction of ECL biosensors for both DNA and proteins by coupling with different biological recognition pairs. For example, using TGA-capped CdTe QDs as DNA labels an ECL method has been developed for DNA assays [88]. t-DNA hybridizes with capture DNA (c-DNA) bound on the electrode and then the amino modified probe DNA, yielding sandwich hybrids on the electrode. Afterward, MPA-capped CdTe QDs are labeled to the amino group end of the sandwich hybrids, and ECL emission of the QD-labeled DNA hybrids is measured for DNA sensing.

A QD-based ECL biosensor for lysozyme has been developed by forming aptamer–lysozyme bioaffinity complexes at the Au electrode [89]. The free probe is hybridized with the 5′-biotin-modified c-DNA oligonucleotide to form double-stranded DNA. Avidin QDs are bound to the hybridized c-DNA through the biotin–avidin system. The ECL signal of the biosensor is responsive to the amount of QDs bonded to the c-DNA oligonucleotides, which is indirectly inversely proportional to the combined target protein. Another QD-based ECL biosensor for protein using thrombin as an example has also been proposed by immobilizing a thiol-terminated aptamer with 15 nucleotides (probe I) on a Au electrode to capture thrombin, which then hybridizes with a 5′-biotin-modified aptamer (29 nucleotides, probe II) to form a sandwich-type structure. Streptavidin-modified QDs (avidin QDs) are bound to probe II to produce an ECL signal [90].

12.3.4.4.1 Signal amplification for QD-based ECL biosensing

In ECL sensing procedures, enhanced ECL intensity is preferred for their high sensitivity, especially when the ECL intensity is not efficient for sensing purposes. Nanomaterials can be used as catalysts, electronic conductors and carriers of labels to obtain the amplified ECL detection signal [91]. CNTs have been shown to dramatically enhance the ECL intensity of QDs [37], which promotes the application of QDs ECL in sensitive biosensor fabrication [35, 92]. By facilitating the CdTe QDs' oxidation and triggering $O_2^{\bullet -}$ generation, graphene oxide can lead to enhanced ECL for sensing glutathione [93]. Another carbon nanomaterial, NCNTs, has been used for preparation of a nanocomposite of CdSe QDs to enhance cathodic ECL emission [73]. With hydrogen peroxide as a coreactant, the emission from the nanocomposite-modified electrode was five times stronger than that from pure CdSe QDs and three times stronger than that from CdSe QDs composited with CNTs.

An efficient coreactant is one important factor to obtain a sensitive biosensing signal. For example, the anodic ECL emission from MPA-cappled CdTe QDs is very weak and cannot be used for sensing purposes in air-saturated pH 7.5 PBS. However, with a sulfite as a coreactant, greatly enhanced ECL emission can lead to sensitive sensing for analytes [25]. In addition the choice of the QDs' architecture is also important for obtaining an amplified ECL signal [36]. By embedding QDs in polymers to produce nanocomposites for preparation of a signal probe, enhanced ECL emission has been obtained. For example, the nanocomposite of a poly(amidoamine) (PAMAM) dendrimer and CdS QDs shows 55-fold enhanced ECL compared to that from a QD film without a dendrimer [94].

The signal amplification for ECL biosensing has attracted considerable interest. Using nanocomposites of QDs and a dendrimer as labels of canavalin A, sensitive ECL biosensors for cell concentration have been reported [95, 96]. By covalently coupling the nanocomposites to the amino group of a DNA probe for signal amplification of ECL measurements, an ultrahigh sensitive protocol for detection of a near-single DNA molecule has also been developed [97]. The ECL DNA biosensor is prepared by immobilizing a hairpin DNA molecular beacon probe with the thiol group at the 5′ end and the amino group at the 3′ end on a Au NP–modified electrode via the Au–S bond. The specific hybridization with t-DNA opens the beacon loop to expose the amino group for covalently linking to the amino group on the surface of the QD–dendrimer nanocomposite, generating strong and stable ECL in the presence of $S_2O_8^{2-}$ as a coreactant, which is related to t-DNA concentration.

The catalytic function of labels to generate or consume the coreactant or affect the formation of a coreactant has been used for signal amplification. A label is loaded on a nanocarrier to label the binding partner. This strategy has been designed for ultrasensitive immunoassays of protein by DNAzyme-biobarcoded NP tags (Fig. 12.7) [98]. The DNAzyme is formed on Au NPs by the interaction of hemin with guanine-rich ssDNA that is assembled on the antibody-bound Au NP surface. As a label, DNAzyme possesses higher coverage and higher activity than a protein enzyme due to its

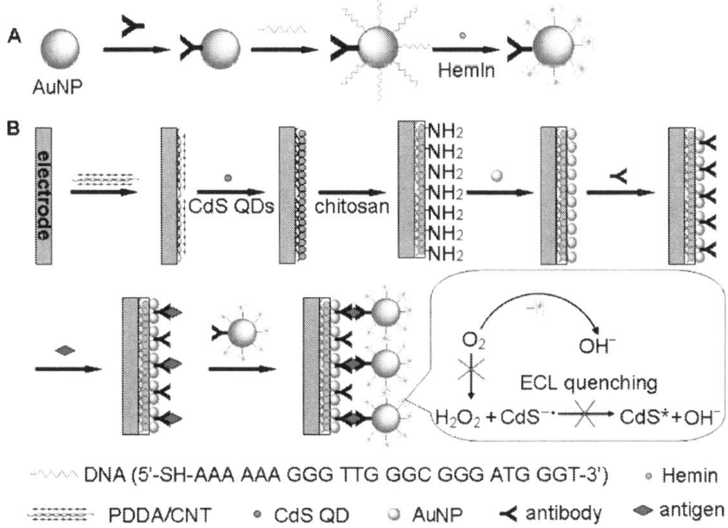

Figure 12.7 Schematic representation of (A) preparation procedure of Ab2-biobarcoded Au NP–G-quadruplex/hemin probe and (B) immunosensor preparation and sandwich-type detection procedure. Reprinted with permission from Ju [98].

smaller size. With a sandwich immunoassay, DNAzyme is captured on the immunosensor surface to catalyze the reduction of dissolved oxygen, the coreactant for cathodic ECL emission, leading to a decrease in ECL intensity. The high loading of DNAzyme for signal recognition events and its strong catalytic activity for quenching ECL emission lead to ultrahigh sensitivity. The detection limit is 4 orders of magnitude lower than that using QD ECL coupled with enzymatic amplification [81], providing a promising potential in clinical diagnosis, especially in point-of-care testing.

12.4 Conclusion

ECL biosensors as powerful tools have been quickly developed for the detection of a wide range of analytes since detailed studies in the mid-1960s. The finding of new, efficient ECL light-emitting molecules or nanomaterials and coreactants, as well as the elucidation of

ECL mechanisms for improving the sensitivity and selectivity of ECL biosensors, is continuous research motivity. The needs in biomedicine, clinic diagnostics, and food and environmental monitoring are promoting the development of the ECL biosensing area.

The improvement of selectivity for ECL biosensing can be achieved by decreasing the overpotential of ECL emission, combining the ECL emitter with specific biological recognition or integrating ECL biosensing with separation techniques. The decrease of overpotential requires a search for new, highly efficient and tunable NP ECL systems (emitters and coreactants) and further understand ECL mechanisms. An efficient approach to enhance the sensitivity of ECL biosensing is to design new signal amplification strategies. Besides the amplification strategies mentioned above, several DNA amplification techniques such as rolling circle amplification, target-induced repeated primer extension, hybridization chain reaction, loop-mediated amplification, t-DNA recycling amplification with endonuclease and exonuclease, and polymerase-based circular strand replacement polymerization can be used to amplify the ECL signal for biosensing of nucleic acids or proteins.

References

1. Dufford, R. T. (1927) Luminescence of grignard compounds in electric and magnetic fields, and related electrical phenomena, *J. Am. Chem. Soc.*, **49**, 1858–1864.
2. Harvey, N. (1929) Luminescence during electrolysis, *J. Phys. Chem.*, **33**, 1456–1459.
3. Richter, M. M. (2004) Electrochemiluminescence, *Chem. Rev.*, **104**, 3003–3036.
4. Miao, W. J. (2008) Electrogenerated chemiluminescence and its biorelated applications, *Chem. Rev.*, **108**, 2506–2553.
5. Zweig, A., Maurer, A. H., Roberts, B. G. (1967) Oxidation, reduction, and electrochemiluminescence of donor-substituted polycyclic aromatic hydrocarbons, *J. Org. Chem.*, **32**, 1322–1329.
6. Tokel, N., Bard, A. J. (1972) Electrogenerated chemiluminescence IX. Electrochemistry and emission from systems containing $Ru(bpy)_3^{2+}$ dichloride, *J. Am. Chem. Soc.*, **94**, 2862–2863.

7. Ding, Z., Quinn, B. M., Haram, S. K. (2002) Electrochemistry and electrogenerated chemiluminescence from silicon nanocrystal quantum dots, *Science*, **296**, 1293–1297.
8. Myung, N., Ding, Z., Bard, A. J. (2002) Electrogenerated chemiluminescence of CdSe nanocrystals, *Nano Lett.*, **2**, 1315–1319.
9. Hercules, D. M. (1964) Chemiluminescence resulting from electrochemically generated species, *Science*, **145**, 808–809.
10. Kuwana, T., Epstein, B., Seo, E. T. (1963) Electrochemical generation of solution luminescence, *J. Phys. Chem.*, **67**, 2243–2244.
11. Zou, G. Z., Ju, H. X. (2004) Electrogenerated chemiluminescence from a CdSe nanocrystal film and its sensing application in aqueous solution, *Anal. Chem.*, **76**, 6871–6876.
12. Jiang, H., Ju, H. X. (2007) Enzyme-quantum dots architecture for highly sensitive electrochemiluminescence biosensing of oxidase substrates, *Chem. Commun.*, 404–406.
13. Santhanam, K. S. V., Bard, A. J. (1965) Chemiluminescence of electrogenerated 9,10-diphenylanthracene anion radical, *J. Am. Chem. Soc.*, **87**, 139–140.
14. Forry, S. P., Wightman, R. M. (2004) *Electrogenerated Chemiluminescence*, ed. A. J. Bard (Dekker, New York), 273.
15. Wehrenberg, B. L., Guyot-Sionnest, P. (2003) Electron and hole injection in PbSe quantum dot films, *J. Am. Chem. Soc.*, **125**, 7806–7807.
16. Haram, S. K., Quinn, B. M., Bard, A. J. (2001) Electrochemistry of CdS nanoparticles: a correlation between optical and electrochemical band gaps, *J. Am. Chem. Soc.*, **123**, 8860–8861.
17. Chang, M. M., Saji, T., Bard, A. J. (1977) Electrogenerated chemiluminescence 30. Electrochemical oxidation of oxalate ion in the presence of luminescers in acetonitrile solutions, *J. Am. Chem. Soc.*, **99**, 5399–5403.
18. Rubinstein, I., Bard, A. J. (1981) Electrogenerated chemi-luminescence. Aqueous ECL systems based on $Ru(bpy)_3^{2+}$ and oxalate or organic-acids, *J. Am. Chem. Soc.*, **103**, 512–516.
19. Miao, W. Choi, J.-P., Bard, A. J. (2002) Electrogenerated chemiluminescence 69. The $Ru(bpy)_3^{2+}$/tri-n-propylamine system revisited; a new route involving TPrA cation radicals, *J. Am. Chem. Soc.*, **124**, 14478–14485.
20. Leland, J. K., Powell, M. J. (1990) Electrogenerated chemiluminescence: an oxidative-reduction type ECL reaction sequence using tripropyl amine, *J. Electrochem. Soc.* **137**, 3127–3131.

21. White, H. S., Bard, A. J. (1982) Electrogenerated chemiluminescence 41. Electrogenerated chemiluminescence and chemiluminescence of the $Ru(bpy)_3^{2+}$-$S_2O_8^{2-}$ system in acetonitrile-water solutions, *J. Am. Chem. Soc.*, **104**, 6891–6895.
22. Myung, N., Lu, X., Johnston, K. P. (2004) Electrogenerated chemiluminescence of Ge nanocrystals, *Nano Lett.*, **4**, 183–185.
23. Myung, N., Bae, Y., Bard, A. J. (2003) Effect of surface passivation on the electrogenerated chemiluminescence of CdSe/ZnSe nanocrystals, *Nano Lett.*, **3**, 1053–1055.
24. Jiang, H., Ju, H. X. (2007) Electrochemiluminescence sensors for scavengers of hydroxyl radical based on its annihilation in CdSe quantum dots film/peroxide system, *Anal. Chem.*, **79**, 6690–6696.
25. Liu, X., Ju, H. X., (2008) Coreactant enhanced anodic electro- chemiluminescence of CdTe quantum dots at low potential for sensitive biosensing amplified by enzymatic cycle, *Anal. Chem.*, **80**, 5377–5382.
26. Liu, X., Jiang, H., Lei, J. P., Ju, H. X. (2007) Anodic electrochemiluminescence of CdTe quantum dots and its energy transfer for detection of catechol derivatives, *Anal. Chem.*, **79**, 8055–8060.
27. Mei, Y. L., Wang, H. S., Li, Y. F. (2010) Electochemiluminescence of CdTe/CdS quantum dots with triproprylamine as coreactant in aqueous solution at a lower potential and its application for highly sensitive and selective detection of Cu^{2+}, *Electroanalysis*, **22**, 155–160.
28. Shen, L. H., Cui, X. X., Qi, H. L. (2007) Electrogenerated chemiluminescence of ZnS nanoparticles in alkaline aqueous solution, *J. Phys. Chem. C*, **111**, 8172–8175.
29. Zheng, L. Y., Chi, Y. W., Dong, Y. Q. (2009) Electrochemiluminescence of water-soluble carbon nanocrystals released electrochemically from graphite, *J. Am. Chem. Soc.*, **131**, 4564–4565.
30. Jie, G. F., Liu, B., Miao, J. J. (2007) Electrogenerated chemiluminescence from CdS nanotubes and its sensing application in aqueous solution, *Talanta*, **71**, 1476–1480.
31. Jie, G. F., Liu, B., Pan, H. C. (2007) CdS nanocrystal-based electrochemiluminescence biosensor for the detection of low-density lipoprotein by increasing sensitivity with gold nanoparticle amplification, *Anal. Chem.*, **79**, 5574–5581.
32. Jie, G. F., Huang, H. P., Sun, X. L. (2008) Electrochemiluminescence of CdSe quantum dots for immunosensing of human prealbumin, *Biosens. Bioelectron.*, **23**, 1896–1899.

33. Han, H. Y., Sheng, Z. G., Liang, J. G. (2007) Electrogenerated chemiluminescence from thiol-capped CdTe quantum dots and its sensing application in aqueous solution, *Anal. Chim. Acta*, **596**, 73 78.

34. Wang, X. F., Xu, J. J., Chen, H. Y. (2008) A new electrochemiluminescence emission of Mn^{2+}-doped ZnS nanocrystals in aqueous solution, *J. Phys. Chem. C*, **112**, 17581–17585.

35. Wang, X. F., Zhou, Y., Xu, J. J. (2009) Signal-on electrochemiluminescence biosensors based on CdS–carbon nanotube nanocomposite for the sensitive detection of choline and acetylcholine, *Adv. Funct. Mater.*, **19**, 1444–1450.

36. Wang, C., Yifeng, E., Fan, L. (2007) Directed assembly of hierarchical CdS nanotube arrays from CdS nanoparticles: enhanced solid state electrochemiluminescence in H_2O_2 solution, *Adv. Mater.*, **19**, 3677–3681.

37. Ding, S. N., Xu, J. J., Chen, H. Y. (2006) Enhanced solid-state electrochemiluminescence of CdS nanocrystals composited with carbon nanotubes in H_2O_2 solution, *Chem. Commun.*, **34**, 3631–3633.

38. Cheng, L., Liu, X., Lei, J., Ju, H. X. (2010) Low-potential ECL sensing based on surface unpassivation of CdTe quantum dots and competition of analyte cation to stabilizer, *Anal. Chem.*, **82**, 3359–3364.

39. Lin, Z., Liu, Y., Chen, G. (2008) TiO_2/Nafion film based electrochemiluminescence for detection of dissolved oxygen, *Electochem. Commun.*, **10**, 1629–1632.

40. Gill, R., Zayats, M., Willner, I. (2008) Semiconductor quantum dots for bioanalysis, *Angew. Chem., Int. Ed.*, **47**, 7602–7625.

41. Bertoncello, P., Forster, R. J. (2009) Nanostructured materials for electrochemiluminescence-based detection methods: recent advances and future perspectives, *Biosens. Bioelectron.*, **24**, 3191–3200.

42. Rubinstein, I., Bard, A. J. (1980) Polymer-films on electrodes 4. Nafion-coated electrodes and electrogenerated chemiluminescence of surface-attached $Ru(bpy)_3{}^{2+}$, *J. Am. Chem. Soc.*, **102**, 6641–6642.

43. Harmer, M. A., Farneth, W. E., Sun, Q. J. (1996) High surface area Nafion resin/silica nanocomposites: a new class of solid acid catalyst, *J. Am. Chem. Soc.*, **118**, 7708–7715.

44. Choi, H. N., Cho, S. H., Lee, W. Y. (2003) Electrogenerated chemiluminescence from tris(2,2'-bipyridyl)ruthenium(II) immobilized in titania-perfluorosulfonated lonomer composite films, *Anal. Chem.*, **75**, 4250–4256.

45. Michel, P. E., van der Wal, P. D., Fiaccabrino, G. C., de Rooij, N. F. (1999) Koudelka-Hep M. Reagentless sensor integrating electrodes, photode-

tector, and immobilized co-substrate for electrochemiluminescence-based assays, *Electroanalysis*, **11**, 136–1367.

46. Zhao, C. Z., Egashira, N., Kurauchi, Y., Ohega, K. (1997) Substrate selectivity of an electrochemiluminescence Pt electrode coated with a $Ru(bpy)_3{}^{2+}$-modified chitosan silica gel membrane, *Anal. Sci.*, **13**, 333–336.
47. Wang, H. Y., Xu, G. B., Dong, S. J. (2001) Electrochemiluminescence of $Ru(bpy)_3{}^{2+}$ immobilized in poly(p-styrenesulfonate)-silica-triton X-100 composite thin-films, *Analyst*, **126**, 1095–1099.
48. Wang, H. Y., Xu, G. B., Dong, S. J. (2003) Electrochemiluminescence sensor using $Ru(bpy)_3{}^{2+}$ immobilized in Eastman-AQ55D-silica composite thin-films, *Anal. Chim. Acta*, **480**, 285–290.
49. Zhuang, Y. F., Ju, H. X. (2004) Study on electrochemiluminesce of $Ru(bpy)_3{}^{2+}$ immobilized in a titania sol-gel membrane, *Electroanalysis*, **16**, 1401–1405.
50. Dutta, P. K., Das, S. K. (1997) Oxidizing properties of zeolite-encapsulated oxobis(2,2'-bipyridine)ruthenium(IV) complexes formed by air oxidation of bis(2,2'-bipyridine)aquaruthenium(II), *J. Am. Chem. Soc.*, **119**, 4311–4312.
51. Zhang, X., Bard, A. J. (1988) Electrogenerated chemiluminescent emission from an organized monolayer of a $Ru(bpy)_3{}^{2+}$-based surfactant on semiconductor and metal-electrodes, *J. Phys. Chem.*, **92**, 5566–5569.
52. Obeng, Y. S., Bard, A. J. (1991) Electrogenerated chemiluminescence 53. Electrochemistry and emission from absorbed monolayers of a $Ru(bpy)_3{}^{2+}$ -based surfactant on gold and tin oxide electrodes, *Langmuir*, **7**, 195–210.
53. Zhuang, Y. F., Zhang, D. M., Ju, H. X. (2005) Sensitive determination of heroin based on electrogenerated chemiluminesce of $Ru(bpy)_3{}^{2+}$ immobilized in zeolite Y modified carbon paste electrode, *Analyst*, **130**, 534–540.
54. Zhuang, Y. F., Ju, H. X. (2005) Determination of reduced nicotinamide adenine dinucleotide based on immobilization of $Ru(bpy)_3{}^{2+}$ in multiwall carbon nanotubes/Nafion composite membrane *Anal. Lett.*, **38**, 2077–2088.
55. Du, Y., Qi, B., Yang, X., Wang, E. (2006) Synthesis of PtNPs/AQ/Ru $(bpy)_3{}^{2+}$ colloid and its application as a sensitive solid-state electrochemiluminescence sensor material, *J. Phys. Chem. B*, **110**, 21662–21666.

56. Noffsinger, J. B., Danielson, N. D. (1987) Generation of chemiluminescence upon reaction of aliphatic-amines with $Ru(bpy)_3{}^{3+}$, *Anal. Chem.*, **59**, 865–868.
57. Li, J. G., Yan, Q. Y., Gao, Y. L., Ju, H. X. (2006) Electrogenerated chemiluminescence detection of amino acids based on precolumn derivatization coupled with capillary electrophoresis separation, *Anal. Chem.*, **78**, 2694–2699.
58. Skotty, D. R., Lee, W. Y., Nieman, T. A. (1996) Determination of dansyl amino acids and oxalate by HPLC with electrogenerated chemiluminescence detection using $Ru(bpy)_3{}^{2+}$ in the mobile phase, *Anal. Chem.*, **68**, 1530–1535.
59. Morita, H., Konishi, M. (2002) Electrogenerated chemiluminescence derivatization reagents for carboxylic acids and amines in high-performance liquid chromatography using $Ru(bpy)_3{}^{2+}$, *Anal. Chem.*, **74**, 1584–1589.
60. Jameison, F., Sanchez, R. I., Dory, L., Leland, J. K., Yost, D., Martin, M. T. (1996) Electrochemiluminescence-based quantitation of classical clinical chemistry analytes, *Anal. Chem.*, **68**, 1298–1302.
61. Leca, B. D., Verdier, A. M., Blum, L. (2001) Screen-printed electrodes as disposable or reusable optical devices for luminol electrochemiluminescence, *Sens. Actuators, B*, **74**, 190–193.
62. Zhu, L. D., Li, Y. X., Tian, F. M., Zhu, G. Y. (2002) Electrochemiluminescent determination of glucose with a sol-gel derived ceramic-carbon composite electrode as a renewable optical fiber biosensor, *Sens. Actuators, B*, **84**, 265–270.
63. Miao, W. J., Bard, A. J. (2004) Electrogenerated chemiluminescence 77. DNA hybridization detection at high amplification with $Ru(bpy)_3{}^{2+}$-containing microspheres, *Anal. Chem.*, **76**, 5379–5386.
64. Lo, W.-Y., Baeumner, A. J. (2007) Evaluation of internal standards in a competitive nucleic acid sequence-based amplification assay, *Anal. Chem.*, **79**, 1386–1392.
65. Debad, J. D., Glezer, E. N., Leland, Sigal, J. K., Wohlsadter, G. B. (2004) *Electrogenerated Chemiluminescence*, ed. A. J. Bard (Dekker, New York), 359.
66. Yin, X.-B., Qi, B., Sun, X., Yang, X., Wang, E. (2005) 4-(Dimethylamino)butyric acid labeling for electrochemiluminescence detection of biological substances by increasing sensitivity with gold nanoparticle amplification, *Anal. Chem.*, **77**, 3525–3530.

67. Miao, W., Bard, A. J. (2004) Electrogenerated chemiluminescence 80. C-reactive protein determination at high amplification with [Ru(bpy)(3)](2+)-containing microspheres, *Anal. Chem.*, **76**, 7109–7113.
68. Zhan, W., Bard, A. J. (2007) Electrogenerated chemiluminescence 83. Immunoassay of human C-reactive protein by using $Ru(bpy)_3{}^{2+}$-encapsulated liposomes as labels, *Anal. Chem.*, **79**, 459–463.
69. Cheng, L. X., Deng, S. Y., Lei, J. P., Ju, H. X. (2012) Disposable electrochemiluminescent biosensor using bidentate-chelated CdTe quantum dots as emitter for sensitive detection of glucose, *Analyst*, **137**, 140–144.
70. Shi, C. G., Xu, J. J., Chen, H. Y. (2007) Electrogenerated chemiluminescence and electrochemical bi-functional sensors for H_2O_2 based on CdS nanocrystals/hemoglobin multilayers, *J. Electroanal. Chem.*, **610**, 186–192.
71. Hu, X., Han, H., Hua, L. (2010) Electrogenerated chemiluminescence of blue emitting ZnSe quantum dots and its biosensing for hydrogen peroxide, *Biosens. Bioelectron.*, **25**, 1843–1846.
72. Liu, X., Lei, J. P., Cheng, L. X., Liu, H., Ju, H. X. (2010) Surface trap of quantum dots by bidentate chelation for low-potential electrochemiluminescent biosensing, *Chem. Eur. J.*, **16**, 10764–10770 (VIP).
73. Guo, L., Liu, X., Hu, Z., Deng, S. Y., Ju, H. X. (2009) Electrochemiluminescence of CdSe quantum dots composited with nitrogen-doped carbon nanotubes, *Electroanalysis*, **21**, 2495–2498.
74. Li, L. L., Liu, H. Y., Shen, Y. Y., Zhang, J. R., Zhu, J. J. (2011) Electrogenerated chemiluminescence of Au nanoclusters for the detection of dopamine, *Anal. Chem.*, **83**, 661–665.
75. Díez, I., Pusa, M., Kulmala, S., et al. (2009) Color tunability and electrochemiluminescence of silver nanoclusters, *Angew. Chem., Int. Ed.*, **48**, 2122–2125.
76. Zhang, R. X., Fan, L. Z., Fang, Y. P., Yang, S. H. (2008) Electrochemical route to the preparation of highly dispersed composites of ZnO/carbon nanotubes with significantly enhanced electrochemiluminescence from ZnO, *J. Mater. Chem.*, **18**, 4964–4970.
77. Liu, X., Hou, Z. T., Zhang, Y. Y., Ju, H. X. (2011) High quantum transfer efficiency of titania dioxide nanotube for low potential electrogenerated chemiluminescence, *Electroanalysis*, **23**, 2629–2632.
78. Liu, X., Zhang, Y. Y., Lei, J. P., Xue, Y. D., Cheng, L. X., Ju, H. X. (2010) Quantum dots based electrochemiluminescent immunosensor

by coupling enzymatic amplification with self-produced coreactant from oxygen reduction, *Anal. Chem.*, **82**(17), 7351–7356.

79. Deng, S. Y., Lei, J. P., Hou, Z. T., et al. (2011) Signal amplification by adsorption-induced catalytic reduction of dissolved oxygen on nitrogen-doped carbon nanotubes for electrochemiluminescent immunoassay, *Chem. Commun.*, **47**, 12107–12109.
80. Zhang, L., Shang, L., Dong, S. (2008) Sensitive and selective determination of Cu2+ by electrochemiluminescence of CdTe quantum dots, *Electrochem. Commun.*, **10**, 1452–1454.
81. Liu, X., Cheng, L. X., Lei, J. P., Ju, H. X. (2008) Dopamine detection based on its quenching effect to anodic electrochemiluminescence of CdSe quantum dots, *Analyst*, **133**, 1161–1163.
82. Shan, Y., Xu, J. J., Chen, H. Y. (2009) Distance-dependent quenching and enhancing of electrochemiluminescence from a CdS: Mn nanocrystal film by Au nanoparticles for highly sensitive detection of DNA, *Chem. Commun.*, **8**, 905–907.
83. Chi, Y. W., Dong, Y. Q., Chen, G. N. (2007) Inhibited $Ru(bpy)_3^{2+}$ electrochemiluminescence related to electrochemical oxidation activity of inhibitors, *Anal. Chem.*, **79**, 4521–4528.
84. Liu, X., Guo, L., Cheng, L. X., Ju, H. X. (2009) Determination of nitrite based on its quenching effect on anodic electrochemiluminescence of CdSe quantum dots, *Talanta*, **78**, 691–694.
85. Jie, G., Zhang, J., Wang, D. (2008) Electrochemiluminescence immunosensor based on CdSe nanocomposites, *Anal. Chem.*, **80**, 4033–4039.
86. Han, E., Ding, L., Lian, H. Z., Ju, H. X. (2010) Cytosensing and dynamic monitoring of cell surface carbohydrate expression by electrochemiluminescence of quantum dots, *Chem. Commun.*, **46**, 5446–5448.
87. Han, E., Ding, L., Jin, S., Ju, H. X. (2011) Electrochemiluminescent biosensing of carbohydrate-CdS nanocomposites for in situ label-free analysis of cell surface carbohydrate, *Biosens. Bioelectron.*, **26**, 2500–2505.
88. Hu, X., Wang, R., Ding, Y. (2010) Electrochemiluminescence of CdTe quantum dots as labels at nanoporous gold leaf electrodes for ultrasensitive DNA analysis, *Talanta*, **80**, 1737–1743.
89. Huang, H., Jie, G., Cui, R., Zhu J. J. (2009) DNA aptamer-based detection of lysozyme by an electrochemiluminescence assay coupled to quantum dots, *Electrochem. Commun.*, **11**, 816–818.

90. Huang, H., Zhu, J. (2009) DNA aptamer-based QDs electrochemiluminescence biosensor for the detection of thrombin, *Biosens. Bioelectron.*, **25**, 927–930.
91. Ju, H. X. (2011) Sensitive biosensing strategy using functional nanomaterials to amplify the detectable signal, *Sci. China Chem.*, **54**, 1202–1217.
92. Hua, L., Han, H., Chen, H. Y. (2009) Enhanced electrochemiluminescence of CdTe quantum dots with carbon nanotube film and its sensing of methimazole, *Electrochim. Acta*, **54**, 1389–1394.
93. Wang, Y., Lu, J., Tang, L. (2009) Graphene oxide amplified electrogenerated chemiluminescence of quantum dots and its selective sensing for glutathione from thiol-containing compounds, *Anal. Chem.*, **81**, 9710–9715.
94. Lu, C., Wang, X. F., Xu, J. J., Chen, H. Y. (2008) Electrochemical modulation of electrogenerated chemiluminescence of CdS nanocomposite, *Electrochem. Commun.*, **10**, 1530–1532.
95. Jie, G., Wang, L., Yuan, J., Zhang, S. (2011) Versatile electrochemiluminescence assays for cancer cells based on dendrimer/CdSe–ZnS–quantum dot nanoclusters, *Anal. Chem.*, **83**, 873–3880.
96. Qian, Z., Bai, H. J., Wang, G. L., Xu, J. J., Chen, H. Y. (2010) A photoelectrochemical sensor based on CdS-polyamidoamine nanocomposite film for cell capture and detection, *Biosens. Bioelectron.*, **25**, 2045–2050.
97. Divsar, F., Ju, H. X. (2011) Electrochemiluminescence detection of near single DNA molecule by using quantum dots-dendrimer nanocomposite for signal amplification, *Chem. Commun.*, **47**, 9879–9881.
98. Lin, D. J., Wu, J., Yan, F., Deng, S. Y., Ju, H. X. (2011) Ultrasensitive immunoassay of protein biomarker for ultrasensitive immunoassay of protein biomarker based on electrochemiluminescent quenching of quantum dots by hemin biobarcoded nanoparticle tags, *Anal Chem.*, **83**, 5214–5221.

Chapter 13

The Self-Powered Biosensors Based on Biofuel Cells

Liu Deng and Shaojun Dong

State Key Laboratory of Electroanalytical Chemistry,
Changchun Institute of Applied Chemistry, Chinese Academy of Sciences, Changchun, Jilin 130022, China
dengliu@csu.edu.cn, dongsj@ciac.jl.cn

With expansion of research in the area of lab-on-a-chip devices, micrototal analysis systems, and point-of-care and home medical testing, the need to miniaturize both transducers and electronic signaling devices has drawn much attention. One of the problems associated with miniaturization and portability of sensors is the power supply. Power supplies, such as batteries, are difficult to miniaturize and require a sensor design that allows for easy replacement or recharging. This review describes the field of self-powered biological sensing, where the sensor itself provides the power for the sensing device. Here we focus on self-powered biosensor devices based on chemical-to-electrochemical energy transformations in biofuel cell (BFC) elements. BFCs are categorized on the basis of biocatalyst applied, such as microbial fuel cells, mitochondrial BFCs, and enzymatic BFCs. All three types of BFCs have been proposed for self-powered biosensing. This review details

Electrochemical Biosensors
Edited by Serge Cosnier

ISBN 978-981-4411-46-2 (Hardcover), 978-981-4411-47-9 (eBook)
www.panstanford.com

the common strategies for self-powered biosensors and discusses the future of the technology.

13.1 Introduction

Electrochemical biosensors with selectivity, rapidity, sensitivity, simplicity, and easy maintenance have a wide range of applications in the area of clinical diagnostics, forensic chemistry, food quality control, biological warfare detection, etc. The development of miniaturized and portable biosensor devices has been spurred by the desire to produce micrototal analysis systems, low-cost point-of-care diagnostics, and environmental monitoring devices [1, 2]. One of the greatest problems associated with biosensor miniaturization is the power supply since power supplies are difficult to miniaturize and require easy replacement or recharging. So the strategy of self-powered sensing was introduced [3–5]. Compared to the conventional biosensor, the analyte presence in a self-powered sensor produces sufficient energy for signaling. Therefore, self-powered sensors do not need an external power source. Recently, a novel concept of self-powered biosensor devices based on biofuel cells (BFCs) was developed [6]. The BFC is similar to conventional fuel cells in that it converts the fuel chemical energy into electrical energy, except that the precious metal catalysts are replaced with a biological catalyst [7, 8]. The biocatalyst utilized in a BFC can be adjusted by modulators (inhibitors or activators). The BFC power output is related to the biocatalyst activity as well as the substrate. So the BFC can be regarded as a sensitive and selective method for the detection modulator or substrate because of the amplifying nature imparted by substrate turnover and the specificity of biological recognition. In fact, the low electrical current of BFCs has advantages in sensing processes since the redox transformation of interfering agents at the electrode can be eliminated. BFCs are categorized on the basis of the biocatalyst applied. If the biocatalyst is a microorganism, the BFC is considered to be a microbial fuel cell (MFC). If the biocatalyst is an organelle (subcellular component), the BFC is considered to be an organelle or a mitochondrial BFC. If the biocatalyst is an enzyme, the BFC is considered to be an enzymatic biofuel cell (EBFC). All three types of BFCs have been proposed for self-powered biosensing and will be detailed and compared in this review.

13.2 The Self-Powered Biosensor Based on the Enzymatic Biofuel Cell

In 2001, the first self-powered biosensor based on an EBFC for chemical and biological sensing was reported by Willner's group [6]. An EBFC with glucose oxidase (GOx) at the bioanode and cytochrome c oxidase (COx) at the biocathode was designed for glucose self-powered sensing, as shown in Fig. 13.1. In the absence

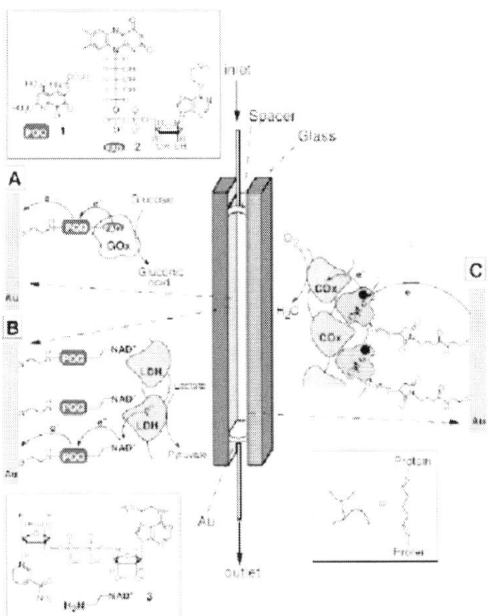

Figure 13.1 Configuration of self-powered BFC-based biosensors composed of (a) pyrroloquinoline quinone-flavin adenine dinucletide/GOx-functionalized anode utilizing a glucose analyte as fuel, (b) pyrroloquinoline quinone-nicotinamide adenine dinucleotide/LDH-functionalized anode utilizing a lactate analyte as fuel, and (c) cytochrome c–/COx-functionalized cathode utilizing O_2 as an oxidizer in combination with both anodes (a or b). Adapted from Ref. [6] with permission from the American Chemical Society.

of glucose, there was no current, voltage, or power, whereas in the presence of glucose, the current, voltage, and power density of the EBFC were as a function of glucose concentration. Another self-powered biosensor developed was a lactate biosensor employing a BFC similar to the previous one with lactate dehydrogenase (LDH) at the bioanode. This type of self-powered sensor could also employ a photo-BFC, as recently described by Tel-Vered et al. instead of a traditional BFC [9].

Another strategy for self-powered biosensors is to detect the analyte by using inhibitor effects in the biosensors' design. Enzymatic inhibitor effects were first used to develop a self-powered cyanide biosensor employing a glucose/air EBFC in our group [10]. We designed an integrated EBFC microchip with both a glucose dehydrogenase bioanode and a laccase biocathode on the chip. Using glucose as fuel, there was current and power output in the presence of the fuel, but the current density and power density decreased with increasing cyanide concentrations. The binding of CN^- onto the T2 Cu of laccase and other multi-Cu oxidases (such as ascorbate oxidase and ceruloplasmin) is well-known, and its negative effect on oxygen reduction is attributed to a perturbed T2/T3 Cu cluster unfavorable toward the internal electron transfer from T1 Cu. Thus the catalysis of laccase was significantly affected by cyanide addition, as shown in Fig. 13.2. Employing an inhibitive effect in self-powered biosensors can expand the analyte's range since an analyte that inhibits the enzyme regardless of redox activity can be detected. Recently our group has developed a self-powered Hg^{2+} sensor by the same methodology [11].

Minteer's group first proposed the use of decoupler effects in a self-powered biosensor's design [12]. A self-powered ethylene-diaminetetraacetic acid (EDTA) biosensor was fabricated on the basis of the activation of a Cu^{2+}-inhibited GOx bioanode (Fig. 13.3). The self-powered biosensor was operated by inhibiting and subsequently activating the GOx-modified electrode in a glucose/O_2 EBFC. As a strong metal chelator, EDTA could reverse the binding of metal ions to the flavin cofactor and reactivate GOx. By using different concentrations of Cu^{2+} to inhibit the bioanode and using a corresponding amount of EDTA to reactivate the enzyme, different amount of EDTA could be detected. This method represents an

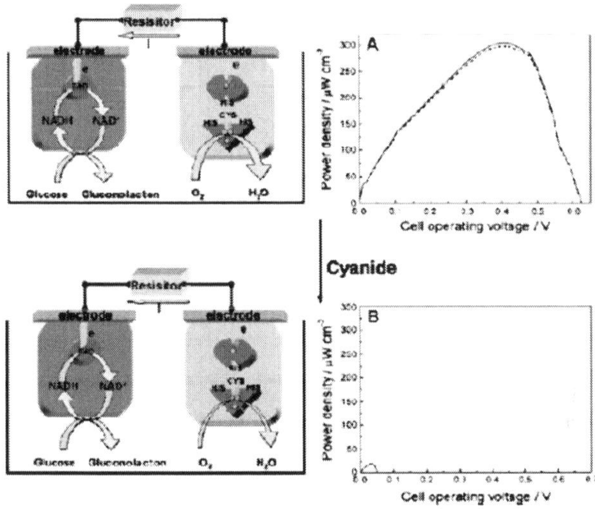

Figure 13.2 Bioelectrocatalysis mechanism of the cyanide inhibitive effect at a laccase-immobilized electrode. (a) Polarization of the uninhibited control and regeneration with 30 min fresh PBS washing of the glucose/O_2 BFC. (B) Polarization of 1 mM cyanide inhibition of the glucose/O_2 BFC. The electrolytes are 10 mM NADH, 40 mM glucose, and 0.2 M PBS (pH 6.5) solution. Adapted from Ref. [10] with permission from the American Chemical Society.

interesting approach for self-powered biosensing since the presence of the analyte caused the sensor to turn on rather than turn off.

Some interesting results were recently reported for EBFCs with power logically controlled by biochemical computing [13–15]. The electrochemical activity of the enzyme-modified electrode could be switched by antigen–antibody recognition [13], pH value alternation [14], oxygen concentration [15], etc., so the EBFC could be controlled by logic operations processing in situ biochemical input signals. The detection mode for traditional biosensors is mainly based on the one-to-one quantitative response between the input signal (target) and the output signal (transducer). Being different from traditional biosensors, logic biosensors based on biocomputing are smart and able to intelligently analyze the relationship between different targets in complex samples according to the Boolean logic operations "programmed" into biocomputing systems.

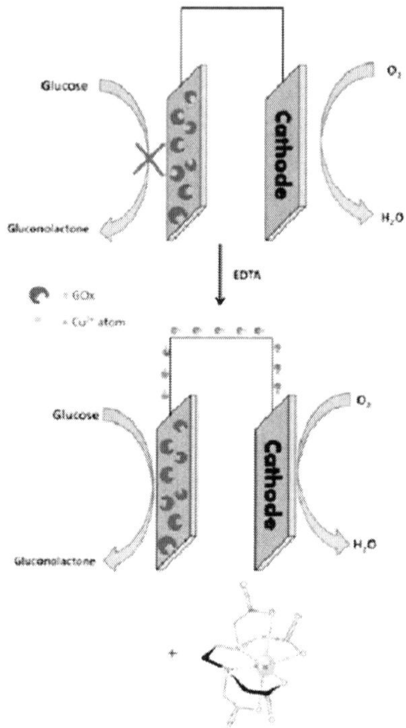

Figure 13.3 Activation of a Cu^{2+}-inhibited GOx self-powered sensor by EDTA. Adapted from Ref. [12] with permission from the American Chemical Society.

Our group constructed self-powered and intelligent logic aptasensors based on the aptamer-controlled power output of EBFCs [16]. As shown in Fig. 13.4, this system was composed of a GOx-/thrombin-binding aptamer-based bioanode and a bilirubin oxidase-/lysozyme-binding aptamer-based biocathode. The electrochemical activity of the modified electrode was switched by the presence or absence of the thrombin (input A) or lysozyme (input B). On the basis of the specificity of the aptamer, the presence of lysozyme could only make the aptamer target recognition occur at the bilirubin oxidase-/lysozyme-binding aptamer-based biocathode, and the thrombin-binding aptamer on the bioanode would only catch thrombin. These proteins on the electrode surface would block the electrode interface, resulting in a power decrease

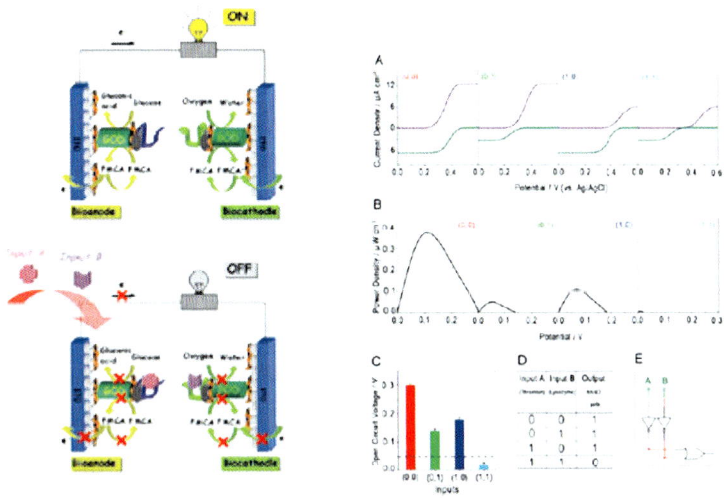

Figure 13.4 Schematic illustration of the assembled aptamer-based BFC logically controlled by biochemical signals that mimic a Boolean NAND logic gate. (A) Polarization curves of the bioanode (violet curves) and biocathode (olive curves) after the input signals **(0, 0)**, **(0, 1)**, **(1, 0)**, and **(1, 1)**. (B) Dependence of the power density on the BFC voltage after the input signals **(0, 0)**, **(0, 1)**, **(1, 0)**, and **(1, 1)**. (C) Bar diagram showing the open-circuit potential of the BFC for different combinations of input signals, derived from Fig. 13.4b. The dashed line shows the threshold (0.05 V). (D) Truth table for a NAND logic gate. (E) Circuit for a NAND logic gate. **Input A**: 8 nM thrombin. **Input B**: 20 nM lysozyme. Adapted from Ref. [16] with permission from the American Chemical Society.

of the assembled EBFC. In the presence of both substrates [input (1, 1)], aptamer target recognition occurred at both the bioanode and the biocathode, and consequently the EBFC was electrochemically mute. The transition between states was used to control electrode activity as NAND logic switches in the system. Through the built-in NAND logic gate, the fabricated EBFCs controlled by aptamer logic systems enabled us to construct self-powered aptasensors that can determine whether the two specific targets are both present in a sample. We further developed a self-powered, reusable aptasensor based on the IMP-Reset gate controlled on-chip EBFC with a similar strategy [17]. This self-powered aptasensor can be used to determine the presence of one specific target in the absence of another target in complex physiological samples (such as human serum) in a single test.

13.3 The Self-Powered Biosensor Based on Mitochondrial Biofuel Cells

As the cell powerhouse, mitochondria contain all enzymes that are needed in biofuel sources' oxidation, such as pyruvate. The mitochondrial BFCs were developed by Minteer's group, which contained two electrodes, a cathode that will reduce oxygen to water and a mitochondrial bioanode that will oxidize pyruvate in a fuel container to carbon dioxide and water [18]. This is the first use of intact and viable mitochondria in BFCs. There are a number of inhibitors (e.g., oligomycin antibiotic) that can stop mitochondrial functioning, which in turn will stop electrical power generation. The mitochondrial metabolic function is shut down by the oligomycin, which is a phosphorylation inhibitor that blocks ATP synthesis by the F_0/F_1 ATPase of the mitochondria. However, this mitochondrial function can be restored by the addition of a decoupler. Minteer's group investigated the unique ability of inhibition and subsequent activation of mitochondria to sense nitroaromatic explosive compounds [19]. Pyruvate was used as fuel in this system. Oligomycin was used to inhibit ATP synthase in the mitochondria, which resulted in a power generation decrease of the mitochondrial BFCs,

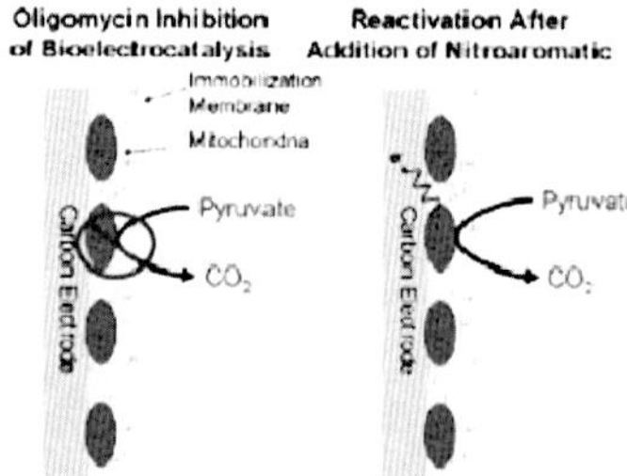

Figure 13.5 Bioelectrocatalysis mechanism of oligomycin inhibition of pyruvate oxidation at a mitochondria-modified electrode and nitroaromatic decoupling of a mitochondria-modified electrode. Adapted from Ref. [19] with permission from the American Chemical Society.

as shown in Fig. 13.5. This self-powered nitroaromatic biosensor would not produce significant power in the absence of the nitroaromatic. When nitroaromatic compounds were added to the cell, the mitochondria were decoupled from the inhibitors and allowed for the oxidation of pyruvate to carbon dioxide, producing a significant power improvement that was used to signal the explosive.

13.4 The Self-Powered Biosensor Based on Microbial Fuel Cells

The extracellular electron transfer of microorganisms at the electrode surface is crucial for a self-powered biosensor based on an MFC. The direct electron transfer (DET) of bacteria toward the electrode is quite attractive in an electrochemical biosensor, which is sensitive to changes in the metabolic status of the cellular biocatalyst and simplifies the construction procedure. Several mechanisms have been proposed for DET in microorganisms [20]. First, a number of proteins in the cytoplasmic membrane, periplasm, and outer membrane (OM) could act as a "molecular wire" and play an important role in the electron transfer process. The role of c-type cytochromes in the extracellular electron transfer was investigated [21]. The biochemical and genetic characterization demonstrated that OmcA, OmcB, and MtrC were exposed on the cell surface and were thought to participate directly in the electron transfer to the electrode surface. MtrcAB and OmcZ were shown to obtain electrons from the host electron transport chain and pass through the membrane, and MtrcA and MtrcB were helpful in this process [22]. Second, some bacteria can produce soluble electron shuttles for the extracellular electron transfer. Rabaey et al. demonstrated that phenazine production by a strain of *Pseudomonas aeruginosa* stimulated electron transfer [23]. The hydroquinone released from the electrochemically evolved *E. coli* was assumed to be self-excreted mediators [24]. These endogenous redox mediators can serve as a reversible terminal electron acceptor to transfer electrons from the bacterial cell to the electrode surface. Third, the conductive nanowires, identified in *Geobacter sulfurreducens* PCA, *Shewanella oneidensis* MR-1, *Cyanobacterium synechocystis*

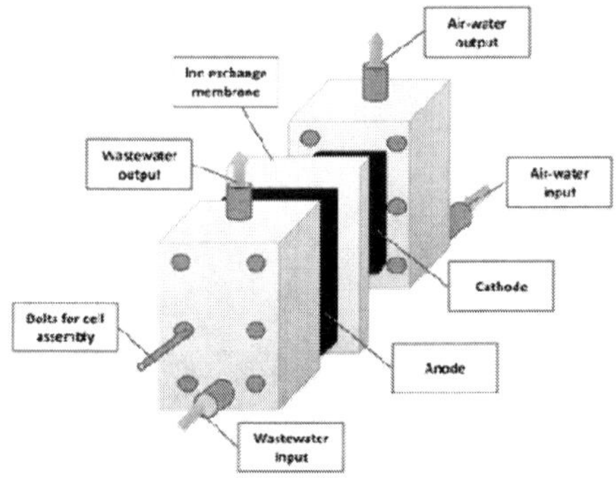

Figure 13.6 Schematic diagram of the two-chamber MFC-type BOD biosensor.

PCC6803, etc., appeared to be directly involved in extracellular electron transfer [25–27].

The first microbial self-powered biosensor was constructed on the basis of a two-chamber MFC [28]. Two electrodes (anode and cathode) were each placed in two chambers joined by a proton exchange membrane (PEM) (Fig. 13.6). The electricity generated from the MFC was directly proportional to the strength of the wastewater. This observation suggested the possibility to use it as a biochemical oxygen demand (BOD) sensor. From then on, various kinds of two-chamber microbial self-powered biosensors were developed [29, 30]. At a low BOD range, the current values increased with the BOD value linearly. However, a high BOD concentration required a long response time. The self-powered BOD biosensor showed long-term operational stability without extra maintenance for over five years [28], far longer in service life span than other types of BOD sensors reported in the literature, demonstrating the applicability of this system to real treatment effluents.

The main disadvantage of self-powered biosensors based on a two-chamber MFC is that the cathode must be aerated to provide oxygen. In addition, reducing their cost is essential. For example, PEMs such as Nafion® are quite expensive. To make a more compact and simple system with reduced cost of operation, Lorenzo et al. tested a single-chamber microbial fuel cell (SCMFC) with an air cathode as a BOD biosensor [31]. The air cathode MFC provided potential advantages over the two-chamber system because aeration, recycling, and catholyte regeneration were not required. The self-powered biosensor performance was evaluated in terms of measurement range, response time, reproducibility, and operational stability. When artificial wastewater was used as fuel, the biosensor output had a linear relationship with a BOD concentration of up to 350 ppm. The biosensor with an SCMFC was used as self-powered sensing for glucose [32]. The sensor response was linear against the concentration of glucose for up to 25 g L^{-1}. The detection limit was found as 0.025 g L^{-1}. This monitoring mode can be applied to real-time BOD determinations for either surface water, secondary effluents, or diluted high-BOD wastewater samples.

To obtain an accurate BOD value, in the presence of electron acceptors of higher redox potential (nitrate and oxygen), respiratory inhibitors were applied to eliminate the influence of these electron acceptors [33]. The current generation from microorganisms was inhibited by inhibitors of NADH dehydrogenase, coenzyme Q, and quinol–cytochrome b oxidoreductase but not by terminal oxidase inhibitors. The azide and cyanide are known to inhibit the activity of the terminal enzyme of the respiration chain. The use of respiratory inhibitors, such as azide and cyanide, was applied for accurate BOD measurement of environmental samples containing nitrate and/or oxygen. These results could be useful in operating self-powered BOD sensors.

The toxic substances have an inhibitory effect on the metabolism of microorganisms and the transfer rate of electrons to the electrode. Kim et al. reported a self-powered biomonitoring system based on MFCs for detecting the inflow of toxic substances [34]. When toxic substances (an organophosphorus compound, Pb, Hg, and polychlorinated biphenyl) were added to the MFC, rapid decreases in the current were observed. The inhibition ratios caused by inflow of these toxic substances (1 mg L^{-1}) were 61%, 46%, 28%, and 38%, respectively. These systems were able to detect the toxins.

13.5 Outlook

The urgent need of lab-on-a-chip devices and point-of-care and home medical testing have brought about the expansion of research in the development of miniaturization and portability of biosensor devices. In one growing area of biosensor devices, self-powered biosensors represent an interesting way. The BFCs that were applied as a self-powered biosensors platform in previous reports produced relatively low open-circuit potentials, current densities, and power densities. So the development of self-powered biosensors based on BFCs for real applications will be devoted to improve BFC power output as well as electrical engineering to circuitry design.

Direct electron transfer mechanisms or optimum mediators were usually employed to minimize the biocatalyst overpotential at the electrode surface. DET is very attractive and would be important to simplify and miniaturize BFCs. Nanomaterials are extensively applied to improve the electron transfer of enzymes. Recent studies have experimentally demonstrated the feasibility of increasing the electron transfer of bacteria using nanomaterials. The bioparticles Pd^0 could act as highly active catalysts for electrode reactions of *Desulfovibrio desulfuricans* [35]. Wu et al. presented evidence that Pd nanoparticles bound to microbes may participate in the electron transfer process (Fig. 13.7). The OM c-type cytochromes wire efficiency of *S. oneidensis* MR-1 could be improved greatly by Fe_3O_4/Au nanocomposites [36]. The electron propagation from OM c-type cytochromes could go along the Fe_3O_4/Au nanocomposite–assembled nanostructure instead of the traditional electron transfer between adjacent bacteria. The incorporation of biocompatible conductive nanomaterials into biofilms opens up a new avenue to accelerate the evolution of self-powered biosensors based on BFC applications.

To develop a stable self-powered biosensor, a matrix for biocatalyst immobilization in BFCs is a crucial factor to retain the activity of biocatalysts for a long period. The ideal biosensor platform should provide a favorable microenvironment to maintain biocatalyst activity and minimize kinetic barriers of the substrate and the product. Using an optimized electrode material with a high surface area and good electrocatalytic properties is one of the most efficient approaches to improve the performance of self-powered biosensors. For example, the unique nanostructured polyaniline (PANI)/mesoporous TiO_2 composite with uniform nanopore distribution and high specific surface area was synthesized, which

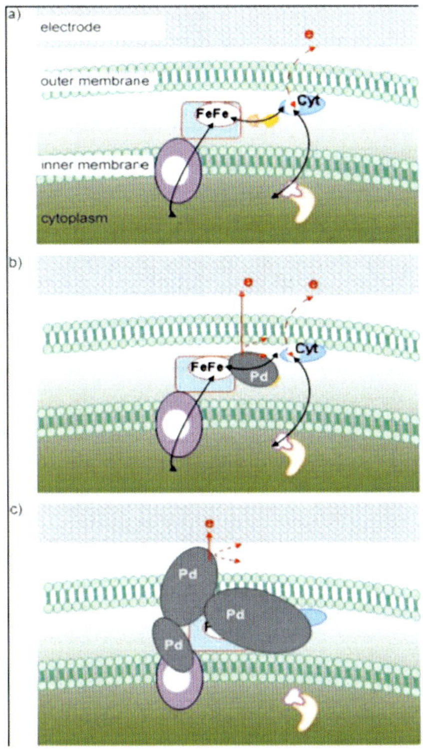

Figure 13.7 Hypothesized pathways for the extracellular electron transfer chain between the cell and the electrode (a) in the absence of Pd^0 (via periplasmic cytochromes and hydrogenases), (b) in the presence of Pd^0 at a low loading, and (c) in the presence of Pd^0 at a high loading. (FeFe) is an iron-only hydrogenase, Cyt is a c-type cytochrome, and areas shaded gray are Pd^0 nanoparticles. Adapted from Ref. [35] with permission from Wiley-VCH.

exhibited excellent bio- and electrocatalytic performance [37]. This electrode material increased the electrode-specific surface area and enhanced the charge transfer capability and consequently caused considerable electrochemical activity improvement. Our group synthesized a carbon fiber mat from silk fiber [38]. This silk-derived carbon fiber mat not only allowed the negatively charged nanomaterials to assemble on the carbon fiber surface, which could accelerate electrical communication between microbes and the

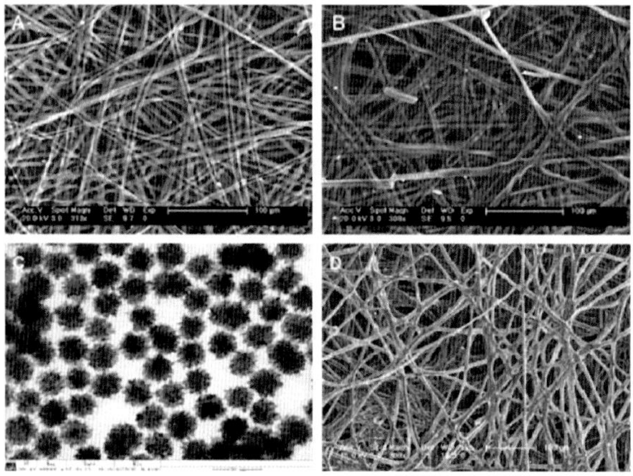

Figure 13.8 SEM micrographs for silk fibers (a), silk-derived carbon fibers (b), TEM micrograph (c) for Au@Pt urchilike nanoparticles, and Au@Pt urchilike nanoparticles-deposited silk-derived carbon fibers (d). Adapted from Ref. [38] with permission from Elsevier.

electrode surface, but also provided the quinone-containing constituents to mimic natural microbial electron acceptors (Fig. 13.8). The combination of nanotechnology, biochemistry, and electronics will provide a wider scope for the design of advanced self-powered biosensors.

Acknowledgments

This research was supported by the National Natural Science Foundation of China (Nos. 21105126, 21075116, 20935003, and 20820103037) and the 973 Project (2009CB930100, 2011CB911002, and 2010CB933600).

References

1. Zhang, Q., Xu, J. J., Chen, H. Y. (2006) Glucose microfluidic biosensors based on immobilizing glucose oxidase in poly(dimethylsiloxane) electrophoretic microchips, *J. Chromatogr. A*, **1135**, 122–126.

2. Moore, C. M., Minteer, S. D., Martin, R. S. (2005) Microchip-based ethanol/oxygen biofuel cell, *Lab. Chip.*, **5**, 218–225.
3. Hansen, B. J., Liu, Y., Yang, R., Wang, Z. L. (2010) Hybrid nanogenerator for concurrently harvesting biomechanical and biochemical energy, *ACS Nano*, **4**, 3647–3652.
4. Qin, Y., Wang, X, Wang, Z. L. (2008) Microfibre–nanowire hybrid structure for energy scavenging, *Nature*, **451**, 809–813
5. Pan, C., Fang, Y., Wu, H., et al. (2010) Generating electricity from biofluid with a nanowire-based biofuel cell for self-powered nanodevices, *Adv. Mater.*, **22**, 5388–5392.
6. Katz, E., Buckmann, A. F., Willner, I. (2001) Self-powered enzyme-based biosensors, *J. Am. Chem. Soc.*, **123**, 10752–10753.
7. Tasca, F., Gorton, L., Harreither, W., Haltrich, D., Ludwig, R., Noll, G. (2008) Highly efficient and versatile anodes for biofuel cells based on cellobiose dehydrogenase from *Myriococcum thermophilum*, *J. Phys. Chem. C*, **112**, 13668–13673.
8. Deng, L., Wang, F. A., Chen, H. J., Shang, L., Wang, T., Dong, S. J. (2008) A biofuel cell with enhanced performance by multilayer biocatalyst immobilized on highly ordered macroporous electrode, *Biosens. Bioelectron.*, **24**, 329–333.
9. Tel-Vered, R., Yildiz, H. B., Yan, Y. M., Willner, I. (2010) Plugging into enzymes with light: photonic "wiring" of enzymes with electrodes for photobiofuel cells, *Small*, **6**, 1593–1597.
10. Deng, L., Chen, C., Zhou, M., Guo, S., Wang, E., Dong, S. (2010) Integrated self-powered microchip biosensor for endogenous biological cyanide, *Anal. Chem.*, **82**, 4283–4287.
11. Wen, D., Deng, L., Dong, S. J. (2011) Self-powered sensor for trace Hg^{2+} detection, *Anal. Chem.*, **83**, 3968–3972.
12. Matthew, T. M., Shelley, D. M. (2011) Inhibition and activation of glucose oxidase bioanodes for use in a self-powered EDTA sensor, *Anal. Chem.*, **83**, 5436–5441.
13. Amir, L., Tam, T. K., Pita, M., Meijler, M. M., Alfonta, L., Katz, E. (2009) Biofuel cell controlled by enzyme logic systems, *J. Am. Chem. Soc.*, **131**, 826–832.
14. Strack, G., Ornatska, M., Pita, M., Katz, E. (2008) Biocomputing security system: concatenated enzyme-based logic gates operating as a biomolecular keypad lock, *J. Am. Chem. Soc.*, **130**, 4234–4235.

15. Tam, T. K., Strack, G., Pita, M., Katz, E. (2009) Biofuel cell logically controlled by antigen–antibody recognition: towards immune-regulated bioelectronic devices, *J. Am. Chem. Soc.*, **131**, 11670–11672.
16. Zhou, M., Du, Y., Chen, C., et al. (2010) Aptamer-controlled biofuel cells in logic systems and used as self-powered and intelligent logic aptasensors, *J. Am. Chem. Soc.*, **132**, 2172–2174.
17. Zhou, M., Chen, C. G., Du, Y., et al. (2010) An IMP-reset gate-based reusable and self-powered "smart" logic aptasensor on a microfluidic biofuel cell, *Lab. Chip.*, **10**, 2932–2936.
18. Sokic-Lazic, D., Minteer, S. D. (2008) Citric acid cycle biomimic on a carbon electrode, *Biosens. Bioelectron.*, **24**, 939–944.
19. Germain, M., Arechederra, R. L., Minteer, S. D. (2008) Nitroaromatic actuation of mitochondrial bioelectrocatalysis for self-powered explosive sensors, *J. Am. Chem. Soc.*, **130**, 15272–15273.
20. Lovely, D. R. (2008) The microbe electric: conversion of organic matter to electricity, *Curr. Opin. Biotechnol.*, **19**, 564–571.
21. Hartshorne, R. S., Reardon, C. L., Ross, D., et al. (2009) Characterization of an electron conduit between bacteria and the extracellular environment, *Proc. Natl. Acad. Sci. U. S. A.*, **106**, 22169–22174.
22. Bouhenni, R. A., Gary, J. V., Biffinger, J. C., et al. (2010) The role of *Shewanella oneidensis* MR-1 outer surface structures in extracellular electron transfer, *Electroanalysis*, **22**, 856–864.
23. Rabaey, K., Boon, N., Hofte, M., Verstraete, W. (2005) Microbial phenazine production enhances electron transfer in biofuel cells, *Environ. Sci. Technol.*, **39**, 3401–3408.
24. Qiao, Y., Li, C. M., Bao, S. J., Lu, Z., Hong, Y. (2008) Direct electrochemistry and electrocatalytic mechanism of evolved *Escherichia coli* cells in microbial fuel cells, *Chem. Commun.*, 1290–1292.
25. Reguera, G., McCarthy, K. D., Mehta, T., Nicoll, J. S., Tuominen, M. T., Lovley, D. R. (2005) Extracellular electron transfer via microbial nanowires, *Nature*, **435**, 1098–1101.
26. Gorby, Y. A., Yanina, S., McLean, J. S., et al. (2006) Electrically conductive bacterial nanowires produced by *Shewanella oneidensis* strain MR-1 and other microorganisms, *Proc. Natl. Acad. Sci. U. S. A.*, **103**, 11358–11363.
27. Bretschger, O., Obraztsova, A., Sturm, C. A., et al. (2007) Current production and metal oxide reduction by *Shewanella oneidensis* MR-1 wild type and mutants, *Appl. Environ. Microbiol.*, **73**, 7003–7012.

28. Kim, B. H., Chang, I. S., Gil, G. C., Park, H. S., Kim, H. J. (2003) Novel BOD (biological oxygen demand) sensor using mediator-less microbial fuel cell, *Biotech. Lett.*, **25**, 541–545.
29. Kim, M., Hyun, M. S., Gaddb, G. M., Kim, H. J. (2007) A novel biomonitoring system using microbial fuel cells, *J. Environ. Monit.*, **9**, 1323–1328.
30. Chang, I. S., Jang, J. K., Gil, G. C., et al. (2004) Continuous determination of biochemical oxygen demand using microbial fuel cell type biosensor, *Biosens. Bioelectron.*, **19**, 607–613.
31. Lorenzo, M. D., Curtis, T. P., Head, I. M., Scott, K. (2009) A single-chamber microbial fuel cell as a biosensor for wastewaters, *Water Res.*, **43**, 3145–3154.
32. Kumlanghan, A., Liu, J., Thavarungkul, P., Kanatharana, P., Mattiasson, B. (2007) Sandwich electrochemical immunoassay for the detection of Staphylococcal enterotoxin B based on immobilized thiolated antibodies, *Biosens. Bioelectron.*, **22**, 2939–2944.
33. Chang, I. S., Moon, H., Jang, J. K., Kim, B. H. (2005) Improvement of a microbial fuel cell performance as a BOD sensor using respiratory inhibitors, *Biosens. Bioelectron.*, **20**, 1856–1859.
34. Kim, M., Hyun, M. S., Gaddb, G. M., Kim, H. J. (2007) A novel biomonitoring system using microbial fuel cells, *J. Environ. Monit.*, **9**, 1323–1328.
35. Wu, X. E., Zhao, F., Rahunen, N., et al. (2011) A role for microbial palladium nanoparticles in extracellular electron transfer, *Angew. Chem., Int. Ed.*, **50**, 427–430.
36. Deng, L., Guo, S. J., Liu, Z. J., et al. (2010) To boost c-type cytochrome wire efficiency of electrogenic bacteria with Fe_3O_4/Au nanocomposites, *Chem. Commun.*, **46**, 7172–7174.
37. Qiao, Y., Li, C. M., Bao, S. J., Lu, Z., Hong, Y. (2008) Direct electrochemistry and electrocatalytic mechanism of evolved Escherichia coli cells in microbial fuel cells, *Chem Commun.*, **44**, 1290–1292.
38. Deng, L., Guo, S. J., Zhou, M., Liu, L., Liu, C., Dong, S. J. (2010) A silk derived carbon fiber mat modified with Au@Pt urchilike nanoparticles: a new platform as electrochemical microbial biosensor, *Biosens. Bioelectron.*, **25**, 2189–2193.

Index

α-1-fetoprotein 164
α-D-glucose 104, 263, 268, 270, 272
α -fetoprotein 234
α-fetoprotein 164, 166, 170, 296
α-glucosidase 271–272
adenosine 82–83, 104–105, 228
adenosine deaminase 104, 228
adrenaline 15–16, 64, 97, 107
adsorption enrichment 57, 64
adsorptive stripping voltammetry 53, 56–57, 59, 61, 63, 68
alkaline phosphatase 83, 106–107, 115, 142, 227, 325
amperometric detection 25, 29, 32, 98, 101, 106–108, 115, 215, 233, 237
amperometric detection of H_2O_2 108, 117
amperometric foot 19–20
amperometric signal 97, 110–111, 113
amperometric spikes 10–11, 19, 21–22, 27, 31
amperometric transduction 95, 98, 105, 111–112, 117
amperometry 5–6, 13, 15, 18–19, 21–22, 25–28, 35, 39, 41, 223, 226, 281, 292
  patch 27
analytes, solution-phase 86
analytical devices 93, 223, 235, 248
anodic photocurrent 311, 322, 329–330
antibodies 80, 111–114, 140, 162–165, 171–174, 176, 178–180, 224, 227, 229–230, 289–290, 295–299, 301–304, 349–351, 354–355
  enzyme-labeled 227, 231, 233
  immobilization of 297, 299
antigen–antibody reactions 139–141, 224–225
aptamer geometry 83
aptamer hairpin 84
aptamers 75–87, 93, 140, 224, 376
  nucleic acid 76
ascorbic acid 23–24, 131–132, 134, 136
aspartame 257–258

β-D-glucose 104, 251, 268, 270, 272, 324
BFCs
  see biofuel cells
  mitochondrial 371–372, 378
bienzyme electrode 106, 109
bioanode 373–374, 376–377
bioassays, enzyme-based electrochemical 142
biocatalyst 324, 327, 331, 371–372, 382
biocathode, aptamer-based 376
biochemical oxygen demand (BOD) 380

biochemical reactions,
 transduction of 309–310, 312, 314, 316, 318, 320, 322, 324, 326, 328, 330, 332
biochemical systems 314
bioelectrocatalysis 375, 378
bioelectrocatalytic reaction 229–230
bioelectrode 94, 100
biofuel cells (BFCs) 108, 371–378, 380, 382, 384
biomolecular sensing
 monosaccharides 130–131, 133, 135, 137, 139, 141, 143
biomolecules 25, 27, 99, 104, 111, 117, 129, 200, 226, 230, 232, 298, 328, 357
biosensing 87, 128, 200, 289, 342–343, 346, 348, 356, 362
 self-powered 371, 373, 375
biosensing electrode 132
biosensor analysis 277
biosensor designs 109
biosensor devices 382
biosensor sensitivity 251, 268, 270, 278
biosensor signal 110–111
biosensors
 amperometric 94, 97–101, 105, 108–109, 117–118, 253
 aptamer-based 75–76, 86
 CNT-based 129
 CNT-based electrochemical 127
 conductometric 248, 250, 252, 255, 257–258, 261–263, 266, 268, 270–275, 277–278, 280–281
 conductometric formaldehyde 259–260
 conductometric glucose 251
 conductometric nitrate 266–268
 conductometric urea 254
 conductometric urease 276
 DNA 142, 197, 205, 210–211, 213, 215, 217
 DNA hybridization 210, 212, 235
 ECL 339, 345, 347–349, 351, 353, 355, 357–359, 361–362
 electrochemical DNA 201–202, 216, 240
 enzyme 93, 97, 257
 enzyme-based 134
 enzyme conductometric 247
 L-arginine 256
 maltose conductometric 272
 nonaligned CNT 129–130
 glucose 95, 101, 131–133, 135, 250–251, 290
 nonenzyme glucose 134
 self-powered 372–375, 377, 379–380, 382
 self-powered based on biofuel cells 371–372, 374, 376, 378, 380, 382, 384
 traditional 375
BOD, *see* biochemical oxygen demand
bovine serum albumin (BSA) 141, 170, 172, 257–258, 261, 264, 266, 297, 301, 350
BSA
 see bovine serum albumin
 non-native 258

C-reaction protein detection 232
C reactive protein (CRP) 140
cadmium 94
carbon fiber microelectrodes 11, 16, 40
carbon fiber ultramicroelectrodes 1, 5
carbon microelectrode arrays 31

carbon nanotubes (CNTs) 39, 101–102, 117–118, 125–146, 170, 172, 230, 240, 299–300, 347, 354, 359–360
carbon paste electrodes (CPE) 54–55, 66, 68, 165, 173, 348
carcinoembryonic antigen 164, 176
  detection of 229
carcinoma antigen 164
catechol 23, 106–108
catechol derivatives detection 356
catecholamines 6, 11, 15–17, 20, 22, 28
cathodic photocurrents 317, 323–324, 330
CdS nanoparticles 200, 212, 320–322, 330, 354
CdSe quantum dots 354, 357, 359–360
CdTe quantum dots 212, 310
chemical oxygen demand (COD) 261, 380
chemical speciation 67
chemical-to-electrochemical energy transformations 371
chemical vapor deposition (CVD) 128, 130, 302
chemisorption 67, 317
chitosan 102, 118, 169, 172–173, 175, 346
chromaffin cells 7, 16–20, 22–23, 25, 28–29, 32, 36, 38–39
  adrenal 23
CNT synthesis 128
CNTs
  see carbon nanotubes
  covalent functionalization of 129
  functionalized/hybrid 126
  nitrogen-doped 354
cobalt 327
COD, *see* chemical oxygen demand
conductivity 5, 32, 165, 206, 228, 233–234, 240, 250, 252, 257, 280–281, 347
CPE, *see* carbon paste electrodes
CRP, *see* C reactive protein
CVD, *see* chemical vapor deposition
cyanide 100, 280–281, 375, 381
cytochrome 264, 329–330, 373
  oxidized 328, 330
cytoplasm 5–6, 26

D-gluconic acid 251, 268, 270, 272
dabycil 205, 211–212
dendrimer 173, 360
desisopropylatrazine 278–279
DET, *see* direct electron transfer
detection
  analyte 241, 314
  biowarfare agent 338
  electrocatalytic 131
  label-free 302
  luminescent 337–338
  multiplexed 202
  protein 138, 262, 354
  selective 242
  single-nucleotide polymorphism 235
detection antibodies 162–164, 176, 178
  labeled 163, 178
differential pulse voltammetry (DPV) 63, 139, 170, 203, 209
digoxin 206, 297–298
dihydroflavins 108–109
dihydronicotinamide adenine dinucleotide (NADH) 105–106, 109, 135–136, 139, 322–324, 347
direct electron transfer (DET) 98, 102–103, 118, 132, 327, 329, 331, 379, 382
diuron 164, 279
diuron detection 278

DME, *see* dropping mercury electrode
DNA, genomic 208
DNA analyses 196, 345
DNA analyte detection limit 242
DNA detection 141, 200, 206, 211, 216–217, 235, 239
DNA hybridization 55, 114, 141–143, 212, 223, 237, 350
  bioelectronic detection of 236
DNA hybridization detection 235, 238
DNA molecules 196–197, 199–200, 202
DNA polymerase 198–199
DNA probe 115, 142–143, 240, 360
DNAzyme 360–361
dopamine 15–16, 22, 63–64, 97, 134, 136
DPV, *see* differential pulse voltammetry
dropping mercury electrode (DME) 61

ECL, *see* electrochemiluminesce
ECL biosensing 339, 341, 343, 345–346, 348–350, 354, 357–358, 360, 362
  principles of 339, 341, 343
ECL coreactant 341, 345, 347–348
ECL detection 141, 338–339, 346, 351
ECL emission 339, 342–344, 346, 348–349, 351, 353–357, 359, 362
ECL emitters 337–339, 345, 350, 362
ECL luminophore 341–342
EIS, *see* electrochemical impedance spectroscopy
electro-oxidation 106, 116, 133, 135
electroactive enzyme labels 223
electroactive indicator 142, 235
electroactive species 10, 12–13, 25, 94, 98, 112, 224
electroactivity 12, 32, 202
electroanalytical chemistry 125, 371
electrochemical aptasensors 81, 83, 85, 87
electrochemical assays 83, 126
electrochemical biosensing applications 162
electrochemical detection 16, 26, 142, 211, 215–217, 316
  sandwich-type 231
  selective 238
electrochemical detection of DNA 213
electrochemical detection of exocytosis 1–2, 4, 6, 8, 10, 12–40
electrochemical detector array 32
electrochemical immunoassays 162, 164–165, 180, 229–230
electrochemical impedance spectroscopy (EIS) 82, 145, 231–232, 238–240
electrochemical labels 82, 208
electrochemical oxidation 23, 97, 180, 341–342, 348
electrochemical polymerization 99, 130, 132, 172
electrochemical probes 16, 31
electrochemical stripping analysis 55
electrochemical transduction 87, 93
electrochemiluminesce (ECL) 141, 337–341, 343–344, 348–349, 351, 353–354, 357, 359
electrochemistry, direct protein 327, 330

electrodes
amperometric enzyme 93–94, 96, 98, 100, 102, 104, 106, 108, 110, 112, 114, 116
antimony film 55
bismuth 54
bismuth film 54, 56, 65
boron-doped nanocrystalline diamond 32–33
carbon 62, 290
carbon fiber 28, 39
CNT electrode 136, 138, 141
conductometric 266
disk 13–14
enzymatic 134
enzyme 96–97, 112, 132, 327
enzyme-modified 375
enzyme–clay 100
gold 63, 86, 209, 296–297, 320, 359
hybrid 132
interdigitated 250, 296
mercury electrodes 54, 65–66, 68
millimetric 12–14
mitochondria-modified 378
monoenzyme 106
nonmercury 54, 65–66
paste 67, 142
platinum 66, 105–106, 112, 117–118, 226, 253
quantum dot 309, 323, 326, 330
saturated calomel 66, 95, 172
screen-printed 297
screen-printed carbon 140, 203, 290, 300, 348, 352
electrodeposition 17, 101, 134, 172, 298, 301
electroenzymatic labeling 223–242
electroenzymatic transduction 116
electroenzymes 238
electrogenerated chemiluminescence 338
electrografting 208
electron acceptors 319, 324, 381
electron microscopy 21
electron transfer
extracellular 379–380
heterogeneous 142, 328
internal 102, 374
interparticle 315
electron transfer kinetics 99, 138, 232, 294, 351, 357
electron transfer resistance 82, 206, 232, 239, 294, 303
electronic transfer resistance 239
electropolymer 101, 346
electropolymerization 301
electrostatic attraction 171–172, 174
electrotransducers 200, 208
modifiers of 200, 208
enzymatic biofuel cells 108, 373, 375, 377
enzyme acetylcholine esterase 322
enzyme glucose oxidase 250–251
enzyme labeling 111–113, 115
enzyme wiring 98
enzymes
biotinylated 115–116
immobilized 98, 103, 107, 233–234, 253
epinephrine 10, 63, 137
ethanol 108, 348
eutrophication 263–264
exocytosis 1–2, 4–40
electrochemical detection of 1, 30
electrochemical measurements of 36
vesicular 2, 9
exocytotic secretion 2–3, 5, 7, 9, 11

FAD, *see* flavin adenine dinucleotide
FET, *see* field effect transistor
field effect transistor (FET) 82, 126, 139
flavin adenine dinucleotide (FAD) 95, 102–103, 108
flavin reductase 108–109
flavins 108
formaldehyde 259, 261

GC, *see* glassy carbon
GC electrode 132–133
GCEs, *see* glassy carbon electrodes
glassy carbon (GC) 54–55, 65, 112, 118, 125, 138, 165, 170–173, 178, 197, 205, 208, 211–213, 225, 299
glassy carbon electrodes (GCEs) 54, 63, 118, 171–172, 205, 208, 212–213, 299
glucose biosensing 131
glucose dehydrogenase 102, 107, 135, 324
glucose detection 95, 105, 131, 133, 324, 352
  selective 100, 106
glucose oxidase 95, 130–131, 175–176, 215, 227, 251, 268, 270, 272–273, 277, 290, 325, 327, 373
glutamate 15, 25, 97, 105, 107–108, 139, 330
gold 54, 77, 112, 161–162, 164–165, 172–177, 180, 215, 240, 251–252, 297, 300, 329
gold nanoparticles 133, 200, 202–210, 212–214, 216, 316, 350, 356, 360
  electrodeposited 173
gold nanospheres, hollow 209–210
gold ultramicroelectrodes 66
graphene 117, 125–126, 146, 172, 174, 214, 328
graphite electrode 115, 351

H_2O_2 detection 112, 351
hanging mercury drop electrode (HMDE) 54, 65, 67–68
heavy metals 54–56, 110, 249, 276–277
hepatitis 115, 143, 164
HMDE, *see* hanging mercury drop electrode
horseradish peroxidase (HRP) 83, 115, 133, 140, 170, 175, 210, 213–215, 229–230, 233, 236, 238, 241, 280, 354
HRP, *see* horseradish peroxidase
HRP-labeled secondary antibodies 230, 233
HSA, *see* human serum albumin
human serum 378
human serum albumin (HSA) 140, 142, 164, 297

IgG antibodies 139–140, 164
immunoassays, sandwich-type 349–351
immunocomplexes 163, 174–175, 177, 228, 234, 357
immunodetection 107
immunoreaction 112, 227, 350
immunosensor arrays 300–301, 303
immunosensors 93–94, 111, 114, 126, 140, 223–234, 236, 238, 240, 242, 290, 295–298, 300, 302–303, 354
  amperometric 112, 114, 228
  conductometric 233–234

electrochemical 161–162, 164, 166, 168, 170, 172, 174, 176, 178, 180, 223–226, 232, 235, 295, 299
electrochemical enzyme 229
impedance 233, 289–290, 292, 294–298, 300, 302, 304
impedimetric 231–232, 296, 300
label-free 297, 357
label-free electrochemical 139
potentiometric 226–228
ultrasensitive electrochemical 230
impedimetric measurements 206, 208, 210–211
indium tin oxide (ITO) 32, 34–35, 82, 202, 205, 208, 322, 355
indium tin oxide electrode 202, 205, 355
insulin 16, 23
ion-selective electrode (ISE) 137, 226, 236–237
ion-sensitive field-effect transistor (ISFET) 82, 226
ISE, *see* ion-selective electrode
ISFET, *see* ion-sensitive field-effect transistor
ITO, *see* indium tin oxide

L-arginine 255
L-arginine detection 255–256
L-cysteine 138–139, 344
labels
catalytic 202, 204
electroactive 83, 200
laccase 97, 100, 107, 110, 374
lactose 102, 269–270
LAPSs, *see* light-addressable potentiometric sensors
light-addressable potentiometric sensors (LAPSs) 228, 314
low-density lipoprotein detection 357

magnetic beads 142, 174, 179, 200, 350
maltose 104, 134, 263, 270–273
maltose phosphorylase 104, 263
mercury 54, 65, 69, 94
mercury film electrode (MFE) 54, 65, 68, 212
MFC, *see* microbial fuel cell
MFE, *see* mercury film electrode
microbial fuel cell (MFC) 371–372, 379–381
microelectrode arrays 66, 130, 300, 302
microelectrodes 2, 41, 66, 104, 130, 206, 300, 303
microspheres, latex 203–204
micrototal analysis systems 371–372
mitochondria 378–379
multiwalled carbon nanotubes (MWCNTs) 126, 131–132, 137, 140–142, 354
mutarotase 104, 268, 270, 272
MWCNTs, *see* multiwalled carbon nanotubes

NADH
see dihydronicotinamide adenine dinucleotide
electrocatalytic oxidation of 135, 322
NADH detection 322
Nafion 17, 131–132, 136, 264–266, 346, 380
nanocomposites 170, 359–360, 382
nanoelectrodes 31, 255, 300
nanolabels, electroenzymatic 223
neomycin 85–86

neurons 2, 4, 22
neurotransmitters 2–3, 5, 17, 19–23, 28, 97, 104–105, 107, 136
nonenzymatic electrocatalytic system 134
noradrenaline 15–16, 18
nucleic acids 54–55, 77, 80, 82, 143, 200, 202, 211, 224, 348, 362

oxidation
- electrocatalytic 139
- enzymatic 101–102, 107–109

oxidation of phenolic compounds 97, 325

PCR, *see* polymerase chain reaction
phenyl phosphate 106–107
photoswitchable bioelectrocatalytic system 329
platinum microelectrodes 32–33
polyelectrolytes 102, 326
polymerase chain reaction (PCR) 78–79, 115, 197–200, 206, 239
polyphenol oxidase 106, 112–113
potentiometric detection 237
probe, electroactive 162–163, 174
prostate-specific antigen (PSA) 139, 164, 302
PSA, *see* prostate-specific antigen

QD electrode 314, 316, 322, 324–329, 331
QDs, *see* quantum dots
quantum dots (QDs) 211–212, 309–324, 326–332, 338–339, 351–355, 357–360
quenching effect 351, 355

redox chemistry 340, 343
redox mediators 99–101, 107, 313
redox protein 329
RNA 202, 238
RNA aptamers 78–79

salicylate 105–106
SAM, *see* self-assembled monolayer
sandwich immunocomplexes 175, 178
scanning electrochemical microscopy 24
SCMFC, *see* single-chamber microbial fuel cell
screen-printed carbon electrode (SPCE) 140, 203, 207, 216, 290, 300, 348, 352
SELEX, *see* systematic evolution of ligands by exponential enrichment
self-assembled monolayer (SAM) 232, 320, 346
semiconductor nanoparticles 211, 230
sensors
- amperometric 115, 138
- amperometric DNA 115–116, 237
- bienzyme 110, 280
- chemical 248
- DNA 94, 111, 114, 116, 141, 143, 199–202, 210, 216, 223, 225, 235, 242
- electrocatalytic glucose 134
- electrochemical 126, 138, 144, 196, 236, 291
- electrochemical-based protein 138
- electrochemical DNA 143, 145, 196, 199, 224, 235
- electronics 126
- glucose 131, 134, 250

photobioelectrochemical 326, 331
photoelectrochemical 330
QD-based ECL 359
self-powered 372, 374, 376
serotonin 10, 15–17, 136–138
signal amplification 113, 115, 225–226, 228–229, 237–238, 241–242, 346, 350, 360
signal amplifiers 206–208
signal antibodies 224, 228–229
silver 54, 65, 173, 177–179, 206
silver nanoparticles 161–162, 164–166, 168, 170, 172–180, 200, 210
single-chamber microbial fuel cell (SCMFC) 380–381
single-nucleotide polymorphisms (SNPs) 195, 201–202, 205, 217
single-walled carbon nanotubes (SWCNTs) 126, 128–129, 132, 139–140, 144, 240
SNARE proteins 21
SNPs, *see* single-nucleotide polymorphisms
solar cells 312–313
SPCE, *see* screen-printed carbon electrode
stripping analysis 68
stripping voltammetry 54, 61, 63, 179
substrates
dehydrogenase 109, 347
oxidase 348, 351–352
surface antigen 164
surface chemistry approach 38
SWCNTs, *see* single-walled carbon nanotubes
systematic evolution of ligands by exponential enrichment (SELEX) 75, 77–78

thionine 139, 172–173, 229–230
thrombin 82–83, 85–86, 359, 376
thrombin detection 86
TIRF, *see* total internal reflection fluorescence
total internal reflection fluorescence (TIRF) 6–7
transferrin 298–299
tyrosinase 97, 107, 136, 279, 356

ultramicroelectrodes 1, 9, 12, 17, 66
ultrasensitive electrochemical detection 242
unlabeled analyte antigen 350
urea 253–255, 276
urea analysis 249, 253
urea determination 252–253
urease 227–228, 249, 252–256, 276
uric acid 98, 131–132, 134, 136

water 12, 94, 100, 103, 106, 108, 110, 129, 263, 266, 273, 277, 297, 300, 378
river 94, 261, 263
water molecules 56
water samples 262, 267

zeolite 346